# Sheep Husbandry and Diseases

## Sixth Edition

Allan Fraser DSc, MD
and
John T. Stamp DSc, MRCVS, FRSE

Revised by
J. M. M. Cunningham
and
John T. Stamp

COLLINS

Collins Professional and Technical Books
William Collins Sons & Co. Ltd
8 Grafton Street, London W1X 3LA

First published in Great Britain by
Crosby Lockwood & Son Ltd 1949
Second Edition 1951
Third Revised Edition with new section 1957
Fourth Edition 1961
Fifth Edition 1968
Sixth Edition published by Collins
Professional and Technical Books 1987

Distributed in the United States of America
by Sheridan House, Inc.

Copyright © Allan Fraser and John T. Stamp 1949,
1951, 1957, 1961, 1968
Copyright © Allan Fraser, J. M. M. Cunningham and John T.
Stamp 1987

*British Library Cataloguing in Publication Data*
Fraser, Allan
  Sheep husbandry and diseases. —
  6th ed.
  1. Sheep
  I. Title.  II. Stamp, John T.
  III. Cunningham, J.M.M.
  636.3    SF375

ISBN 0–00–383272–4

Printed and bound in Great Britain by
Mackays of Chatham, Kent

# Contents

# Preface to the Sixth Edition

Sheep are becoming more and more important in the agricultural economy, not only in Britain, but in Europe generally. If they are to compete successfully, flockmasters and shepherds will have to be well informed of present-day sheep husbandry, including modern methods of disease control. Gone are the days when traditional methods alone sufficed and this book makes this abundantly clear by supplying all the information that is essential to make sheep farming ever increasingly productive. Professor Ian Cunningham, Principal of the West of Scotland College of Agriculture and one of the two authors of this book, has for many years been closely associated with, and responsible for, the new knowledge of sheep husbandry, productivity, genetics and sheep breeding. He has written for flockmasters, in a very understandable manner, not only about scientific advances but linked this with a basic knowledge of sheep farming. Dr John Stamp, retired director of the Animal Diseases Research Institute, Edinburgh, a world authority of diseases of sheep and a former editor along with the late Dr Allan Fraser of previous editions of this book, has set out clearly up-to-date methods of disease control. Thanks are given to Dr W. Martin, formerly Director of the Animal Diseases Research Institute for his considerable help in the writing of the disease section.

It should perhaps be explained that references in the husbandry section of the book are up to date and helpful to flockmasters and shepherds but are absent in the disease section since veterinary technical references are not appropriate in a book of this type. Veterinary surgeons are more than willing to give further advice when this is required.

# 1 The World's Sheep Trade

The distribution of sheep throughout the world is very unequal. Some countries have few or none while there is a dense sheep population in others. The reasons for these localised concentrations are geographical, historical, and commercial.

There are certain geographical factors that exclude sheep husbandry – others that favour it. Thus there are few sheep in either equatorial or sub-arctic regions. There are many sheep on islands and in coastal regions. There are many sheep on the lands that fringe the continental deserts.

Nichols[1] gave the following limiting conditions for sheep husbandry: mean temperature, 28–77°F(–2–25°C); rainfall, 8–115mm per month; relative humidity, between 55 and 70 per cent at the higher limit of temperature, and between 65 and 91 per cent at the lower temperature. There is a close association between a 50–100mm annual rainfall and the density of sheep population.

Historical and commercial factors are of no less importance so that in different parts of the world sheep differ widely both in type and commercial purpose.

To take Europe first – there are three main areas of sheep concentration: (1) Britain; (2) Spain; and (3) South-East Europe.

(1) The position in Britain is of special interest. Here is a country densely populated and highly industrialised, where the standard of living is relatively high and agricultural production intensive and yet sheep remain an important part of animal husbandry.

Although statistics of sheep population fluctuate from year to year and can never be very accurate for any one year, it can be stated in round figures that, whereas Britain contains some 25 million sheep, France in contrast holds only 11 million with the total in the European Economic Community (EEC) 12 being around 92.0 million head.

In order to understand the greater density of sheep population in Britain, it is necessary to go a little way back into history.

Before the discovery and colonisation of new continents, Europe had to clothe herself and, owing to the coldness of the climate and the absence of

1

cotton and synthetic fibre as competitors, the main source of clothing was wool. Sheep were kept primarily for wool. Other domestic animals could be exploited for food, the cow and goat for dairy products, cattle and swine for meat. The only animal that could provide clothing without the necessity of first killing it was the easily plucked or shorn sheep. Therefore, before the development of the wool-producing industry of the southern hemisphere, the demand for wool in Europe was so great that the countries highly populated by sheep gained fortunes by them. As far as wool production was concerned, England and Spain were in medieval days the Australasia of their times. The export of raw wool from England to Europe was once of first importance to the country's wealth, preceding the development of a trade in manufactured woollens by some centuries. Lord Ernle writes:[2]

Wool was the chief source of the wealth of traders and of the revenues of the Crown. It controlled the foreign policy of England, supplied the sinews of our wars, built and adorned our churches and private houses'.

That the Lord Chancellor sits on a woolsack is no meaningless ceremonial, but a symbol that the wealth and majesty of England once rested, like the Lord Chancellor, on a solid and elastic foundation of sheeps' wool.*

England, from being an exporter of raw wool, became a manufacturer of her own wool products. The value and extent of manufacture outrunning both the relative importance and the resources of the national sheep industry, England, from being a noted exporter of wool, in the end became the world's greatest wool importer.

The increase of manufacture, the Industrial Revolution, the growth and urban concentration of population, led to a further important change in the country's sheep. A people whose traditional meat was roast beef, who were rapidly increasing in number and with expanding purchasing power, cut off by the sea and without modern means of meat storage by refrigeration, turned to the country's flocks as a further source of an expanding meat consumption. From the old shortwool and longwool flocks were rapidly and successfully evolved the English breeds of mutton sheep, the first breeds of sheep in the world to be especially designed for the provision of mutton, and which, for that purpose, retain their pre-eminence throughout the world today. Had it not been for the diversion of the main sheep enterprise of England from wool to mutton during the late eighteenth and early

*Up to the reign of Edward III of England the Woolsack was actually filled with English wool. From then, until 1940, it contained horsehair. In that year the woolsack was renovated, and, at the suggestion of Sir Ian Clunies Ross of Australia, was stuffed with wool from the British Dominions.

nineteenth centuries, it is probable that, subsequent to the development of wool production in the southern hemisphere, the sheep population of Britain, like that of other industrial European countries, would have suffered a rapid decline.

(2) Leaving the British Isles for the moment, let us turn to another country with a great sheep history – Spain. Spain was the first certain known home of the Merino sheep, which more than any other breed had been bred for the special, indeed almost exclusive purpose of fine wool production. Spain and England, were, in fact, the two great sources of raw wool in Europe in medieval and in later times; indeed, right up to the nineteenth century. English and Spanish wools, like their respective countries of origin, were long in keen and bitter rivalry. Since the nature of that rivalry is little understood it is well to recognise that in its short and long wool England had two separate and distinct products. The long wool never came into serious competition with Spanish wool. It was English short wool that suffered from the competition of Spanish wool, also short, and used for similar purposes in manufacture, yet definitely finer and more delicate in its fibre. The distinction was well stated by Lord Ernle:[3]

'In England, long wool was employed mainly for worsted fabrics, but also to give strength and firmness to cloth. Abroad, it was eagerly bought in its raw state for both purposes. In long wool, or combing wool, England had practically a monopoly of the markets, and to it the export trade of raw material was almost exclusively confined. Short wool, on the other hand, was used for broad-cloth. In its raw state it had a formidable rival abroad in the fleeces of the Spanish Merino. Only in the manufactured state did it compete with Flemish and French fabrics on the continent, and often found itself unable, owing to the excellence of Merino wool and the skill of foreign weavers, to maintain its hold on the home market.'

For centuries Spain maintained a virtual monopoly of production of Merino wool, although from the end of the eighteenth century onwards, Merino sheep spread to other European countries, including England, and then to the wool-producing countries of the southern hemisphere – South America, South Africa, Australia, New Zealand – and to the United States. Just as England may be claimed to be the original home, the fountain head, of mutton breeds of sheep, so Spain was the source of finewool breeds, now spread to the farthest corners of the habitable globe. Unlike England, however, Spain has not retained its pre-eminence as the home of finewoolled Merino sheep. That pre-eminence went to Australia, which in 1928 passed legislation to control the efflux of its finest sheep to rival wool-producing countries. Yet Spain remains a country with areas of dense sheep concentration, although the majority of the sheep are not Merinos, and the Merino sheep persisting there have been outshone by their descendants in newer lands.

(3) In South-East Europe – Italy with 8.4 million sheep and the Balkan countries, notably Greece, with 9 million – the density of sheep population is due to rather different causes. The sheep are derived neither from English mutton nor from Spanish finewool breeds. They are sheep of a more primitive type, associated with peasants and subsistence peasant farming. They are triple-purpose sheep producing wool, milk, and mutton, and their importance in hilly districts of the more economically primitive European countries may be considerable. Hollecek-Holleschowitz,[4] writing in 1936, says that the general regression of European sheep-breeding in the later nineteenth century, due to the importation of cheaply produced wool from the southern hemisphere, did not affect the stability of the triple-purpose, Zackel, sheep population of South-East Europe. That it was so is understandable, since the peasant's sheep in South-East Europe has traditionally been maintained for family subsistence rather than commerce. This is changing as flocks of triple-purpose breeds are replaced by specialised dairy-type breeds and as transhumance systems, dependent on the utilisation of natural grasslands, give way to modified systems of increased intensity. There are some 40 million sheep in East Germany, Poland, Hungary, Rumania and Bulgaria, 60 per cent being in the latter two countries. Radical changes have been effected in these countries with large, more specialised flocks becoming more common.

Stretching across the continents of the southern hemisphere, between 20° and 60° of south latitude, there are four great and important areas of sheep concentration. These are: (1) Argentina and Uruguay in South America, (2) South Africa, (3) Australia, and (4) New Zealand. In all four areas the sheep sheep industry has had much the same general economic history: the introduction of wool-producing Merino sheep at various periods between the middle of the eighteenth and the middle of the nineteenth centuries, the development of an export trade in wool, tallow and hides, the switch-over towards the more profitable export trade in mutton and lamb made possible by Tellier's invention of refrigeration, put to practical purpose on a large commercial scale about the year 1882. These newer centres of the world's sheep population, favoured by cheap, abundant land and favourable climatic conditions, now contain about half the sheep population of the world. Their development has had two important effects on the sheep population of Europe. First, the abundant importation of cheaply produced wool of the first quality checked the expansion of Merino sheep over Europe, resulting in a steady decline of wool production in industrialised European countries. Secondly, the export of frozen lamb of first quality, directed almost exclusively at the meat markets of Britain, resulted in some profound changes in the economy of the sheep industry of that country in order to meet the new competition. Thus, although the sheep population of Britain remained relatively stable, those areas which, by reason of natural infertility and high altitude, were only suited to the production of mature wether mutton suffered a substantial decrease in sheep stocks.

The development of the sheep industry in the southern hemisphere merits further discussion, not only because of its intrinsic interest, but because of the information it affords as to factors that have affected the world trade in sheep products.

(1) Argentina and Uruguay. Sheep are not native to the South American continent and appear to have been introduced by the Spanish during the colonial days in the seventeenth century. Later, towards the end of the eighteenth and the early nineteenth centuries, Merinos were introduced from Spain, France and Germany and interbred with local types. At a later stage the Australian Merino was also imported and it became extremely popular, but it has given way to the Corriedale, which now outnumbers all other breeds.

At one time these counties were significant sources of mutton and lamb to the British market, but it is as leading producers and exporters of wool where their sheep industries are now important.

(2) South Africa. The original and native sheep of South Africa were the leggy, hair-bearing, fat-rumped and fat-tailed types of Eastern sheep indigenous to hot dry climates and the primitive husbandry systems of Africa and Asia. Importations of European sheep began as early as 1654, but the seeds of the modern sheep industry of South Africa were sown in 1789 when the King of Spain gave a flock of Merinos to the Prince of Orange. Some of these animals were then sent to the Cape of Good Hope, which was then owned by the Dutch.

The sheep industry of South Africa has grown greatly during the present century. Between 1904 and 1955 the population of woolled sheep increased from 11 million to over 30 million, and numbers remained at a similar level in 1979. The majority of the sheep are Merinos and 80 per cent of the wool produced is derived from this breed. Wool production is one of the country's biggest agricultural industries, with an output of 108 million kg in 1978/79 of which around 90 per cent or more is exported, with an increasing amount being processed – scoured, before sale. The EEC countries are now the major buyers.

(3) Australia. For many years, Australia was the world's greatest sheep country but according to figures published by the International Wool Secretariat[5] numbers declined from 170 million in 1965 to 135 million in 1978, so that it is now the second largest after the Soviet Union, with 140 million sheep.

It is the largest producer (709 million kg in 1978/79) and exporter (596 million kg) of fine wools. Almost one-half of the continent's sheep are concentrated in its south-eastern corner, in New South Wales.

Taking Australia as a whole it may be said that, despite refrigeration and energetic efforts to build up an export trade in frozen meat, mutton and lamb have remained secondary to wool as a source of the country's wealth.

(4) New Zealand is the fourth area of the southern hemisphere in which a close concentration of sheep occurs and its sheep history is both more recent and also rather different from that of Australia.

The first sheep were imported from New South Wales in 1834. They were Merinos, and up to the introduction of refrigeration in 1882, this was the predominant breed of the country, the only important export product being Merino wool. With the coming of refrigeration, the New Zealand sheep industry switched very quickly over from wool to lamb production as the main form of sheep enterprise. This change was made possible by the use and further importation of English mutton breeds to cross with the Merino.

The climate and pasture of New Zealand are more favourable to sheep breeding and fat-lamb production than are the far more arid and infertile grazings of Australia or of South Africa. The early concentration on production of fine wool was due entirely to the ease of transporting it overseas. When, by refrigeration, mutton and lamb were able to survive a sea passage the New Zealand sheep industry developed the production of mutton and lamb, particularly lamb, for export, a trade in which the dominion has, so far, found no serious rival. It is noteworthy that New Zealand, in developing her fat-lamb industry, did not sacrifice her wool exports. Whereas in 1978 all sheep meat accounted for some 26.2 per cent of the value of her total exports, the corresponding figure for her wool clip was 18.1 per cent. The wool nowadays, however, is crossbred wool, only some 3 per cent being Merino. The country is the third largest of the world's wool producers (320 million kg), with over 90 per cent being exported, more than half of which is scoured. The Romney, introduced from England in the late 1850s, along with its crosses is the most important breed, these producing a fairly coarse and relatively strong crossbred class of wool, much of which is used in carpet manufacture.

In addition to the areas of greatest sheep concentration already discussed, three other countries, chiefly because of their large size, contain many sheep, thus ranking high among the sheep-producing countries of the world. They are: (1) the United States, (2) India and (3) the Soviet Union.

(1) The United States. Although the early settlers in America brought sheep with them from their respective countries, and although there have been sheep there for more than 300 years, improved breeds were not introduced until the nineteenth century. Following the Embargo Act of 1807 which prevented the entry of foreign wool, Merino sheep were imported. A sheep boom between 1808 and 1816 followed. Further developments were the gradual retreat of sheep westwards, drawn there by the cheap grazing of undeveloped lands and the increase of population and more intensive farming in the eastern states. Sheep have ever been frontier stock. American sheep of the western states still contain much Merino blood, even though, within more recent times, wool in the whole American sheep industry has become a by-product of lamb. The reason for this retention of Merino blood is the close flocking instinct of the Merino which makes herding on open ranges much more practicable, and because the Merino is the type of sheep most suitable to semi-arid climates. In the eastern states, however, sheep play a minor part in the economy of the cultivation of the mixed farm, and English mutton blood prevails.

The sheep population of the United States has shown a substantial decrease from 51 million in 1939 to 33 million in 1960 and 12.8 million in 1976, a change which appears likely to be permanent. According to figures published by the Meat and Livestock Commission,[6] mutton and lamb production dropped by 13 per cent in 1978 compared with the previous year and a further substantial decline is forecast. Mutton and lamb have never been popular among American meat consumers, who have always preferred beef and pork, and the woollen textile industry of America, like that of Britain, has become dependent for its raw material on wool imported from other countries.

(2) India and Pakistan, together forming a vast continent of agricultural village communities, might be expected to contain a concentration of sheep similar to that of the peasant communities of South-East Europe. Hot, dry climates, however, while acceptable to the wool-bearing Merino and often proving suitable to wool production on an expensive scale, do not favour the production of mutton and milk. Regions where heat and heavy rainfall are combined, as in Burma and Assam, are notoriously unsuitable to sheep of

any kind. In any event, sheep are not of first importance in Indian animal husbandry. There is only 1 sheep to every 17 bovines (cattle and buffaloes). The standard of sheep husbandry is low and wool produced is predominantly of carpet class. India and Pakistan are exporters of wool, although not of mutton or lamb.

(3) The Soviet Union. Before the Bolshevik Revolution, the Soviet Union supported a large sheep population, clearly divisible into two separate classes. Triple-purpose, coarse-woolled sheep were held by the peasantry. Flocks of Merino-type sheep were the property of the large landowners. Following the Revolution, the general policy adopted was to upgrade the coarse-woolled sheep by crossing with the confiscated Merinos, aiming to bring the country's wool up to higher textile standards. Large importations of fine-woolled sheep, mainly from Australia, and the adoption of artificial insemination on a very wide scale, were designed to hasten this development. Because of the exceptionally severe winter climate in so much of the Soviet Union, both wool and sheep skins are of particular importance in providing clothing.

During the Second World War and early postwar years both sheep population and wool production suffered a severe recession. That loss has not been regained. The Soviet Union now has the world's largest flock of sheep, 141 million. With its huge sheep population and production of 462 million kg wool, second only to Australia, it still imports 127 million kg of wool, to make it one of the leading importers. In recent times, in addition to wool, mutton production is also being encouraged.

Having briefly considered the outline of production in the main sheep-rearing countries, it is now possible to discuss the world sheep trade as a whole. It is divisible into two main parts: (1) The wool trade, (2) The trade in mutton and lamb.

(1) Wool is extremely variable and diverse in type and further complexity is introduced by the common practice of blending or mixing wools during manufacture. A single piece of cloth may contain wool from several continents.

Wool may be broadly divided into three main classes: (a) Merino, (b) crossbred, and (c) carpet wool.

It should be noted that the term 'crossbred' used in this connection means that the wool so designated is of medium fineness, coarser than Merino wool. It does not imply that the sheep from which such wool is shorn are crossbred. Indeed, many pedigree sheep, including those of most English mutton breeds, would fall within this category.

(a) Merino wool is produced mainly in four countries: Australia, South Africa, Argentina and the Soviet Union. Together they produce 90 per cent of the world's Merino wool, Australia making much the largest contribution.
(b) Crossbred wool. The four main producers of 'crossbred' wools are New Zealand, Argentina, Australia and the Soviet union.
(c) Carpet wools are produced mainly in Eastern countries, particularly the Soviet Union, China, India and the Balkan states. Of the virgin wool which is exported, about 86 per cent grows on sheep in the southern hemisphere and is shipped to the textile industries in the northern hemisphere.

Australia is the world's greatest wool-producing country, and the contribution of other countries to the world's wool trade is shown in Table 1.1.

**Table 1.1** World wool production and exports (1978/79)

(million kg)

| Country | Production | Export |
|---|---|---|
| Australia | 709 | 596 |
| Soviet Union | 462 | — |
| New Zealand | 320 | 262 |
| Argentina | 173 | 90 |
| South Africa | 108 | 65 |
| Uruguay | 63 | 33 |
| United Kingdom | 49 | 19 |
| United States | 46 | — |
| World | 2567 | 1234 |

After Watson[7]

The diversity of wool production and the fundamental differences in value and utilisation are such that a country may produce certain classes of wool unused in, or surplus to, the requirements of its own textile industry. An example is the carpet wool of the Scottish Blackface mountain sheep, which finds its main market not in English factories but in those of Italy and the United States. There is consequently a great coming and going of wool cargoes all over the world.

All of the industrialised countries are importers of raw wool. The industrialisation of Japan and more recently Korea has led to a rapid and considerable increase in wool imports into these countries. Indeed Japan, where the textile industry began modestly in 1879, is now the biggest buyer of wool, a position long held by Britain.

In 1978 the comparative imports in millions of kg were: Japan, 202; Britain, 149; the Soviet Union, 127; and France, 114.[8] However, of the major industrial nations, it is West Germany which has the highest use of wool per person.

Most of the wool produced in Australia, New Zealand and South Africa is sold at public auction by wool brokers, who sell won behalf of the farmer. Sale is by sample and certificate. A sample is extracted from the bale and details of the wool fibre thickness (in microns) and the yield after scouring are recorded on a certificate, which is attached to the bale.

The continuance of a world trade in wool, such as exists today, depends upon the success of the wool producer and woollen manufacturer in meeting the competition of other fibres used in the manufacture of human clothing.

Throughout the twentieth century the demand for clothing has been increasing and this has been greater than can be supplied by the production of wool and cotton. Man-made fibres have been developed to meet this increasing demand but many of these are derived from oil-based products which have increased in price.

To promote the use of wool and to carry out research and development on the use of wool, the International Wool Secretariat (IWS), based in London, and now with offices in 30 countries, was set up in 1937. The celebrated woolmark was introduced by the IWS in 1964 as a symbol of quality, which assures the buyer that the product is manufactured from virgin wool. A set of quality control systems associated with the woolmark ensures that the product meets certain technical requirements, such as resistance to rubbing and snagging, fabric strength, and colours remaining fast to light and liquid.

The woolblend mark introduced in 1971 indicates blended fabrics with a content of more than 55 per cent wool. This mark is recognised by millions of people throughout the world.

(2) The world's trade in *mutton and lamb* is more limited than that of wool. Total sheepmeat production has been estimated at 6.8 billion tonnes and in 1985 world exports of sheepmeat were around 750,000 tonnes.[10] Although the EEC of 12 countries has only 6 per cent of the world sheep stocks it produces 11.0 per cent of the total sheepmeat and contributes to one-third of world trade.

Between them Australia and New Zealand account for 80 per cent of world trade exports and Britain, France and Japan represent 40 per cent of all sheepmeat imports. Indeed, the flow of lamb from New Zealand to Britain in itself contributes to 20 per cent of world trade but it is thought that this flow may gradually decrease.[9] Of significance is the developing trade to the Arab oil-producing countries, while some Latin American countries, such as Peru and some in South-East Asia, Malaysia, Singapore and Hong Kong are increasing imports.

In the sheepmeat trade there are two commodities – lamb and mutton. The major supplier of the first is New Zealand, and Britain and to some extent France are the main demand countries. Traditionally, consumption of lamb in Britain has been high, but is presently declining, according to figures published by the European Communities Commission,[10] having been 11.3 kg per capita in 1960 and dropping to 7.0 kg in 1984. In most areas of the world consumption of sheepmeat is low and fairly stable at 1–3 kg per head per annum but, as is shown in Table 1.2, much higher levels of consumption occur in some countries.

Table 1.2. Consumption of sheepmeat

|  | Per Head Consumption kg |
| --- | --- |
| New Zealand | 80.0 |
| United Kingdom | 7.3 |
| Republic of Ireland | 10.5 |
| France | 3.7 |
| Greece | 13.4 |

The predominant supply country for mutton is Australia with Japan as the major importer.

The EEC only produces on average 60 per cent of its consumption and is therefore dependent on imports to a large extent. Britain is the chief Community importer, mainly frozen lamb from New Zealand, this having dropped by 37 per cent over the past decade. Consumption levels have been rising in France and West Germany. Nonetheless, the future prospects of the world's trade in lamb depends to a large extent on a continued and high consumption by the urban population of Britain. Should that consumption substantially decrease, the prosperity of the New Zealand industry could be seriously impaired.

## References

(1)  Nichols, J. E. (1932) *A Study of Empire Wool Production* (Leeds).
(2)  Ernle (Lord) (1961) *English Farming, Past and Present*, Heinemann.
(3)  Ernle, *English Farming*.
(4)  Hollecek-Holleschowitz, C. (1936) 'The Zackel sheep of the Sylvanian

Carpathians, Ruthenia', *Z. Schafz.* **25**, 194 and 221; *A.B.A.* **4**, 423.
(5) International Wool Secretariat (1978) *Wool Facts*, (September).
(6) Meat and Livestock Commission (1979) *International Market Survey*. Economic Information Service. Meat and Livestock Commission (Bletchley).
(7) Watson, T. (1979) *International Wool Secretariat* (London).
(8) Watson, *International Wool Secretariat*.
(9) Wilson, P. N., and Brigstocke, T. (1978/79) 'The sheep market: meat and wool as it affects UK farmers', *Farm Management* **3**, no. 10, 465.
(10) European Communities Commission (1971) *Sheepmeat*: CAP Newsletter Division for Agricultural Information; European Communities Commission (Brussels).

# 2 Breeds and Breeding

The origins of the domestication of the sheep, which belong to the genus *Ovis*, is lost in the mists of time. It is thought that sheep may have been domesticated about 10000 B.C., following the dog. It is intriguing to consider the extent to which the dog contributed to the process of domestication, as is also the possible extent to which early man in his search of food would follow the wild sheep on their seasonal migration from the winter grazing areas in the valleys to the high mountain pastures. Wild herbivores, including feral sheep, still show the same behaviour. As sources of meat, skins and wool, sheep would be much valued. Indeed, the transhumance system, utilising two main pastoral zones, is known to have existed from prehistoric times and it remains today in many parts of the world as described by Demururen[1]. The largest concentration of migratory systems is in Asia but it is also found in Africa, south-east Europe and in North America.

From an examination of the remains of sheep from the Quartenary deposits and comparison of these with skeletons of existing primitive wild types and with improved breeds, Ewart Cossar[2] concluded that all three old-world kinds of wild sheep – the Argali (*Ovis aminon*), Urial (*Ovignec*) and Mouflon (*O. musimom*) – contributed to forming our domestic breeds. The British breeds have probably all sprung from two wild Asiatic types, the Urial and Mouflon, whereas the Asiatic groups of modern sheep probably stem from the Argali.

One wild type, the Bighorn, which is found in the Rocky Mountains of the United States, has not been domesticated and there is no evidence that it has ever been used in the development of a domesticated breed.

The Soay is possibly the best example of a primitive type in Britain and its double coat of long hairy fibres with an undercoat of short fine wool is probably typical of the original types.

The breeds, crosses and varieties of sheep are untold in numbers but the historical development and relationships of the important types have been investigated in detail by Michael Ryder[3] whose publication is a massive work of scholarship.

Owen[4] has listed the world's main groups of sheep as shown in Table 2.1

13

and Ryder[5] has illustrated the relationship between some of the more important breeds (Fig. 2.1).

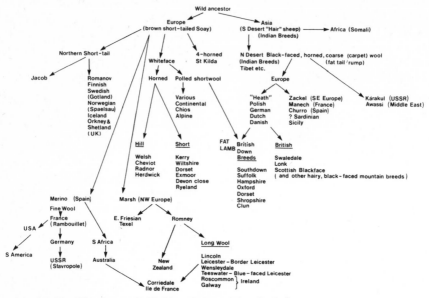

*Fig. 2.1* Classification and affinities of sheep (after Ryder[3]).

Over the centuries, breeding and selection have given rise to breed types significantly different in their utilitarian value. At the one extreme, the Merino with its very dense mass of fine wool fibres has lost all vestiges of the primitive types, while these are still apparent in many of the hardier types adapted to mountain environments. Sheep have been developed as multi-purpose animals, some primarily for milk for processing into cheese and yoghurt and other breeds for superior carcass attributes.

**Table 2.1** List of main world sheep groups

| | |
|---|---|
| (1) Groups of worldwide distribution and influence | Examples of constituent breeds |
| Merino | Australian Merino, Caucasian Merino, German mutton Merino, Precoce, Rambouillet, Russian Merino, Sopravissana. |
| Longwool | Bluefaced Leicester, Border Leicester, Devon Longwood, English Leicester, Galway, Ile de France, Corriedale (Merino cross) Lincoln, Roscommon, Teeswater, Wensleydale. |
| Down | Dorset Down, Hampshire, Oxford Down, South Down, Suffolk |
| Romney | Lowicz, New Zealand Romney, Romney (Kent) |

Table 2.1 (contd.)

| | |
|---|---|
| (2) Regional groups of historical and widespread importance | |
| Northern Short-tailed (N. Europe) | Finnsheep, North Ronaldsay, Romanov, Shetland (includes primitive multihorned, e.g. Soay, Jacob) |
| Marsh (N.W. Europe) | East Friesian, Oldenburg, Texel |
| Heath (N.W. Europe) | Danish Landrace, Dutch Heath, German Heath, Polish Heath |
| Zackel (S.E. Europe) | Albanian Zackel, Bulgarian Zackel, Greek Zackel, Racka, Sumava, Turcana, Valachian, Voloshian |
| Tsigai (S.E. Europe) (Merino derivative) | Azou Tsigai, Romanian Tsigai, Ruda, Russian Tsigai |
| Karakul (Soviet Union) | Arabi, German Karakul, Malich |
| Fat-rumped (Soviet Union) | Chuntuk, Kalmyk, Karanogai, Kazakh |
| Fat-tailed (Asia, Europe, Africa) | African long fat-tail, barbary fat-tail, Caucasian fat-tailed, Han-yang, Hu-yang, Mongolian, Tan-vang, Tung-yang |
| Indian (India and Pakistan) | Bikaneri, Chanothar, Gujarati, Hassan, South India Hairy, Thal, Waziri |

| | |
|---|---|
| (3) Local groups of varying importance | |
| Scottish Blackfaced hill (UK) | Lonk, Scottish Blackface, Swaledale |
| British Whitefaced hill (UK) | Cheviot, Herdwick, Llyn, Radnor, Welsh Mountain |
| Awassi (Middle East) | Awassi, Israeli Improved, N'emi, Shafali |
| Somali (Africa) | Adali, Blackhead Persian, Toposa |
| Churro (Spain & Portugal) | Braganca, Castilian, Galician, Mancha, Segura |
| Sardinian (Italy) | Large Lowland Sardinian, Medium Hill Sardinian, Small Mountain Sardinian |
| Irish (Eire) | Galway, Roscommon |
| French (France) | Berrichon, Bluefaced Maine, Causses, Contentin, Lacaune, Prealpes du Sod |
| Tibetan (Asian) | Northern Tibetan, Southern Tibetan, Steppe Tibetan |
| Lop-eared Alpine (Alps) | Bergamo, Biella, French Alpine, Swiss White Alpine, Tyrol Mountain |
| Chios (Greece & Turkey) | Greek Chios, Turkish Chios |
| Dahman (South Central Morocco) | Dahman |
| British Whiteface shortwools (S.W. Britain) | Devon Closewool, Dorset Horn, Exmoor Horn, Kerry Hill, Ryeland, Wiltshire Horn |

To clarify discussion it may help to skip whole volumes of history and enquire as to how breeds have originated in more recent times. Of the five best-known sheep breeds in Scotland, the origins of three, the Border Leicester, the Suffolk and the Oxford, are known. The Border Leicester is almost certainly an inbred crossbred between the English Leicester and the Cheviot. It is known and admitted that the Suffolk is an inbred crossbred between the Southdown and Norfolk Horn. It is also known and admitted that the Oxford is an inbred crossbred between the Hampshire and the Cotswold. That, indeed, appears to be the most rapid and successful method of establishing a new sheep breed. No doubt during the first few generations there is greater variation and a necessity for heavy culling, but the increased variation gives more scope for selection. In any event it is quite clear from a critical consideration of breed histories that practically every so-called 'pure' breed of sheep existing today is crossbred in its recent or remote origin. Thus all the English longwools were crossed with the Improved Leicester while all the English shortwools were crossed with the Southdown. The modern 'Peppin' – a type of Australian Merino was, according to some Australian authorities, formed by the introduction of English longwool blood into the original Merino, the admitted origin of both the Corriedale and Polwarth. Is there, then, such a thing as a purebred sheep in the world, or rather, to put it another way, have the essential attributes of a pure breed been properly defined?

These are not the accepted views of most breeding associations and societies. Breed societies, unless the facts are very clearly against them, like to make the following assumptions about the breed they sponsor: (1) An ancient origin; (2) No admixture with other breeds at any stage of its history; and (3) Progressive improvement by careful selection and mating of individuals within the breed.

In certain instances, facts are stretched rather far to fit these theories. To suggest, for example, that the Border Leicester has arisen by selection within the English Leicester breed and without any admixture with the Cheviot is to assault the intelligence of anyone who knows anything of Scottish sheep. There is a certain false sanctity about ancient lineage and uncontaminated breeding. It is the essence of so many sagas and epics that it might well be regarded as the earliest form of propaganda of a ruling class. Mankind, and particularly English mankind, is inclined to transfer the prejudices of an originally Norman aristocracy to legends of the breeding of English sheep.

The fact is that there is no reason to suppose that either ancient origin or avoidance of crossing, *in themselves*, confer any commercial advantage upon a breed. It must also remain doubtful, in view of genetical critique, whether selection within the breed by customary methods can, in fact, lead to its progressive advance. Of course, it is very easy to waste time and space in attacking the pretensions of breed societies. They are in no sense dispassionate advocates of the productive improvement of the world's sheep. They have a certain breed to sell and customers to convince, and no salesman can be expected to take an unbiased view of his own products. The purpose of breed points is also frequently misunderstood. Obviously it makes no earthly difference to the mutton or wool a sheep produces whether or not a ram's horns have one coil or two, or whether its top-knot of wool stops between its ears or extends to its eyebrows, but these points, quite unrelated to production, may be valuable trademarks when it comes to a sale.

There are really too many breeds, either to describe or to discuss their characteristics. However, some merit mention because of their numerical importance or also because of their influence in the development of other breeds.

## Merino sheep

The word 'Merino' is Spanish, meaning originally an inspector of sheep walks, but it came to be applied to a particularly valuable breed of sheep. Between the years 1500 and 1700 the breeding of a fine-woolled sheep, probably based on North African types introduced by the Moors, was developed. Spain established an international wool trade and maintained a jealously guarded monopoly by placing an embargo on the export of the breed. Eventually this broke down due to the generosity of Spanish kings to their royal neighbours in France, Austria and England. It was thereafter only a matter of time until the breed had spread across the seas to other continents, arriving in Australia in 1797.

The original Merinos were small in size with an average greasy fleece weight of 2.5 kg, a length of 5–7.5 cm, with a yield of about 40 per cent. The Merino breed never established a dominant position in Britain and indeed failed in competition with native breeds. The full and detailed history of the Merino has been described by Carter.[6] This failure is usually interpreted as due to the British climate, soil or husbandry having proved unsuited to Merino sheep or its wool. It proved unsuited to the hill areas in Scotland and this may have been due to the deficiencies of a fine fleece in wet conditions and a lack of hardiness leading to a high mortality.

The very characters which rendered the breed unsuited to the Scottish Highlands were destined to prove, on the contrary, of supreme value in the development of the sheep industries in the semi-arid areas of the New World. (Plate 2.1.)

In the eighteenth century wool, particularly fine wool, was one of the most valuable agricultural products in proportion to its weight. Furthermore, it was easy to store, classify for sale and transport and in distant lands did not compete with meat in the era before refrigeration.

It followed that the Merino was the first important sheep colonist of all new countries. America, both North and South, Australia, New Zealand and South Africa all found it an early source of prosperity and foreign exchange and the breed still remains supreme in Australia and South Africa.

*Plate 2.1* Spanish Merino ewes. (Photograph courtesy of Spanish Embassy)

The history of the Australian Merino is complicated with breeders having used types from many countries as explained by Cox.[7] Australia has bred Merinos for over 180 years with a history of significant achievement. Whereas a purebred Spanish Merino ewe in 1800 might cut on average 1.8 kg greasy and around 1.4 kg. scoured, today the best studs would average 6–8 kg.

It is not only in quantity that striking progress has been achieved. What of the quality? Cox,[8] in his *Evaluation of the Australian Merino* writes:

'Density, a virtue derived from the Saxony blood, has been retained. In length of staple great progress has been made, the length of the medium and fine-woolled types has been doubled. With wool of the 64s to 70s type, 7.5–12 centimetres is easily obtained today in the best Australian flocks. As for colour, the brightness and colour obtained is beyond comparison with the wool of a hundred years ago. Elasticity, always one of the attributes of the Saxony Merino wool, has been retained.'[8]

The supreme excellence of the best Australian wool for the textile purposes of today is unquestioned. Medium-fine wool of good combing length is in better demand, except for special purposes, than the shorter and definitely finer wools the Saxon breeders produced. There is no doubt that the Saxon wools of their time were finer, approaching 100s count. By comparison, wool of 64s and 70s wool is only of medium fineness. That is not to say that Australian breeders could not (for indeed when they wished to do so they did) have produced wool equally as fine as the finest Saxony of course. They

*Plate 2.2* Australian Merino ewes. (Photograph courtesy of Australian News and Information Bureau)

found, however, that wool in greater quantity and of medium fineness paid better. The only argument is how that change has been produced. (Plate 2.2.)

Taking Australian wools as a whole, one finds that over a century there has been a great increase in fleece weight when compared with the original Merino, and that there has been a noteworthy increase in staple length and a perceptible loss of fineness. To what are these changes due? It is customary to read of their being credited to the genius of the Australian breeder and to the suitability of the Australian climate and pasture for fine-wool production, the assumption being that there has been no admixture with other breeds of sheep, during the long history of Australian Merino breeding. Is it not more probable – or as has been suggested when discussing Peppin sheep – that there was at some time or other, perhaps on several occasions, a cross with English longwool breeds? Such a cross would at once have raised fleece weight and staple length with some necessary diminution of fineness of fibre, leaving only fleece density to be fostered and preserved by subsequent and careful selection. Whatever the methods used by stud-breeders in Australia their success is amply proved by the demand for their stock.

Although the first sheep in North America were actually British breeds, the early history of sheep-raising is identified with the Merino which diverged into characteristically different types such as the Vermont. This strain which had excessively wrinkled skin was exported to Australia where this trait was eventually more or less eliminated due to the practical difficulties in shearing such sheep.

The Rambouillet or French Merino can be traced to Spanish Merinos imported into France but there is considerable doubt about its subsequent history which has been traced in considerable detail by Allan Fraser in earlier editions of this book. The principal stronghold of the Rambouillet is in the United States, where it is regarded as a dual-purpose breed and is frequently crossed with rams of the mutton breeds. The breed is also found in France, Germany, the Soviet Union and South America.

The Merino was the first sheep breed to be established in the great pastoral industry of New Zealand. The introduction of refrigeration completely altered the predominant breeds, with British breeds, notably the Romney and Southdown, coming into their own. It is only in the high country in the South Island where the breed is to be found today.

One of New Zealand's own contributions to the world's sheep breeds is the Corriedale. In essence this breed is an inbred halfbred, between the Merino and English longwools, mainly Leicester and Lincoln.

It was an old custom of Merino breeders, both in Australia and New Zealand, to cross the older ewes with longwool rams and to sell the halfbred progeny to farmers in the more fertile areas.

Another derivative of the Merino and a longwool type is the Polworth. This breed is the product of a Merino × Lincoln bred back to the Merino and this three-quarter Merino was fixed as a type. The breed has a heavy dense fleece of 60/64s quality.

In the United States the tendency has been to breed the Merino away from its original type to produce several so-called 'Compromise Breeds'. The derivation of two: Columbia – Lincoln ram × Rambouillet ewe, and Targee – Rambouillet × Lincoln Corriedale, are illustrative of the role of the Merino in the development of new breeds in another continent.

In general the Merino is not particularly milky nor is it a prolific breed but the emergence of a new strain in Australia – the Booroola – is considerably more prolific.

Most of the lamb produced in Australia is from first-cross ewes, which traditionally are Merino Border Leicester. This cross has a high reproductive rate but has a defined breeding season. New breeds to replace the Border Leicester are being developed as well as self-replacing breeds of high fecundity and year-round joining ability.

**English longwools**

In his *Communications to the Board of Agriculture*, Dr Parry writes:[9]

'The long wool of this island is its peculiar treasure, indispensable to its manufacturers, unrivalled, probably, by that of other European countries and best grown on rich and deep land.'

If short, fine wool was once upon a time the pride of Spain, long coarse wool was the basis of England's industrial prosperity. Indeed, long wool was, in medieval times, an English monopoly and her most valuable export. What of the sheep that produced that wool? They were bred and reared in the rich, flat lands of England – in Leicestershire, Lincolnshire, on the rolling uplands of the Cotswold hills, on the bleak Romney marshes swept by the sea-spray. The origin and history of England's longwool sheep are lost in the mists of her pastoral history. How did men who knew nothing of genetical science, of nutritional science, of veterinary science; with pastures unimproved, roots non-existent and cake undreamt of, produce breeds that have stamped an English seal on the faces of sheep of three continents, so that the fame of their counties of origin, of Leicester, of Lincoln, of Romney, have been carried to distant corners of the farthest islands of the Antipodes? These are questions of the deepest interest to the biologist no less than to the farmer. Unsolved though they are and unsolved though they may ever be, yet surely they should make humble those who in their arrogant modernity would seek to scorn the splendid triumphs of their fathers' breeding.

There are, however, certain facts that are known or can be reasonably deduced about the early English longwool sheep. They must have been bred for wool almost exclusively, as the early Spanish Merino sheep were bred for wool exclusively, and they must have been extremely slow in reaching full maturity. These facts are certain, since two of Bakewell's acknowledged aims in longwool improvement were better mutton and mutton maturing at an earlier age. To judge from their descendants, they must have been heavy and cumbrous sheep. They were certainly profitable, for the most beautiful villages in England – those in the Cotswold hills – were built on the large profits of long wool. They must have existed for many centuries in the isolation of their native counties, for unlike Spain, England had no great flocks of migratory sheep. Perhaps more through that isolation than by any conscious selection within the breed, sheep became adapted to the soil they grazed on, until it might be said with justice that they became imbued with the very essence of that soil itself.

Definite history of longwools dates from the time of Robert Bakewell who improved the Leicester long wool breed, producing thereby the Improved

English Leicester – and because of Bakewell's Improved English Leicester modified all the other longwool breeds of England and therefore of the world. I shall deal first with that old and famous breed, although other longwools, notably the Lincoln and the Romney Marsh, in the end achieved a greater world importance.

## The Leicester

The breed owes its origin to one man, Robert Bakewell (1725–95) whose book and achievements are chronicled in Pawson's[10] book on Robert Bakewell.

Prior to Bakewell's time the Leicester appears to have been a long-legged, lanky type of animal with few of the properties looked for by the butcher but kept almost entirely for its wool. The ideal sheep Bakewell[11] proposed to produce by improvement was 'possessed of the most perfect symmetry, with the greatest aptitude to fatten, and rather smaller in size than the sheep generally bred'.

The methods used to attain this aim are open to argument in that some consider he introduced new types as a consequence of his travels, and once having established the necessary range of variation, he may well have inbred. He did, however, avoid one profound error into which many breeders of much more modern times have fallen. He did not profess to judge the breeding potentials of sires on external appearance alone.

Following the scriptural injunction 'By their fruits ye shall know them', he assessed the breeding value of the sires he used by the most sensible and reliable method – the observed quality of the lambs they left. It was almost certainly for this purpose that he introduced the practice of letting rather than of selling his rams. By so doing he kept control of a much wider range of potentially valuable sires. Should one of his let rams leave especially good lambs he could recall it and use it as a sire in his own flock at Dishley Grange.

Many modern breeders do differently. They presume, by consideration of pedigree and a ram's external appearance alone, to be able to select him from his fellows as a future flock sire, selling his apparently less desirable brethren to others. Perhaps because of some early and striking example in his sheep-breeding experience Bakewell realised how rare of occurrence, how impossible of detection apart from breeding performance, is a genuinely outstanding sire. By his novel system of letting, Bakewell ensured that when and if the ideal sire to his purpose arose it should be his in the future to have and to hold.

Having detected and retained the ideal sire for his purpose, Bakewell made full use of its exceptional breeding powers. He employed, and may have invented, the expedient of teasers – aproned rams of small value which, running with the flock, detected the ewes that were in heat, or to use the quaint old shepherds' word were 'blythesome' – so that ewes in their proper season could be brought to service by selected sires.

The new breed secured a widespread influence and vast success being used to improve practically every other breed of English longwool including the Lincoln and Romney Marsh. The modern breed of sheep is large and produces a fleece of fine lustrous wool weighing from 7 to 9 kg.

## Border Leicester

Undoubtedly derived from the Leicester in the Scottish border area, this breed became predominant as the sire of first-cross lambs from hill ewes, particularly Cheviot, Blackface, Welsh and other types. The Cheviot cross, the Scottish Halfbred, formed for many years the traditional breeding flock on the majority of lowland Scottish farms.

This breed is to be found in many countries throughout the world but other breeds, including modern 'synthetics', are presenting it with a challenge to the extent that it is declining in importance. (Plate 2.3.)

## Bluefaced Leicester

This is a breed of relatively recent origin appearing in the Hexham area in the North of England within the last four decades. Probably developed from the

Plate 2.3 Border Leicester ram. (Photograph from *Farmer & Stock-Breeder*)

crossing of Border Leicester and Wensleydale (See Hall and Baxter[12]) it has had a remarkable expansion for use as a crossing sire with the Swaledale to produce mules, and with the Blackface, the Greyface or the Mule.

The breed is rather bare-skinned, that is, it has a short rather open fleece, and it has never been particularly renowned for its hardiness. (See Plate 2.4.)

*Plate 2.4* Lincoln ewes. (Photograph © Charles Reid)

### Kent or Romney Marsh

The unique conditions of the marsh area in Kent with its cold and harsh winter climate yet luxuriant and lush summer pasture are unique.

Bred to use pasture rather than arable products, the breed has found favour in numerous countries and New Zealand in particular has made the breed her own. In that country the breed differs considerably from the English strain, being better fleeced, lighter boned and low-set and active. (Plate 2.5)

From the New Zealand Romney two new breeds have emerged, the Perendale and the Coopworth. The first is based on crossing with the Cheviot and it inherits the activity of the latter with the fleece quality and conformation of the former. The second, the Coopworth, is a Leicester-cross Romney.

The historical importance and influence of the English longwool was immense. The need for breeds of this type has steadily declined but they will maintain a role as sire-producing breeds for Merino crossing or in providing rams for crossing with hill breeds in Britain only as long as the crosses have attributes required by the commercial sheep producers of finished lamb. If

*Plate 2.5* New Zealand Romney ram. (Photograph via High Commissioner for New Zealand)

new types emerge which produce crossbred sheep more suited to requirements then these will be favoured and fancy points will be of little value.


## English shortwools

While England for long held a virtual monopoly of longwool production, the wools of her more numerous shortwoolled breeds were in competition with the produce of the Spanish Merino. Some of the English shortwools, grown on upland pastures, were fine indeed but there is no evidence that they ever equalled Merino wool in fineness of fibre, in density, or in softness of handling. Indeed, far from succeeding in displacing Spanish shortwools from European markets, English short wool had difficulty in maintaining its position against Merino wool in the market at home. Nevertheless, in their earlier history, the importance of English shortwool breeds lay in wool rather than in mutton. If a medieval shepherd could shear 1 kg of fine wool off a shortwool sheep he thought such production fit occasion for a feast. As the Clown remarked in *The Winter's Tale* before the rogue Autolycus robbed him:

'Let me see – every 'leven wether tods', by which he implied that 11 wether sheep yielded a tod or 13 kg bringing out the average fleece weight of a wether at about 1.2 kg of wool.

What Bakewell did for the longwool John Ellman (1753–1832) of Glynde of Lewes in Sussex was to do for the shortwools. With an expanding industrial meat market to cater for, Ellerman used the shortwool types of uplands and downs to lay the foundations of the Southdown breeds. Very little is known about Ellerman's breeding methods but it is said that in paying close attention to carcass quality, which included maximum development of the cuts preferred by consumers and minimum waste, he took counsel of butchers in fixing the ideal towards which he aimed. What was accomplished was a profound change in the proportions and symmetry, from the old Heath breed of Sussex to the modern Southdown.

One of the striking features of the breed is its small size and its use has declined. However, the Southdown was to be incorporated in all the numerous Down breeds that have ever arisen, including the Suffolk, Shropshire, Oxford, Hampshire and Dorset.

## Down breeds

### Suffolk

The breed is an inbred crossbred from the old Norfolk breed and the Southdown. It is pre-eminently a meat breed and its main function is as a terminal sire for crossing with other breeds and crosses. Its adaptability in producing lean carcasses at a range of weights and age, combined with good growth rate of its progeny, has brought it to the position of being the most widely used terminal sire in Britain.

The breed is very characteristic in appearance with its jet black face and legs, head free of wool and a short fine fleece. (Plate 2.6.)

### Dorset Down

This breed is smaller than the Suffolk but has found a niche in Britain as a terminal sire for the production of early maturing milk lamb from grassland flocks. It has also been used successfully for crossing with hill breeds. (Plate 2.7.)

### Oxford Down

This is the largest sheep of all Down breeds and it was very popular when heavy lamb, carcass weight of 50 kg or more, was acceptable either as lamb or hogget. Its popularity has declined.

*Plate 2.6* Suffolk ram. (Photograph from *Farmer & Stock-Breeder*)

## *Future of Down breeds*

Originally these breeds formed the commercial farm flocks of their native areas where the folding method was an essential part of the farm management system.

As terminal sires it is inevitable that the Down breeds will be compared with each other and with imported breeds, as is now taking place in Britain and other countries.

*Plate 2.7* Dorset Horn shearling ewes. (Photograph from *Farmer & Stock-Breeder*)

Breed comparisons are difficult in that types of breeds may differ significantly between countries while the ewe breeds on which terminal sires are used are fundamentally different. Furthermore, systems used and type and quality of carcasses needed to meet consumer requirements may vary between and within countries. In Britain, the traditional demand has been for lightweight lamb in London and the South, whereas a carcass some 4–5 kg heavier is acceptable in the North of England and Scotland. Recently, new cutting techniques and methods of presentation have established a demand for heavy (21–25 kg), but lean, carcasses.

One of the breeds introduced into Britain is the Texel which has been shown to have a higher lean content in the carcass.

Commercial breeders select sires depending on the ewe type and their market objectives but the main attributes of sires of slaughter lambs should be growth rate, lean meat content of lamb carcasses and their conformation. Fertility should not, however, be overlooked, since the first task of the ram is to get ewes in-lamb.

Some of the facts which may influence the choice of a terminal sire are listed in Table 2.2.

**Table 2.2** Terminal sire breeds

| Breed | Adult weight (kg) | Growth rate of cross-lambs (g/d) | | Carcass characteristics | |
| --- | --- | --- | --- | --- | --- |
| | | 0–12 wks | 0–slaughter | % Lean | Lean:Bone ratio |
| Oxford | 107 | 294 | 248 | 55 | 3.4 |
| Suffolk | 91 | 284 | 238 | 55 | 3.6 |
| Dorset Down | 77 | 277 | 218 | 53 | 3.4 |
| Texel | 89 | 260 | 218 | 59 | 3.8 |
| Oldenburg | | 279 | 218 | 56 | 3.6 |
| Ile de France | | 274 | 216 | 54 | 3.4 |

There is another breed which merits mention, namely the Dorset Horn and its polled derivative. The character of this breed which distinguishes it from others in Britain is its extended breeding season. It has good fecundity producing up to 180 per cent of lambs and within recent years the conformation of the breed has been improved.

## Hill Breeds

In many countries there is a progression both of breed and of production level from the least to the more fertile areas. If sheep-rearing is to be successful these two factors, breeds and fertility, have to be matched and adaptability to climate factors should also be coupled with these.

The capacity for rapid growth and early maturity in a breed cannot be exploited unless there is food of an adequate amount and quality at all seasons.

While in certain countries – Australia and South Africa for example – the level of sheep production is regulated largely by rainfall; in Britain the overriding factor is mainly altitude or type of pasture. In general, the productive capacity of the land decreases as the valleys give place to the hills and these in turn to the mountains.

As there are breeds suited to such lowland pastures so also there are breeds adapted to the harsh conditions of climate and nutrition in the hills.

The hill breeds of Britain may be divided into two groups: (1) White-faced breeds with crossbred wool; and (2) Dark-faced breeds with carpet wool.

The first group contains the Welsh Mountain, the Cheviot breeds, the Herdwick and the Exmoor; the second group the Scottish Blackface, the Swaledale, Dalesbred, Rough Fell and Lonk.

## Welsh Mountain

This, the hill breed of Wales, is Britain's smallest sheep. It is an extremely active animal and the ewes appear to milk well. It is not unusual to see Welsh ewes being suckled by cross-lambs of a relatively young age already grown to a size greater than their dam. There are several types of Welsh Mountain sheep: hill, improved, Senni Bridge. (Plate 2.8.)

*Plate 2.8* Welsh mountain ewes. (Photograph from *Farmer & Stock-Breeder*)

*Cheviot*

Another white-faced hill breed which is native to the hills of that name. Of larger size than the Welsh Mountain, the Cheviot is probably rather less hardy, thriving best on green hills of lower elevation and in districts of more moderate rainfall. At one time the Cheviot was a horned breed and showed a good deal of tan on face and legs. Now, horns do not occur in ewes at all and only occasionally in rams and the face and legs should be white. There is no more beautiful sheep than the Cheviot as found on its native hills. It has a carriage, alertness, and keenness of expression unsurpassed by any other breed.

The Cheviot spread northwards over Scotland until it reached the shores of the Pentland Firth. Then a recrudescence of the Blackface cleft the Cheviot territory in twain, so that the Cheviot became concentrated on the hills of Scotland's extremities – the Cheviot hills in the south, the Sutherland hills in the far north. In these two disconnected areas the Cheviot has developed into two varieties – indeed, into two separate breeds – the South Country Cheviot and the North Country Cheviot.

*Plate 2.9* Cheviot shearling ram. (Photograph from *Farmer & Stock-Breeder*)

The Cheviot or, as it is sometimes less correctly called, the South Country Cheviot to distinguish it from its northern cousin, is the smaller breed. Its main territory today is in the Scottish Border country, especially on the hills which gave the breed its name. The two main sale centres are Hawick to the east and Lockerbie to the west and there is a tendency towards a further breed subdivision in that the Lockerbie type, whilst rather less stylish, is considered by many to be a more genuine hill sheep, hardier and less under the influence of fashionable breeders. The relative merits of the two types is a matter of opinion but the tendency to a separation into two types is a definite fact. (Plates 2.9 and 2.10.)

Both surplus ewe lambs and cast ewes have a special value because of the popularity of the Scottish Halfbred ewe derived by crossing the Border Leicester ram with Cheviot ewes of either South Country or North Country type. The wether lambs, following a prolonged spell of disfavour, are now once again in growing demand. The South Country Cheviot wether lamb is small yet late-maturing and will keep on growing for some time before showing much sign of fattening. Probably it reaches its best mutton quality when 9–10 months old when, in the early spring it should kill out at 11–22 kg deadweight, still relatively small in the joint and not over-fat. Under modern market conditions it seems, therefore, to be particularly well-suited for winter feeding.

*Plate 2.10* North Country Cheviot ram. (Photograph lent by North Country Cheviot Sheep Society)

## North Country Cheviot

Cheviot sheep came first to the northernmost counties of Scotland, Caithness and Sutherland, in the year 1792 when Sir John Sinclair of Ulbster brought 500 ewes from the Cheviot hills to his farm of Langwell in Caithness. It was Sir John Sinclair who gave the name 'Cheviot' to the breed previously known as 'long hill sheep' in contrast to the Scottish Blackfaces of that time generally known as the 'short hill sheep' or 'Linton'. Since 1792 until today Cheviot sheep have continued to be the only important sheep breed in Caithness and Sutherland. With the reduction in Cheviot territory in favour of the Scottish Blackface, the Northern Cheviot flocks became isolated from those on the Borders and a divergence first into two distinct types and now into two recognised breeds has been the result.

Just as in the case of its South Country cousin this breed has developed into two types. The sheep bred in Sutherland, with a sale centre at Lairg, are more truly hill sheep than the type which has developed in Caithness which is larger and has spread throughout Scotland mainly in the upland areas. The Cheviot breeds crossed with the Border Leicester produce the famous Scottish Halfbred.

### Scottish Blackface

This is the most numerous and widely distributed hill breed although other breeds, notably the Swaledale, have been gaining ground in recent years. Indeed, some strains of Blackface have a considerable amount of Swaledale blood and other types in them. It is a horned, dark-faced breed with a carpet wool fleece. There are distinct types within the breed and it is wool type, amongst other characteristics, which distinguishes them. Wool varies from relatively short and fine, the 'bare-skinned' type from Galloway, to the longer-woolled 14–16 cm, of the Perth type. The breed has the capability to produce satisfactory slaughter lambs off the dam or alternatively store-lambs can be kept and finished on aftermath, forage or root crops to give carcasses of 14–17 kg. The breed is extremely prolific with up to a 175 per cent lamb crop being attainable, with 140 per cent being more common, under good conditions, but on most hills the nutritional environment will limit production to around 80–90 per cent lambs weaned. (Plate 2.11.)

On the better hills ewes are frequently crossed, more recently with the Bluefaced as well as the Border Leicester, to produce the Greyface.

Comparison of breeds needs to be carefully interpreted and should take into account the environmental and other conditions under which these are made. Under favourable upland conditions for true hill breeds Weiner and Hayter[13] compared Scottish Blackface, Cheviot and Welsh ewes. The Scottish Blackface gave the highest lifetime output, producing more lambs per ewe with the heaviest weaning weights. However the Welsh, which had a mature weight some 30 per cent less, did not lag behind on overall output when this was related to maternal weight.

All the crosses between these breeds showed an advantage but this varied considerably; for example, litter size showed no advantage, in the Scottish

*Plate 2.11* Scottish Blackface shearling ram – Newton Stewart type. (Photograph lent by Blackface Sheep Breeders' Association)

Blackface – Welsh Mountain cross, but a 10 per cent improvement in the Scottish Blackface-cross Cheviot. Turner[14] suggests possible differences in specific combining ability or nicking, a feature which has been suggested may occur in a wide range of crossbreds.

Under hill conditions in Wales, Purser[15] undertook a comparison of the Scottish Blackface and the Welsh Mountain and the crossbred of these breeds.

This work clearly illustrates that it is not only production for the individual ewe which matters but that other criteria are also important. In this instance, stocking rate of the Scottish Blackface was around 20 per cent

**Table 2.3**

Lambing performance, growth, carcass characteristics and production of two hill breeds and crossbred

|  | Welsh Mountain | Scottish Blackface | Crossbreds |
|---|---|---|---|
| Ewe liveweight (kg) | 35.3 | 47.4 | 42.2 |
| Ewe fleece weight (kg) | 1.20 | 1.45 | 1.36 |
| Lambs weaned/100 ewes mated | 111 | 123 | 122 |
| Weaning weight/lamb (kg) | 20.1 | 26.0 | 23.4 |
| Age at slaughter (days) | 222 | 200 | 193 |
| Carcass weight (kg) | 12.3 | 14.3 | 13.3 |
| Killing-out percentage | 45.2 | 42.0 | 43.6 |
| Production/100 kg ewe lamb carcass | 28.3 | 28.0 | 29.1 |
| Production/h lamb carcass | 19.8 | 21.2 | 21.5 |

lower and so output per unit area becomes relevant in assessing the production levels and economics of a hill farm. Other important aspects include labour as related to ewe numbers, headage payments and market outlets.

In 1954 a breeding experiment based on selection for long and short cannon bone length in Scottish Blackfaces was set up by Purser.[16] The initial aim was to establish the relationship between this trait and conformation, growth and carcass quality. By 1979 the difference in average cannon bone length between the short and long lines was 25 per cent. At the same liveweight, short-cannon-bone lambs were fatter. But one of the most striking differences that have emerged between the two flocks is in the prolificacy of the ewes and the survival of their lambs. The flocks were originally compared on a hard hill farm but cast-for-age ewes were transferred to more favourable conditions leading to an even greater disparity in performance (see table 2.4), with an increase of 25 per cent in meat production per ewe from long cannon bone types.

During the period of this experiment the Blackface breed, by whatever means, has changed significantly, becoming much longer in the leg even though bone thickness is still important.

It is of interest to speculate that many of the breeds with long thin bones (Finish Landrace is an extreme) are highly fertile, whereas breeds with short thick bone structures tend to have a poorer reproductive rate.

---

**Table 2.4**

Performance of Scottish Blackface sheep with different cannon bone lengths

Cannon length Lamb production

| | Lambs born (%) | | Lamb losses | | Weaning weight (kg) | | Carcass composition (30kg) | |
|--------|-----|-----|-----|-----|------|------|---------|----------|
| | U | H | U | H | U | H | Fat (%) | Lean (%) |
| Short | 92 | 95 | 10 | 16 | 35.4 | 22.1 | 28 | 55 |
| Medium | 103 | 97 | 5 | 12 | 36.1 | 23.0 | 25 | 58 |
| Long | 119 | 103 | 2 | 11 | 37.1 | 23.2 | 24 | 59 |

U = Upland    H = Hill

After Purser[17]

---

## Swaledale

Although related to the Scottish Blackface the breed is larger in the body, stands higher on its legs and is lighter on the forequarters and narrow in the back. From the point of view of mutton conformation it is not such a good sheep. The ewes are good milkers and excellent foragers. The rapid expansion of the Swaledale from a breed of purely local interest in Yorkshire and Durham to its present position as the predominant hill breed in the North of England and in many districts of the Scottish Highlands is one of the most interesting features of recent hill sheep-farming history. The breed has always had a reputation for producing excellent crossbred lambs, traditionally with the Teeswater, to produce the Masham and more recently the mule mainly from the Bluefaced Leicester. (Plate 2.12.)

Those breeds described are numerically the main hill breeds but there are others including the Dalesbred, Rough Fell, Derbyshire Gritstone, the Lonk, Exmoor Horn, Radnor, Herdwick, and Shetland all of which have a niche and possess valuable qualities.

*Plate 2.12* Swaledale ram. (Photograph from *Farmer & Stock-Breeder*)

## Grassland sheep

In every sheep country there is a tendency for what has been termed 'stratification' of the industry to develop. Less fertile areas – infertile because of altitude or aridity – support the more independent breeds and types of sheep, managed on an extensive scale with but little shepherding. These extensively managed flocks, be they on range or hill, are a source of breeding stock for more fertile areas. Very often the first cross of hill or range sheep is used to exploit more cultivated grassland. This is so in Britain, where the principal source of grassland sheep lies in the surplus ewe lambs and cast ewes from hill flocks, or in the first crosses by mutton rams from ewes of hill breeding. These recruits from the hills have proved the most satisfactory type of sheep to keep on temporary leys or permanent lowland pastures. One essential reason for this lies in the latent milking capacity of ewes of hill breeding, rather better perhaps than that of breeds bred more specifically for mutton under the folding systems of arable farms.

There are, however, certain breeds of sheep with a good deal of the hill breeds in their ancestry which give promise of proving eminently satisfactory where a self-contained flock suitable for grassland is for any reason preferred to a flying flock. Two such breeds deserving of special mention are the Clun Forest and Kerry Hill.

## New Breeds

The history of sheep-breeding is such that the value of and demand for some breeds decline while others have a period of ascendancy. Most breeds have traits which are valuable and the preservation of breeds against the day of changing needs may well be a justifiable and sensible policy.

Those breeds enjoying present popularity will find competition from breeds imported to bring desirable genes, and only stringent health regulations limit this trade as in countries such as New Zealand and Iceland. New technology, such as embryo storage and transfer techniques, may well be a means of overcoming these disease problems.

Meanwhile, we have seen the introduction into Britain of the Texel, Charolais, East Friesland, Blue de Maine, Oldenburg, Ile de France and Vendeen, to list but a few, with no doubt others to follow.

Parallel to this we have had the development of new breeds. The Colbred, which could be described as an ovine cocktail of Border Leicester, Clun, Dorset Horn and East Friesian, was developed by Oscar Colborn with the aim of producing a breed with high prolificacy and exceptional milk yield.

The Cadzow 'Improver' was another type designed to occupy the same role as the Colbred that is in direct competition with the longwool-type sires. A recent type is the Cambridge developed with a high reproduction rate for crossing. To be successful new breeds for crossing on hill ewes will need to produce a more uniform, a more prolific and a more milky crossbred ewe with good mothering abilities, yet with adequate conformation and growth potential to produce a first-rate carcass from a cross with a terminal sire. Neither of these breeds nor the more recent Animal Breeding Research Organization (ABRO) Damline have been so markedly superior that their future is assured.

The traditional meat type or terminal sire breeds are also under assault. The Texel, with its propensity for high lean content in the carcass, may find a role in the stratified British system. Indeed, the attraction of several of the breeds now being imported is their ability to produce lean carcasses with good development of the valuable cuts.

At Worlaby in Lincolnshire, Henry Fell[18] has developed the Meatlinc breed. In combining the Suffolk, Dorset Down, Ile de France and Berrichan du Cher he has combined what he suggests, 'sounds like a glorious and totally illogical mess of potage'. If the objective of producing a breed with good growth rate, heavily fleshed, particularly in the hind leg, and without excess fat, with the ability to provide finished lamb at a wide range of slaughter weights and equally with the capacity to stamp this on its crossbred progeny is achieved, then the market-place will be the final arbiter since this is what it is looking for.

The British sheep industry owes much to the Bakewells and Ellermans of the past and today's pioneers may also find themselves in the hall of fame. There are too many breeds to describe in detail in a text of this nature. In the previous edition of this book (1968) many breeds were discussed in detail and reference should also be made to *British Sheep*[19] and *Sheep of the World*[20] for those seeking a description of modern breeds.

## Milch sheep

It seems as though sheep were originally triple-purpose animals, producing wool, mutton and milk, and although today the vast majority are kept for their wool or mutton, or for both, in earlier times, mutton, nowadays the most valuable sheep product, was the least important of the three articles for which sheep were bred.

Wool, certainly, was of more value and in all probability so was milk. The sheep, like the cow and the goat, in more primitive communities is considered a dairy animal. It was so in the old days in the Scottish Highlands. Thus in 1726, Burt[21] wrote:

'Children share the milk with the calves, lambs, and kids; for they milk the dams of them all, which keeps the young so lean that when sold in the Low-country they are chiefly used, as they tell me, to make soups withall; and when a side of any one of these kinds hangs up in our market the least disagreeable part of the sight is the transparency of the ribs.'

In Britain the sheep has long since ceased to be of any importance as a dairy animal. Recently there has been a resurgence of interest in milking sheep for the production of processed products, mainly cheese. Indeed, a new breed the British Milksheep, developed from British and European breeds, has a high yield capability with the best individuals producing up to 700 litres in a full lactation.

It is in the countries of southern Europe and those around the Mediterranean that ewe milk is important, primarily for manufacture into cheese of various types.

There are numerous breeds or types, many of relatively local distribution. In those breeds developed specifically for dairy qualities average yields of 600–700 litres are obtained. Breeds within this category include the East Friesian, Awassi, Chios, Laucaune, Sardinian and Sicilian breeds. (Plate 2.13.)

*Plate 2.13* East Fresian milking ewe. (Photograph lent by The Netherlands Embassy)

The milk breeds can have a considerably extended lactation with the capability of milking for up to 180 to well over 200 days. In many peasant communities however, yields are low and lactation is little more than 3–3.5 months. In some cases lambs may be allowed to suckle for several weeks whereas in others they may be removed at a young age.

*Fur-bearing sheep*

In many breeds of sheep the birth-coat of the lamb differs from that of the adult sheep in both type and quality of fleece.

In the Karakul breed this difference occurs to a marked degree with the birth-coat of the lamb being composed of tight curls of hair usually black in colour with an exquisite gloss, while the adult fleece is brown, coarse, wiry and of carpet class. The Karakul lamb skin is the raw material for the trade in Persian and Astrakan furs.

Another breed valued for the skin, but more so from more mature animals, is the Gottland breed developed in Sweden and found throughout Scandinavia. The wool is beautifully curled, is glossy and in the ideal types has a uniform grey appearance which may vary in shade.

*Eastern sheep breeds*

Throughout Asia and North Africa there are numerous sheep breeds specially adapted to dry or desert climates. Generally their conformation as mutton sheep is of the poorest and the quality of the meat unacceptable to European palates. The wool is usually coarse and hairy and in some types it is mainly hair. Many breeds have the ability to store fat and the desert types do so in their tail or rump. One of the best-known is the Blackhead Persian which has spread to several countries. It crosses well with British meat breeds and the cross with the Dorset has been fixed as a breed – the Dorper.

**Pure-breeding and breed improvement**

A premium may be obtained when selling sheep for breeding. But what is it that the purchaser is looking for? It can be, of course, a hundred and one things depending on the breed, its characteristic points and the particular interests of the breeders.

There is tremendous importance attached to the head in judging many types of sheep. A common observation made is, 'Why all this fuss about a sheep's head? Nobody eats sheep's heads nowadays.' There is an answer to this query. When a sheepman is buying a pen of ewe lambs, for example, he isn't thinking about the mutton they'll make, but about the lambs they will leave. The question is how will the sheep breed? The head compared to the body, being less affected by environmental influences such as nutrition, is held to be a safer guide to the animal's breeding. The head again is an indication of health. Sheep dull in the eye, drooping in the ear, puffy over the cheeks are seldom healthy or thriving. On the contrary, the keen, bright eye, the pricked ear, and the clean hair and fine mobile skin over the cheek bones are all signs of vitality and health.

So, also it is with the fleece or 'coat'. Unless the fleece is of the right type for the breed this will be a fault. In the Border Leicester, for example, a loose fluffy coat is a fault as is also excessive curl or 'pirl' at the staple tips. The general character of a fleece may be indicative of the type of breeding both of purebreds and crossbreds.

As with the head, the fleece also gives an indication of health, or the reverse; natural bloom on a sheep's coat is a sign of thrift. A thriving sheep will have a gloss in its coat which is apparent down to the skin when the fleece is parted.

Many breeders also give careful consideration to the question of bone. The bones of the legs should be strong, straight, smooth and flat. Since bone is inedible the question is again frequently put as to why strong bone is an advantage in an animal whose products are mutton and wool. It has been argued that bone is a mineral capital which is drawn upon while an animal is milking. This, however, is a controversial issue and the current trend for extremely strong-boned sheep in some strains of the Suffolk breed in Britain as opposed to the original finer-bred type is an issue for argument amongst breeders with scientists on the sidelines having little objective information to offer. It may not be without significance that the Texel, which has a high lean content in the carcass, is not particularly heavy-boned.

After head, fleece and bone comes action and movement or the lack of it. In a flock of sheep this can convey much valuable information as to health, previous treatment and general condition.

Many breeders also attach considerable importance to overall conformation, be it length, width and depth of body, spring of rib, length of neck and carriage of head, development of hind-quarter or some other aspect of relevance to the breed.

In addition, the traditional pedigree breeder attaches importance to continuity in breeding to type. Pre-potency is the outcome of long pedigree and ancient blood and if a ram's parents have been bred for generations with particular objectives in view there will be a greater certainty that when he becomes a sire he will reproduce these qualities. What the flockmaster requires is that the ram must possess certain qualities such as conformation, fleece, mutton or breed characters which he desires and that the ram will be pre-potent and pass on these desirable qualities to his progeny.

The question at issue is, whether it is pedigree as such and the uniformity that may be achieved in this way, with its emphasis on the so-called fancy breed points, or the more rigorous use of scientific principles that chart the way ahead. The scientist would argue that selection should concentrate much more on those traits or characters which are of economic importance, such as growth rates, milk yield, carcass quality, reproductive rate, ewe survival, wool yield and quality, or relevance to the particular breed.

The breeding of sheep, as of all other classes of livestock, stands at the crossroads where genetical science and the art of the practical breeder meet.

## Improvement of sheep

Flockmasters have to decide about the best way by which improvement may be achieved. In doing so they must take into account the system in which the sheep will perform. In many instances it may be that it will be management factors such as nutrition that are the limiting factors so that the intrinsic merit of the breed is not expressed. This occurs with hill sheep, for example. The Scottish Blackface may produce a lamb crop of 80–90 per cent on the hill yet the cast ewes moved to a better environment will later produce 140 per cent or more. Improved breeding may only be evident if the environmental factors allow this to be expressed and so breeding and management go hand in hand, states Russel.[22]

In any flock of sheep there will be some variation in any one trait and this may differ from the variation found in the breed generally or a random sample from it. This variation will be due to differences in breeding or genotype as well as differences due to environmental factors such as feeding, age and management generally and is usually described as 'phenotypic variation'. This latter is often thought erroneously to refer to the appearance of an animal. In the case of some characters, for example face colour, phenotype will have a large genetic component, but many performance traits such as growth rate and milk yield are influenced by both breeding and management.

In breed improvement what is really important is the extent to which the variations which occur are due to genetic differences that are inherited. The term 'heritability' is used to describe that part of the phenotype variation which is due to genetic differences.

Many of the characters which give breeds their distinctive appearances such as face and leg colour, presence or absence of horns are highly inherited and in some cases are influenced by a few gene pairs.

However, several of the traits of interest to the commercial sheep-breeder, in many instances have a relatively low to medium heritability such as reproductive traits and liveweight gain. Others, however, including most of the wool traits, are highly heritable explaining why selection for fleece characteristics can be effective. Some examples taken from Rae[23] who has comprehensively tabulated heritability estimates are shown in Table 2.5.

Practical breeders are well aware that selection for one trait may well have an influence on others and so geneticists have sought to establish correlative effects.

**Table 2.5** Heritability estimates

| Trait | Range |
|---|---|
| *Reproductive traits* | |
| Number of lamb born | |
| 2 years old | 0–0.20 |
| 3 years old | 0–0.40 |
| Weight of lamb weaned | 0–0.20 |
| *Liveweights* | |
| Weaning weight | 0.10–0.30 |
| Rate of gain | |
| Birth to weaning | 0.15–0.40 |
| Post weaning | 0.20–0.50 |
| *Carcass traits* | |
| Slaughtered at constant weight | |
| Lean percentage | 0.25–0.40 |
| Depth of fat (12 ribs) | 0.30–0.50 |
| *Fleece traits* | |
| Greasy fleece weight | 0.30–0.40 |
| Average fibre diameter | 0.30–0.50 |
| *Milk yield* | |
| First lactation yield | 0.15–0.50 |

In an early selection experiment for twinning rate in the New Zealand Romney, fleece production declined, although it is now known that simultaneous improvements in both traits can occur (see Clark[24]).

Those traits which are influenced by environment have led geneticists to look for means of indirect measurement which may be less influenced by such factors and are therefore better indicators of genetic merit. For example, testes size in the ram may also indicate genetic merit for female performance. Whatever the selection method used, a fundamental requirement is an efficient recording system with emphasis on objective measurement rather than subjective wherever that is possible. Croston *et al.*[25] have described how many countries have established national sheep-recording schemes to assist in the evaluation and selection of breeding animals. This report emphasises the importance of testing and selecting under field conditions. Inevitably this leads to the need for flexibility in evaluation procedures to meet the varied selection objectives appropriate to a range and diversity of production systems. This is not to say that there are not circumstances when the reduction of environmental effects may not be of considerable benefit. Early weaning combined with artifical rearing may be an effective method of selecting potential sires for growth rate but the technique is not without its limitations as Owen et al.[26] have shown.

The development of selection plans is a task for the specialists. Clear objectives have to be established and methods used which allow extensive and complex data to be simplified. Adjustments are frequently needed for factors such as sex, age of dam, type of birth, age at measurement of a trait such as growth rate of lambs. Account has to be taken of many other aspects of genetic theory and those who wish to delve further should consult the work of Helen Newton Turner,[27,28] Rae[29] and Robertson[30] as well as the contemporary scientific literature.

## Group breeding schemes

As the science of population genetics has developed during the twentieth century means have been needed to apply this new knowledge in methods for sheep improvement. The concept of the Group Breeding Scheme (GBS), first proposed by Professor A. L. Rae of Massey University in New Zealand, was quickly seized upon by sheep-breeders in that country and the first scheme was established in 1967 by a group of Romney breeders. There are now some 30 GBSs in New Zealand, involving around some 30 farmers contributing about 20 per cent of the rams in the national flock. Breeding schemes have been developed in Australia, South Africa and more recently in Britain. The Australian Merino Society operates what is probably the largest GBS in the world. The scale of operation is gigantic and this now involves 700 participants with a nucleus flock of 14000 ewes. This flock receives the best 1 per cent of ewes from 46 Ram Breeding Cooperatives (RBCs). The RBCs with a total of 100000 ewes in turn receive the best 1 per cent from 2 million two-tooth ewes screened annually for performance (growth rate and wool yield), states Lucas.[31]

The principles of GBSs are essentially fairly simple in that they involve the collaboration of a number of breeders, who screen their flocks to identify superior females. These sheep are submitted to a nucleus flock where they are further tested in a common environment. The nucleus flock also selects superior males and females and the degree of recruitment from within this flock, or from the base flocks, is a matter for discussion. Rapid generation turnover, which may be less acceptable in a commercial flock, may be justified to enhance genetic improvement in a nucleus flock. The nucleus flock is the main ram-breeding flock for all members, so breeding objectives have to be clearly stipulated and agreed.

Many systems incorporate some form of ram progeny-testing especially for recruits to the nucleus flock. Schemes vary widely in their size and goals and the latter will depend to an extent on the breed and the aims of the breeders. Geneticists should be involved in the formulation of plans since they can assist in developing a suitable selection index which will incorporate those traits that are measurable and can weigh them relative to their economic importance, heritability and other factors.

The extent to which structural factors such as teeth, jaws, feet, legs, udders, etc. are used as a basis of elimination or as a basis of selection, in addition to an index, is a matter for agreement. Appraisal of this type will reduce the selection pressure that would be achieved by use of an index alone.

The breeding advantage of a GBS is that the rate of improvement will be some 10–15 per cent greater than in an individual flock and inbreeding will be reduced. Both of these advantages can be attributed to the numerical scale of the operation. The development of GBS methods and their structural features have been discussed by Rae[32] and Jackson and Turner.[33]

The first scheme to be developed in Britain was one formed by a group of Welsh Mountain sheep-breeders. This scheme, popularly known as CAMRA, serves to illustrate some of the aspects of such schemes. There are 10 members with around 6000 ewes which are screened by the Meat and Livestock Commission (MLC) individual ewe index over a 3-year period. Members contribute 5 ewes annually to the nucleus flock.

The selection index developed by CAMRA for the nucleus flock incorporates 5 factors:

(1) The adjusted lamb 12-week weight.
(2) Performance of sires in terms of the mean value of his progeny group.
(3) The dam's performance as a mother.
(4) The dam's body weight at 18 months. This is taken prior to mating and is used to avoid using subsequent weights which may be affected by barrenness or twinning.
(5) The dam's litter size. In practice this is given little weight in the index, being for a hill breed.

All of this information is combined into one number to indicate overall merit and with 100 as average. Rams are selected from the top 25 per cent on index, this being reduced to 20 on the basis of appraisal in terms of conformation, structural features and breed type.

A further aspect is that ram lambs are performance-tested for body weight at 18 months, with the best being retained for the nucleus flock for 1–2 years, thereafter returning to the base flocks while other rams are available at various stages.

The aim is that all replacement rams used in member flocks should be products of the scheme GBS (see Barker et al.[34]).

Group Breeding Schemes have been developed for other breeds in Britain, including the Cambridge, Romney Marsh, Lleyn and Speckleface.

Since GBSs require agreement amongst people, usually with different views and attitudes to breeding, experience shows the need to formalise arrangements, usually with a legal basis, to ensure smoother running and continuity. At the end of the day, the sheep produced must meet the needs of commercial markets and must also be superior to stock bred by other methods.

## Crossbreeding

The practice of crossbreeding is probably more prevalent in sheep than in any other domestic animal. It is usually the result of a planned policy based on a variety of sensible considerations.

In two authoritive articles, Rae[35] has comprehensively discussed the practice of crossbreeding and its role. He divides the subject into two clearly separated sections:

(1) Crossbreeding in the formation of breeds and for grading up to a superior breed.
(2) Crossbreeding to utilise the benefits of heterosis and to produce sheep suited to a range of environmental conditions in the production of lamb and mutton.

The discussion that follows here is concerned only with the second of these aspects.

Crossbred sheep frequently have the important asset of increased vigour relative to their parents and this has been technically termed 'heterosis' or 'hybrid vigour'. This effect can be seen in the better performance of the individual, for example in growth rate and in the survival of crossbred lambs. Sheridan[36] points out that maternal heterosis, which occurs when crossbred dams are used, leads to improved fertility with fewer barren ewes and better prolificacy (larger litters) so that the benefit in reproductive performance in particular can be of economic importance. Rae[37] reminds us that it is sometimes thought that the greater the differences in type between breeds so then will heterosis be increased, but there is very little evidence to support this view.

Heterosis may not always occur. Terrill[38], in discussing crossbreeding in the United States, observed that under extremely good conditions of feeding and husbandry where the purebreds are well adapted to the conditions little or no heterosis may be found.

Mitter[39] also points out that commercial crossbreeding has been of particular benefit in countries with a wide range of environmental conditions which requires different types of sheep.

The very definite advantages of the first-generation hybrid are utilised in the successive crosses between different breeds, which is the basis of the economic structure called 'stratification' governing the production programme of several sheep countries. Stratification and the reasons for it are very clearly illustrated by the system of sheep-breeding prevailing in Scotland, although it is common in Britain (See Read[40])

In Britain, the foundation stocks and the basis of the whole system are the two hill breeds, Blackface and Cheviot. To produce sheep more suited to the slightly greater fertility and milder climate of the uplands, both Blackface and Cheviot are crossed with the Border Leicester. By this cross the Halfbred (Border Leicester × Cheviot), and the Greyface (Border Leicester × Blackface), are produced. On lowland pastures either of these Border Leicester crosses are again crossed with one or other of the Down breeds – usually Suffolk or Oxford – to give Down-cross lambs, all of which, both wether and ewe, are destined for slaughter. This scheme of Scottish sheep stratification can be best illustrated diagrammatically:

| *Blackface ewe* | *Cheviot ewe* |
|---|---|
| (1) × Border Leicester ram | × Border Leicester ram |
| Greyface ewe | Halfbred ewe |
| (2) × Down ram | × Down ram |
| Fattening Down-cross lamb | Fattening Down-cross lamb |

In addition to securing a fresh infusion of hybrid vigour at each successive stage, this stratification system ensures a workable compromise between the limitations the land imposes and the qualities the market requires. (Plate 2.14.)

Thus, save quite exceptionally, it is clearly impossible to produce a first-quality fat lamb off a barren hillside. All that the land's fertility will support is the slow growth rate and slow reproduction rate of hill breeds of sheep. Suppose, however, that the hill is more fertile or, by pioneer reclamation, has been joined to the fringe of the cultivated uplands, then it is possible to cross the hill ewes with the Border Leicester, thereby producing a much more valuable class of lamb. The Greyface lamb usually makes more money than a Blackface lamb, and a Halfbred lamb more than a Cheviot lamb.

The ewe side of the Border Leicester – hill crossbred has proved an excellent grassland sheep. The ewe inherits a good deal of the independent

*Plate 2.14* Halfbred ewe lambs – Border Leicester X Scottish Blackface. (Photograph lent by the Society of Border Leicester Sheep Breeders)

habit and the milking ability of hill sheep and some of the weight and fertility of the Border Leicester. These ewes crossed with Down rams bear lambs of excellent mutton conformation and quality.

Thus, by a judicious system of successive first-crosses the final result, on good lowland pasture, is a combination of the hill breed's constitution, the Border Leicester's fertility, the Down sheep's mutton, expressed in the form of twin lambs of Down type drawing abundant milk from a mother of hill descent.

The Bluefaced Leicester has become more popular in recent years in preference to the Border Leicester, particularly for crossing with the Blackface, Swaledale and other horned hill breeds. ABRO has produced a new synthetic breed, the Damline which has been compared with the more traditional breeds as sires of crossbred ewes.

Although the Damline sires appeared to transmit higher fertility and earlier sexual maturity to their crossbred daughters than the Border Leicester or Bluefaced Leicester a higher level of postnatal mortality reduced this advantage. When production was assessed on the basis of total weight of lamb at 10 weeks per 100 ewes mated in a study by Cameron *et al.*,[41] the estimated weights of lamb produced were 2829, 2639, and 2363 kg for the Bluefaced Leicester, Damline and Border Leicester respectively. This does not take into account the fact that the ewe mating weight at 31 months of age was 62.6, 53.5 and 58.2 kg respectively, for the crossbred ewes in the order of the sires above. It also illustrates the complexity of comparisons.

Other sires are being used for crossing with hill breeds, including the East

Friesland which has been shown in trials of crossbred ewes to produce larger litters which are heavier at 12 weeks than crossbred ewes from Damline or Border Leicester sires, while the Oldenburg and others are also being used.

Within the last half-century this stratification system has acquired further substantial advantages by the ever-increasing demand for milk lamb – that is to say, lambs that come to slaughter weight while still sucking. To produce such a lamb it is necessary that its parentage supplies both first-class meat potentialities and abundant milk through which these potentialities can be both quickly and effectively expressed.

Now to combine first-class fleshing and milking qualities in the one breed has proved one of the more difficult problems of animal domestication. The dual-purpose sheep is evidently a reasonably practicable proposition where the two purposes are wool and meat, but when they are milk and meat practicability is much more dubious. Despite the many advantages of ideal dual-purpose cattle, breeding continues to diverge towards the two specialised objectives of meat and milk.

There is reason to believe that to combine first-class fleshing and milking ability in the same sheep breed is an equally difficult task. The stratification system of successive first-generation hybrids allows this difficulty to be overcome. The first cross between a ram of mutton type and an ewe of high milk yield is one theoretical answer. The cross between a Down ram and a crossbred ewe is at least one practical and profitable response.

Amongst the Down breeds the Suffolk enjoys considerable popularity in Britain with the Dorset Down and many other types also being used. Choice of a terminal sire depends on the production system and the market to which it is geared. The potential growth rate of lambs sired by the different breeds used tends to be related to their adult bodyweights.

Given the appropriate conditions, the daily gain of Oxford-cross lambs will be superior, with the Suffolk generally showing an advantage over the Oldenburg, Dorset Down, Ile de France or Texel as a terminal sire. However, carcass characteristics are also important and lean content is becoming increasingly valuable and relevant in meeting consumer demands.

It is essential that breeds are compared within the context of the production system – milk lamb, weaned lamb off grass, intensively finished in feedlot or as hoggets. Because of variations between maternal types and the systems used, data from trials may have limited application being relevant only to the area where the trials are conducted.

Benefits can also be obtained from a simple two- or three-way rotational crossing system.

Even though the Scottish Blackface and Swaledale breeds are very similar in type, being adapted to hill conditions, they may be crossed with advantage. Swaledales are generally considered to be better mothers but Blackfaces produce heavier lambs at weaning. First-cross sheep in Cameron et al.'s[42] study were 9.5 per cent more prolific and produced heavier lambs at weaning. The merits of this cross have been widely recognised and Swaledale rams have been widely used in formerly dominant Blackface areas. Usually the cross-ewes are bred to Blackface rams but a simple criss-cross breeding programme is a simple way to retain hybrid vigour.

## Inbreeding

Bakewell was the first animal-breeder, so far as is known, to have used inbreeding as a deliberate method of breed improvement. His contemporaries spoke of the 'constant incestuous intercourse' current in his flock. This practice of inbreeding seems to be the most probable reason for his exceptional success, for by inbreeding it would be possible to secure a greater degree of uniformity within a few generations than by less closely related matings. The reason for the distinctive differences between the many breeds of sheep which developed in England may well be due to close inbreeding rather than any credit which could be assigned to pedigree and selection based upon it.

Inbreeding is often regarded as leading to breed degeneration but it may also be a method of breed purification. Many recessive characters are undesirable, and inbreeding may therefore, by exposing recessives, open the cupboard door to display fully hidden the skeletons hidden in a breed's genetic constitution. It is possible that in some breeds, by using inbreeding so that recessives become homozygous, this practice has led to their redirection. Allan Fraser held the view that given a sound genetic constitution inbreeding can do no harm whatever.

In support of this argument he cited examples of close inbreeding in mankind:

'In biblical times, Jacob wedded his first cousins, Rachel and Leah, while Abraham married his half-sister, and Moses his aunt. The Egyptian Pharaohs and Ptolemies mated with their sisters whenever possible – Cleopatra having been the offspring of six generations of such brother–sister marriages – while she, in her turn, married her younger brother. The ancient Peruvians, too, also believed that the only bride royal enough for a King was his own sister.'

There are sufficient examples of successful inbreeding in animals, both laboratory and domestic – and even in mankind (see Austin[43]). For of Cleopatra was it not written: 'Age cannot wither her, nor custom stale her infinite variety.'?

Why, then, the firm prejudice of many sheep breeders – particularly commercial as contrasted with stud sheep breeders – to any form of inbreeding?

Increasingly, scientific evidence demonstrates that inbreeding has a depressing effect on production and especially so on reproductive performance. This is seen in a drop in numbers of lambs born and in their survival. However, as Ercanbreck[44] points out, inbreeding may affect wool production and it also has an influence on several important traits.

The attraction of inbreeding for the pedigree or stud breeder could be due to increased uniformity which will be obtained even though there will be a trend for segregation into lines which may be distinctively different.

The question as to whether crossing intensively inbred lines would be an effective means of breeding for improved performance has been investigated and notably in a large-scale investigation in Idaho in the United States. Three breeds, the Rambouillet, Columbia and Targhee were inbred to produce some 50 lines with the inevitable decline in production. Crossing inbred lines did little more than lead to a recovery in performance with the line-cross offspring being slightly superior in a few trials and certainly insufficient to compensate for the reduction in performance sustained in establishing the inbred lines (see Lamberson and Thomas[45]). In this experiment, inbred rams were compared with rams from the non-inbred control flock and they sired superior lambs to those from the inbred sires.

There is little convincing evidence, states Dr Parry,[46] that inbreeding will be a potent tool in breeding sheep for improved performance.

Outbreeding, although substantially decreasing the incidence of expressed undesirable recessive characters, will not eliminate them. These recessive characters keep cropping up. Some, by their nature, are easy to detect. Allan Fraser said,

'I have, for example, never handled a flock of sheep without the appearance of an occasional lamb with undescended testicle or parrot mouth, in former times accepting these deformities as inevitable and an unavoidable source of loss.'

It is now known, however, that both these defects are heritable although recessive characters. They can be 'bred out'.

Other heritable defects of sheep are not so readily detected. A very interesting example is that of the grey shiraz lamb coat in Karakul sheep. This type of Karakul pelt is of special economic value, but cannot be bred pure for the reason that lambs homozygous for this type of coat never reach maturity; they all die, apparently because of some imperfection in their digestion. In genetical language the grey shiraz colour cannot be bred pure

*Plate 2.15* Karakul lambs. (Photograph lent by Soviet Weekly, Soviet News & Press Agency, London)

because it is correlated with a lethal factor. The knowledge of the existence of this particular lethal factor in karakul sheep has been made evident because of its correlation with the desirable factor of economic importance. It is known that there are many other lethal and semi-lethal factors in sheep resulting in deformed lambs, weakly lambs, lambs that are doomed to die before maturity. The existence of these factors may be unmasked by inbreeding, and that is undoubtedly one reason why so many practical breeders believe that inbreeding causes degeneration and is something to avoid. (Plate 2.15.)

## References

(1)  Demururen, A. (1982) 'Migratory (transhumance) system' in I. E. Coop, (ed.) *Sheep – Goat Production.* Elsevier. Amsterdam.
(2)  Cossar, Ewart. J. (1913–14) 'Domestic sheep and their wild ancestors' *Trans. High. Agric. Soc. Scotland.*
(3)  Ryder, M. L. (1983) *Sheep–Man*, Duckworth, London.
(4)  Owen, J. B. (1981) 'Exploited animals and sheep,' *Biologist, J. Inst. Biol.*, **28** (3), 153–60.
(5)  Ryder, *Sheep–Man.*
(6)  Carter, H. B. (1964) *His Majesty's Spanish Flock.* Angus–Robertson, Ltd.
(7)  Cox, W. (1936) *The evaluation of the Australian Merino.* (Sydney)

(8)   Cox, *Evaluation of the Australian Merino.*
(9)   Parry, Dr (1906) *Communications to the Board of Agriculture*, Vol. 4.
(10)  Pawson, H. C. (1957) *Robert Bakewell.* Crosby Lockwood.
(11)  Anon. Leicester Sheep Breeders' Association. *History of the Breed.*
(12)  Hall, J. S. and Baxter, F. G. (1950) 'Bluefaced Leicester sheep' *Agriculture*, **56**, 523.
(13)  Weiner, G. and Hayter, Susan (1975) 'Maternal performance in sheep as affected by breed, crossbreeding and other factors', *Anim. Prod.*, **20**, 19–30.
(14)  Turner, H. N. (1969 'Genetic improvement of reproduction rate in sheep,' *Anim. Breed. Abstr.*, **37**, 545–63.
(15)  Purser, A. F. (1981) 'The performance of Welsh Mountain and Scottish Blackface sheep in a Welsh hill environment,' Animal Breeding Research Organization Report, pp. 3–6.
(16)  Purser, A. F. (1980) *Cannon Bone Size and Meat Production in Sheep*, Animal Breeding Research Organization Report, pp. 15–19.
(17)  Purser, *Cannon Bone Size and Meat Production in Sheep.*
(18)  Fell, H. (1979) *Intensive Sheep Management*, Farming Press.
(19)  (1982) National Sheep Association, *British Sheep.*
(20)  Ponting, K. (1980) Sheep of the World, Blandford Press.'
(21)  Burt (Captain) 1726 *Letters from Scotland.*
(22)  Russel, A. J. F. (1978) 'The relative contributions of nutrition and genetics to improvement in the efficiency of sheep production', *Agric., Prog.*, **53**. 92–7.
(23)  Rae, A. L. (1982) *in Breeding Sheep and Goat Production*, ed. I. E. Coop, Elsevier.
(24)  Clark, J. N. (1972) 'Current levels of performance in the Ruakura fertility flock of Romney sheep,' *Proc. N. Z. Soc. Anim. Prod.*, **32**, 99–111.
(25)  Croston, D., Danell, O., Elsen, J. M., Flamant, J. C. Hanrahan, J. P., Jakubeg, V., Nitter, G. and Trodahl, S. (1980), 'A review of sheep recording and evaluation of breeding animals in European countries,' *Livestock Prod. Sci.*, **7**, 373–92.
(26)  Owen, J. B., Brook, Lesley., Read, J. L., Steene, D. E. and Hill, W. C. (1978), 'An evaluation of performance-testing of rams using artificial rearing,' *Anim. Prod.* **27**, 247–59.
(27)  Turner, H. N. (1976) Methods of improving production in characters in importance in Sheep Breeding, e.d. G. J. Tomes, D. E. Robertson, and R. J. Lightfoot, W. Aust. Inst. Technol. pp. 81–99.
(28)  Turner, H. N. (1977) 'Australian sheep breeding research,' *Anim. Breed Abstr.*, **45**, 9–13.
(29)  Rae, *Breeding Sheep.*
(30)  Robertson, D. E. (1983) Defining selection objectives in Sheep Production, (ed.) W. Haresign, Butterworth.
(31)  Lucas, Rev. J. (1982) 'Several breed improvement schemes in New Zealand and Australia,' *J. R. Agric. Soc. (Engl.)*, **143**, 89–97.
(32)  Rae, A. L. (1974) 'The development of group breeding schemes: some theoretical aspects,' Sheep Farming Annual. Massey College, New Zealand, pp. 121–127.
(33)  Jackson, N. and Turner, H. N. (1972) 'Optimum structure of a cooperative nucleus breeding system,' *Proc. Aust. Soc. Anim. Prod.*, **9**, 55–64.

(34) Barker, S. D., Coutts, C. N. and Jones, T. (1983) Group Breeding Schemes. Report., Anim. Breed. Res. Org., pp. 17–22.
(35) Rae, A. L. (1952) 'Crossbreeding of sheep,' *Anim. Breed. Abstr.*, **20**, 197 and 287.
(36) Sheridan, A. K. (1981) 'Crossbreeding and heterosis,' *Anim. Breed. Abstr.*, **49**, 131–44.
(37) Rae, *Breeding Sheep*.
(38) Terrill, C. E. (1974) 'Review and application of research on crossbreeding of sheep in North America,' *Proc. 1st World Congr. Genet. Appl. Anim. Prod. Madrid*, **1**, 765–77.
(39) Mitter, G. (1978) 'Breed utilisation for meat production in sheep,' *Anim. Breed. Abstr.*, **46**, 131–40.
(40) Read, J. L. (1982) Application of Crossbreeding of Sheep in the United Kingdom. *Proc. World Congr. on Sheep and Beef Cattle Breeding.* Dunmore Press.
(41) Cameron, N. D., Smith, C. and Deeble, F. K. (1984) Comparative performance of crossbred ewes from three crossing sire breeds. Report Anim. Breed. Res. Org., pp 5–9.
(42) Cameron, Smith and Keeble, 'Comparative performance of crossbred ewes.'
(43) Austin, H. B. (1943) *The Merino, Past, Present and Probable*, Sydney.
(44) Ercanbreck (1973) 'Heterosis in yearling ewes from inbred lines.' *J. Anim. Sci.*, **36**, 1191.
(45) Lamberson, W. R. and Thomas, D. L. (1984) Effects of inbreeding in sheep – a Review. *Anim. Breed. Abstr.*, **52**, 287–95.
(46) Parry *Communications to the Board of Agriculture.*

# 3 Reproductive Physiology of the Ram

In sheep, as in other mammals, the sexes are born in numbers that are roughly equal. Nevertheless, the species is naturally polygamous, and this polygamy can be directed and guided by the selective breeder. An entire ram can sire hundreds, but castration, by preventing less desirable males from leaving progeny, was one of the great advances in animal breeding. Were it more nearly possible to predict the breeding value of a sire by his individual appearance, that advance would have been greater. Since some impressive-looking males may leave worthless progeny, and some notable breeding sires be less remarkable in appearance, it must have happened repeatedly in breeding history that some of the potentially best breeding males were castrated and some that were less desirable as breeders, left entire. There is no easy solution, since it is clearly impossible to progeny- or performance-test all males born. Castration must therefore continue to be the first step in selective breeding of the male.

## Castration

Castration involves the destruction of testicular function, the testicles being the two organs that lie together in the ram's scrotum or purse. There are several methods of castration. In sheep the old method was to cut open the scrotum and remove both testicles by knife or teeth.

A second method, at one time widely advocated, was to crush the spermatic cords connecting the testicles with the rest of the body, thereby leading to a slow degeneration and eventually a complete atrophy of the two testicles in the unopened scrotum. This operation is performed by a mechanical castrator (named the Burdizzo after its inventor), essentially a pair of heavy and blunt pincers that crush the spermatic cords without cutting through the overlying skin.

A third method, and that now most widely used, is termed 'elastration', in which a strong rubber ring is fitted over the lamb's scrotum and left there

54

until the whole scrotum atrophies and falls off. Recent legislation in Britain has made it an offence to use this method on lambs over one week old.

The final results of castration by any of these methods are similar. The male sheep becomes a wether or wedder (the words are synonymous). The wether loses all sexual power and desire. It cannot leave lambs and has no sexual interest in ewes. Any distinguishing masculine characters in a breed, such as horns, are either entirely suppressed or develop towards the female pattern.

## Wethers or Wedders

Wether sheep have characteristics of their own, and it seems a convenient opportunity to say something here about their economic advantages, under certain conditions, in sheep husbandry. Wethers are much hardier than breeding ewes and for this reason are kept in more adverse environments in various parts of the world.

Wether sheep were maintained in large numbers in the Scottish Highlands but the change in market demand for lambs, as opposed to mature wether mutton nearly a century ago, spelled the doom of this trade. In wool-growing countries, such as Australia and South Africa, wether flocks are still kept on the poorer land for the sake of their valuable wool and their mutton, saleable in the less-exacting home markets of those countries and also favoured in other areas such as the Middle East.

In fat lamb production systems castrates are generally slaughtered at relatively young ages, 3 months to 1 year old or so, and at light weights, 15–30 kg carcass weight. The use of entire animals has been proposed because of the possible pain and distress involved in castration, suggesting that this is an unacceptable mutilation.

Depending on the system of management, there appears, states Wilcox[1] to be but small differences in liveweight gain between castrates and entires up to the weaning stage.

Bradford and Spurlock[2] state that, at best it is usually around 5 per cent improvement in growth rate in favour of entires, but this may increase to 15–20 per cent by the yearling stage.

By normal slaughter stage, entires tend to be leaner, with less subcutaneous fat and an increase in the edible portions of the carcass, although the forequarter may be proportionally heavier. However, in systems where growth is slow, entire rams may have reached puberty before they are fit for slaughter and the carcass thereafter may be less acceptable or completely unsuitable.

## Rigs

In almost every large flock there is a small percentage of rams that cannot
be completely castrated by any of the usual methods because one or both
testicles remain in the abdomen and do not descend in the normal manner
into the scrotum. These, called 'rigs' in Scotland – ridgel or riggald are
variations of the word – are usually marked at castration time by the simple
expedient of leaving their tails uncut. Rigs with both testicles undescended
are never fertile and can leave no lambs, but their mating instincts being
unaffected, they will search out ewes in heat. They can sometimes be
employed usefully as 'teasers', as may rigs with one testicle undescended,
provided the testicle that *has* descended is first removed.

### Dual function of the testicle

Uncastrated male sheep with one or both testicles descended can conceive
lambs. With natural mating a ram may leave a hundred or more lambs in
one season; with methods of artificial insemination (AI) he may leave
thousands. Proper selection of rams in relation to genetic merit is therefore
of first importance to the sheep industry as a whole. However, the
comparative ability of rams to mate is rarely questioned.

In this connection – the actual use of rams in service – there are many
problems and some common misunderstandings. The variation in mating
capacity may be attributable to differences in the quality and quantity of
sperm produced, sexual urge and activity, copulatory efficiency, mating
preferences and other factors, such as age, health and fatness.

To follow the matter clearly it is necessary to consider a few of the
elementary facts concerning the sexual physiology of the male sheep.

The essential sexual organs of the ram are the two testicles the scrotum
contains. The testicle fulfils two separate and distinct, although closely
correlated functions:

(1) It produces the spermatozoa in the coiled tubules of which the testicle is
mainly composed. Spermatozoa – or sperm – when viewed under the
microscope look very like tadpoles with swollen heads and thin flagellant
tails. One, but only one, of the millions of spermatozoa ejected at a service
is destined to unite with the microscopic ovum or egg of the ewe served,
thereby beginning the formation of a lamb. Rarely, spermatozoa may be
absent from the seminal fluid or they may be of abnormal appearance. In
such cases the ram may be sterile. This sterility can be temporary or
permanent, but unless spermatozoa are present in its seminal fluid, a ram –

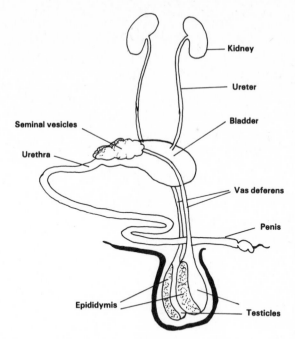

*Fig. 3.1* Reproductive organs of the ram.

be he ever so magnificently masculine in appearance and energetic in his mating instincts – cannot leave a solitary lamb.

(2) In addition to spermatozoa, the testicles produce a hormone or internal secretion of the androgen group called testosterone. This hormone is produced by glandular tissue lying between the coil tubules of the testicle, and this glandular tissue is therefore called the interstitial tissue. The hormone, testosterone, produced by it passes directly into the blood.

It is this hormone that stimulates the development of all the distinctly masculine features and behaviour of the ram. A ram of exceptionally developed masculine appearance or behaviour is one with particularly active testicular interstitial tissue. There is no reason to assume that because of his exceptionally masculine appearance and behaviour he is thereby more likely to leave many lambs. In fact, there is no reason to presume that he will leave any lambs at all. The keenest working ram I ever observed was completely and permanently sterile.

The twofold function of the testicles explains why a rig with undescended testicles may mate energetically with many ewes and yet leave no lambs. Undescended testicles are retained within the abdominal cavity, and spermatozoa never develop in that abnormal situation. The glandular interstitial tissue, on the contrary, can and does, with the result that a rig with undescended testicles may behave like a ram in every way, but to no useful reproductive purpose.

In 16 per cent of the rams examined by Lees[3] testicular abnormalities occurred, either in the form of misshapen testicles, or as being inadequately descended. The rate of growth of the testes in rams is dependent on the stimulus of lutenising hormone produced by the anterior pituitary gland. It has been suggested by Land[4] that testes growth and their eventual size, relative to the body weight of the ram, may be used as an indication of the potential libido and therefore potency of the ram. Puberty in the ram, which may be difficult to define in physiological terms, if regarded as adequate endocrine function and spermatogenesis, so that the ram will seek to mate and conceive offspring, is usually attained between $4\frac{1}{2}$–5 months of age, states Dyrmundsson.[5] However, the age-range within which puberty occurs is extremely wide and was found to be from 99 to 176 days in one study by Dyrmundsson.[6] It seems to be more related to body growth than to chronological age, and appears when the ram has reached around 35–45 per cent of adult body weight, with successful copulation within 40–50 per cent of mature weight. Ram lambs, if to be used in breeding programmes where there are benefits in earlier selection and more rapid generation turnover, should be encouraged to maintain a high growth rate.

Forty years ago, McKenzie and Berliner,[7] in studying sperm production in 8 Shropshire and 8 Hampshire rams over a period of 14 months, found very little correlation between mating behaviour and the production of spermatozoa. For example, they found that mating desire, as expressed by attempts at mating, was not a reliable indicator of sperm-producing activity, for it might be normal in sterile rams and slight in fertile rams. From this evidence it is clear that the common belief that a ram of exceptionally masculine appearance and behaviour is the most likely to leave many lambs is a misconception supported neither by experimental evidence nor by practical observation.

Nevertheless, there is this very definite advantage of a masculine, energetic ram in the field. He will go on searching for ewes in season where a ram of less masculine type, although possessing seminal fluid equally good, is lying at rest. On hill grazings the ram's physical energy – or rather the lack of it – may well prove the limiting factor in the number of ewes he settles and the number of lambs he leaves.

There are also differences in the ability of individual rams to detect oestrus, particularly outwith the normal breeding season, even though this may be satisfactory at the height of the breeding season. It is possible that detection of ewe lambs in oestrus is a special problem with variable performance between rams.

Rams when sold are usually in very high condition, and the question therefore arises as to what effect, if any, this high condition has upon their breeding powers. McKenzie and Berliner[8] investigated this problem, using microscopical examination of the semen as the critical test. They concluded that while a capacity for quick fattening and low breeding power tend to be associated in the same sheep, nevertheless, high condition is not necessarily detrimental to a ram's breeding potential. Here again, however, the question of the ram's ability to seek and to find ewes in heat must be given due consideration.

## Fertility

In sheep the fertility of a flock is more generally limited by the fertility of the ewes than by that of the rams, so that in the past perhaps too little attention was given to the possibilities of a weakness in this respect on the male side of sheep-breeding. Interest in, and technical advances in, AI have together provided much new information on changes in and factors influencing the natural fertility of rams. With periodical collection of semen by the methods employed in AI, it has become possible to make a really satisfactory study of this important problem for the first time.

Rams' semen, collected either in the artificial vagina or by electrical stimulation, can be examined microscopically and its quality judged by the numbers, normality or abnormality, motility, and other qualities of the spermatozoa it contains.

The use of these methods of assessing rams' fertility has provided new information, much of it acquired under experimental conditions, but nevertheless applicable to conditions as met with in the field.

How frequently is a poor lamb crop due to low fertility in the rams? Although most farmers might conclude from experience that infertility is a rare occurrence, Lees[9] has found that 1 ram in 10 has been wholly or almost completely infertile and about 3 times that number may be less than satisfactory for other reasons. A clinical examination for potential fertility has been suggested by Fraser,[10] taking into account not only the general health status of the ram, including freedom from lameness, but also involving an examination of the testes and the semen.

The test includes examining the volume and density of the ejaculate, as well as the motility and normality of the spermatozoa. This service is becoming increasingly available from veterinarians in Britain.

However, it must be remembered that as shown by Robinson,[11] rams producing semen of similar density and motility can vary in fertilising capacity. Thus libido and copulatory efficiency are also important.

Experimental evidence shows that while domestic sheep have not the definite rutting season of the stag, nevertheless rams possess better semen, and are therefore presumably of higher potential at certain times of the year than they are at others. McKenzie and Berliner[12] in their close study of Shropshire and Hampshire rams observed distinct periods of increased sperm-forming activity in both breeds, that for the Shropshire being from October to January and of the Hampshire from August to January. Under American conditions Lees[13] has suggested that the sexual pattern of the ram responds initially to the influence of increasing daylength and declines with reduced daylength. Using scrotal measurements as an indicator of testicular size, which in turn appears to be related positively to spermatogenic activity, it was found that the testes reach their maximum size about 3 months after the longest day. There is therefore a cyclical time-lag in breeding activity associated with daylength.

Studies by Lees[14] have shown that change in the length of the daily photoperiod leads to seasonal changes in semen production and plasma hormone levels in rams.

It has been suggested that the pineal gland mediates photoperiodic effect on reproductive activity causing altered patterns of luteinising hormone (LH) and prolactin secretion.

Some breeds of rams are ineffective during the summer period in Britain, and in studies with three breeds, Kerry Hill, Hampshire and Suffolk, Lees[15] showed that only the Kerry produced autumn-born lambs from Clun ewes. Apart from reduced semen production, the reason for failure to breed may also be a lack of libido and an inability to detect oestrus.

Climatic conditions may also be important. In Australia, Gunn[16] observed seasonal changes in rams' semen correlated with changes in atmospheric temperature. Hot weather (in the range 90–100°F) caused seminal degeneration and cooler weather favoured recovery. This effect of heat, noted both in the Merino and in British breeds, was so striking that Gunn believed it probable that in the hotter districts of Australia rams may become temporarily sterile because of the heat alone. More recent investigations by Dutt and Simpson[17] have confirmed the fact that high environmental temperatures (certainly those above 80°F) lower the quality of rams' semen and reduce their fertility. Dutt and Hamm[18] point out that this effect is most noticeable in unshorn rams.

Gunn[19] found, moreover, that many other accidents to which rams are subject, besides heat stroke, may cause seminal degeneration. These included arsenical dips, blowfly strike, foot rot, abscesses, and wounds.

More generally it can be said perhaps that no sick ram is a fully fertile ram, which might, indeed, have been presumed, but the importance of Gunn's work in this connection lies in showing how very sensitive the rams' sperm-forming capacity is to these morbid influences.

Malnutrition is essentially one form of disease, and it is therefore perhaps a little surprising to find sperm formation in rams to be rather more resistant to errors in diet.

Vitamin A deficiency, which may occur in the field where sheep are grazing withered, drought-stricken herbage, can lead to temporary sterility.[20] With this exception, sperm formation in the ram would seem to be rather independent of nutritional factors. In plainer language, a half-starved ram may continue producing normal sperm. The probable reason is that malnutrition does not cause rise of temperature or fever, and it is to this influence that sperm production seems peculiarly sensitive.

Among other causes of temporary infertility in rams Webster[21] lists change of climate, severe storms, and heavily woolled scrotum. The last point is of particular interest in view of the insistence by so many sheep-breeding societies that the ram's scrotum be closely covered with wool. Since the physiological purpose of the scrotum is the maintenance of a testicular temperature slightly below that of the rest of the body, a heavy covering of wool in that region can hardly be considered as being the best aid to fertility.

Mattner and Braden[22] state that, irrespective of the number of ewes mated the volume, density and number of sperm per ejaculate tends to fall over the first 2–4 days after joining and Lightfoot[23] says that this continues for 2 weeks or so. Lightfoot[24] also states that older rams 2½ years and over, produce more sperm per ejaculate than these of a younger age and the semen of immediate post-pubertal rams tends to be poor, but improves with age.

## Potency

It is essential to make a clear distinction between fertility and potency. Fertility in rams implies the capacity to produce normal sperm in abundance. Potency means his power of mating. The reverse of fertility is infertility; of potency, impotence. A normal ram is both fertile and potent. It may happen, however, that a sterile ram may be fully potent and a fertile ram may be impotent. Impotence may be temporary or permanent, due either to physical or psychological causes.

Artificial insemination techniques have demonstrated what vast quantities of sperm a fertile ram may produce, and how difficult it is to exhaust either his sexual energy or powers of sperm production. McKenzie and

Berliner[24] found that 3 Hampshire rams could perform 24, 20 and 13 services respectively within 9 hours without showing any signs of exhaustion. They also found tremendous variability between individual rams. Hulet et al.,[26] in their studies of the mating behaviour of the ram in single-sire pens, found that the average ram mated naturally 45 times in 6 days.

As the number of ewes per ram is increased, so also does the activity of the ram increase. Successful results have been obtained by Allison[27] in New Zealand when up to 720 ewes have been joined with 4 rams. When around 35 to 70 ewes are run with one ram there appears to be little difference between 2-tooth and older rams (6-tooth) in breeding success, but when numbers were increased to 180 per ram, younger rams (2-tooth) mated with a lower percentage of ewes.

In an extensive study in New Zealand by Quinlivan and Martin,[28] wide differences between rams were observed as regards the conception rate of ewes. Numbers joined had no effect on this other than the proportion of ewes detected in oestrus.

There is some evidence to suggest a direct effect on litter size by the ram. It is understandable that exceptional rams possessed of all the attributes, should lead to higher conception rates and survival of the resulting embryos can be expected, but their identification is perhaps difficult in practice even though different performance is of considerable economic importance.

## Behaviour

When rams are run together as a group before the mating season they establish a hierarchy or dominance order, with the heavier and more active rams asserting this superiority. This pattern was found by Bourke[29] to continue when 2 or more rams are run with the ewe flock, to the extent that an extremely dominant ram may attempt to claim one, or a large proportion of ewes, or alternatively prevent others from mating. Other studies, for example by Tomkins and Bryant[30] have shown little effect of ram dominance or mating behaviour. However, Mattner et al.[31] have shown that competition and aggression may only arise when a sexually active ram is run with one which is relatively inactive, and that, in any event, the number of services by each ram may not necessarily be affected.

The wide range of research which has been undertaken in Australia on reproduction in the ram is the subject of a comprehensive review by Moule.[32]

In practice, paddock size may be important, especially in extensive systems. To overcome the problems of the ram acquiring a harem and so failing to seek out ewes in heat, or because of a lack in libido, fertility or other failure, a generous allowance of 1 ram to 40 ewes is needed. In more intensive systems with greater control and using rams of proven stock-getting ability, there is no reason why 1 ram per 100 ewes should not be used. In contrast, where synchronisation of oestrus is practised, at least 1 ram per 10 ewes becomes necessary. With small flocks it should be possible to share rams if synchronisation can be staggered between farms.

Many systems demand that ewes which have been mated and, indeed, remated can be identified, so that later they may be grouped for management purposes such as late-pregnancy feeding and lambing. The use of sire harness is a convenient means of identifying mating within appropriate periods. Great care is required to ensure a careful fit of the harness, otherwise physical damage to the ram may occur.

**Artificial insemination**

Artificial insemination is the introduction of spermatozoa into the female reproductive tract by means other than natural mating.

There are two accepted methods of securing semen from the ram:

(1) By the use of the artificial vagina which, essentially, is a lubricated rubber sheath ending in a glass flask and surrounded by a jacket containing warm water at the appropriate temperature (45–46°C) and pressure; or,

(2) By electrical stimulation of the lumbar region of the spinal cord. Because of the possible pain or discomfort involved this latter method is not widely favoured.

The artificial vagina, invented by the Russian, Milovanov, in 1931, is the method most widely used. Rams' semen is very much more concentrated than that of the bull. About 1 cc is usually produced at each ejaculation. In the collection of semen there are beneficial effects in using a ewe in standing oestrus, as opposed to a non-oestrus ewe, since it has been shown to increase the volume of ejaculate, giving 17 per cent more sperm.

Conception rate can be affected by dilution rate, depth of freezing and storage time. Dott[33] found that a dilution rate of 1 in 6 is the maximum consistent with an acceptable conception rate (CR). Colas et al.[34] discovered that when skim milk powder diluent is used, a dilution rate of 1 in 3 is more satisfactory. Salamon[35] states that a semen dose containing a minimum of $11–12 \times 10^6$ spermatozoa is desirable to obtain a satisfactory CR.

One of the major obstacles to the widespread use of AI is the short storage life of 'fresh' semen. With present techniques, satisfactory results can be obtained with semen kept for up to 14 hours. Salamon and Robinson[36] have found that retention up to 24 hours can entail a loss of fertilising power of some 15 per cent when it may be used for the second lamb, not the first insemination.

Survival of ram sperm under deep-freezing techniques is also uncertain and unsatisfactory.

In Australia, Salamon[37] found that frozen semen stored for 36 months has given conception rates of around 53 per cent, while similar results using semen frozen in straws have been obtained by Colas and Brice.[38] Doubtless, further research will modify and improve these results, but at the present time the practice is to use 1 mm fresh undiluted semen for each insemination.

Semen, after collection, is inserted into that part of the ewes' genital tract called the cervix where it should be deposited as deep as possible for the best results. The simple operation requires a speculum to dilate the vagina, a graduated syringe to hold and deliver the correct amount of semen, and a head lamp to illuminate the interior of the ewe's genital tract. An alternative is to use a speculum with an inbuilt light source.

The detection of oestrus in the ewe gives rise to a problem in applying AI in sheep-breeding. Apart from her behaviour towards the ram, a ewe shows no reliable indications of heat or oestrus such as is evident in many other mammals. Either teaser rams or a programme of oestrus synchronisation or a combination of the two, may be used.

Teasers (entire rams prevented mechanically from actual mating by the tying of an apron or sack under their bellies) or vasectomised rams (those with the vasa differentia or tubes connecting testicles and penis severed), can be employed to detect the ewes that are in heat. These ewes, thus detected – usually by colour marks left upon them by the sterile rams – are then collected together for AI.

Vasectomised rams are the most satisfactory detectives, the operation of vasectomy being easily performed. The spermatic cord connecting the testicle with the pelvic cavity comprises an artery, a vein, a nerve and a much thicker, tubular structure called the vas deferens, up which the semen passes during ejaculation. Through a skin incision the vas deferens can be isolated from the other essential elements of the spermatic cord and divided. The result is a ram with the full masculinity and behaviour of the entire male, but one that is completely and permanently sterile.

A standardised technique of oestrus synchronisation described by Gordon[39] is now well established. This may be followed by the use of teaser rams to identify ewes in heat or simply by two inseminations at 50–56 hours and 60–65 hours after the withdrawal of the intravaginal pessaries (sponges).

Recent research suggests that one insemination ($400 \times 10^6$ sperm per strain) 56 hours after sponge withdrawal may indeed be superior to natural mating, with 1 ram to 10 ewes in synchronised flocks.

*Advantages of artificial insemination*

The first advantage is that a ram of outstanding value may be used to impregnate a vastly greater number of ewes. With natural mating, even where ewes in heat are brought in to the ram for service, conserving his physical energy by so doing, he can be expected to serve no more than 100–200 in a season. By the use of AI he may serve many thousands. In the Soviet Union, the country where AI in sheep is most widely practised, Stamp[40] reported that 1 imported ram inseminated 34000 ewes. Dutt and Simpson[41] report that in another instance, during a period of 105 days the sperm of 1 ram was used to inseminate 17700 ewes, resulting in the birth of some 18000 lambs. With such possibilities of almost unlimited paternity it clearly of the first importance to ensure that the ram's genotype has been rigidly examined by adequate progeny-testing. Provided this safeguard is employed and the dangers inherent in inbreeding fully recognised, AI can clearly be of the greatest benefit. Sires of outstanding breeding merit are rare. When such are discovered, AI ensures that the exceptional improving and proven sire can be used to the swiftest and fullest extent. Rams of inferior genetic merit need not be used at all. In this sense AI is merely an extension and supplementation of castration.

On the contrary, were rams used extensively by AI without preliminary progeny-testing and on the very uncertain genetic evidence of phenotypic excellence, fashionable pedigree or show records, the results could well be unfortunate. The employment of AI can, in fact, lead quite as rapidly to progress in the wrong direction as in the right one.

The second advantage is where a swift change in breed dominance is desired. That, rather than breed improvement, was the reason underlying the very rapid adoption and early development of AI in the sheep industry of the Soviet Union. Before the Revolution the sheep industry of the Soviet Union was clearly divisible into two unrelated sections. There were some 7 million sheep of Merino type in the big flocks of the landowners, and some 38 million coarse-woolled sheep in the small flocks kept by the peasants. The policy adopted by the Soviet government was a rapid increase in the number of fine-woolled sheep. This was achieved by crossing Merino rams, either native or imported, with the coarse-woolled peasants' sheep. Between 1926 and 1931 importations were extensive. The embargo placed by the Australian government in 1928 prevented the further importation of Australian Merinos by the Soviet government. The difficulty was met by the extension of AI so that the Merino rams available could be used to the fullest extent. During the Second World War the sheep industry of the Soviet Union was decimated, and AI received a further impetus during its restoration.

In the year 1964 it was claimed that over 90 per cent of the ewes in the Soviet Union were given AI, and Roberts *et al.*[42] report that the technique has become so commonplace that the shepherds, following short instructional courses, perform the operation themselves.

The reasons for the rapid adoption of AI by the sheep industry of the Soviet Union are therefore readily understandable. The objective was the mass improvement of the country's sheep within the shortest possible time; the aim, a rapid increase of sheep yielding wool of textile class to provide the raw material of a national, self-sufficient, textile industry in the face of the difficulty and expense of importing sufficient rams of the desired type. All these circumstances favoured and accounted for the tremendous advances both in technique and application of AI in the sheep industry of the Soviet Union.

Artificial insemination could have an important role in intensively managed flocks where synchronised mating, requiring a high proportion of rams, is an integral part of the system. Indeed, AI is being successfully employed in flocks of this type, notably by Robinson at the Rowett Research Institute.

In Britain, AI has had little acceptance in conventional sheep systems, probably because, as shown by Meat and Livestock Commission[43] calculations, the costs are twice as great compared with natural mating. However, potential is considerable and increasing interest is being shown by progressive breeders.

Because of the need for laboratory facilities and trained personnel, as well as cost, AI has only limited application in developing countries, although it could have a profound influence on sheep improvement if semen storage could be solved.

## References

(1)  Wilcox, J. C. (1968) 'Time and method of castration on the performance of fat lambs,' *Exp. Husb.*, **17**, 52-58.
(2)  Bradford, G. E. C. and Spurlock G. M. (1964) 'Effects of castrating lambs on growth and body composition.' *Anim. Prod.*, **6**, 291.
(3)  Lees, J. L. (1978) 'Functional infertility in sheep,' *Vet. Rec.*, **102**, 232-36.
(4)  Land, R. B. (1974) 'Physiological studies and genetic selection for sheep fertility,' *A.B.A.* **42**, 155-58.

(5) Dyrmundsson, O. R. (1973) 'Puberty and early reproductive performance in sheep II Ram lambs,' *A.B.A.* **41**, 419-30.

(6) Dyrmundsson, O. R., and Lees, J. L. (1972) 'Puberal development of Clun Forest ram lambs in relation to time of birth,' *J. Agric. Sci.* Camb., **79**, 269-71.

(7) McKenzie, F. F. and Berliner, V. (1937) 'The reproductive capacity of rams,' *Res. Bull. Mo. Agric. Exper. Stat.*, No. 265.

(8) McKenzie and Berliner, 'Reproductive capacity.'

(9) Lees, J. L. (1969) 'The reproductive pattern and performance of sheep,' *Outl. Agric.*, **6**, 81-88.

(10) Fraser, A. F. (1979) 'Clinical examination of rams for fertility,' *Vet. Rec.*, **87**, 200.

(11) Robinson, T. J., Salamon, S., Moore, N. W. and Smith, J. P. (1967) *'The Control of the Ovarian Cycle in Sheep,'* (ed.) T. J. Robinson, Sydney University Press, p.208.

(12) McKenzie and Berliner, 'Reproductive capacity.'

(13) Lees, 'Reproductive pattern and performance.'

(14) Lees, 'Reproductive pattern and performance.'

(15) Lees, 'Functional infertility.'

(16) Gunn, R. M. C. *et al.* (1942) 'Studies in fertility of sheep: (2) Seminal changes affecting fertility in rams,' *Bull. Coun. Sci. Indust. Res. Aust.*, No. 148, 140.

(17) Dutt, R. H. and Simpson, E. C. (1957) 'Environmental temperature and fertility of Southdown rams early in the breeding season,' *J. Anim. Sci.* Camb., **16**, 136-43.

(18) Dutt, R. H. and Hamm, P. T. (1957) 'Effect of exposure to high environmental temperatures and shearing on semen production of rams in winter.' *J. Anim. Sci.*, Camb., **16**, 328-34.

(19) Gunn, 'Studies in fertility.'

(20) Dutt, B. (1959) 'Effect of vitamin A deficiency on the testes of rams.' *Brit. Vet. J.* **115**, 236-238.

(21) Webster, W. M. (1952) 'Infertility in rams,' Proc. Soc. Anim. Prod., 11th Ann. Conf., 1951, p.62.

(22) Mattner, P. E. and Braden, A. W. H. (1967) 'Studies in flock mating of sheep (2) Fertilisation and pre-natal mortality,' *Aust. J. Expl. Agric. Anim. Husb.*, **7**, 110-16.

(23) Lightfoot, R. J. (1968) 'Studies on the number of ewes joined per ram for flock matings under paddock conditions (2) The effect of mating on semen characteristics,' *Aust. J. Agric. Res.*, **19**, 1043-1057.

(24) Lightfoot, 'Flock matings under paddock conditions.'

(25) McKenzie and Berliner, 'Reproductive capacity.'

(26) Hulet, C. V. *et al.* (1962) Mating behaviour of the ram in the one-sire pen. *J. Anim. Sci.*, **21**, 857-64.

(27) Allison, J. (1970) 'Flock mating in sheep. (iii) Comparison of two-toothed and six-toothed rams joined with different numbers of ewes per ram,' *N.Z. J. Agric. Res.*, **21**, 113-18.

(28) Quinlivan, T. D. and Martin, C. A. (1972) 'Survey observations on the mating performance of sheep in New Zealand Romney stud flocks.' *Aust. J. Agric. Res.*, **23**, 309-20.

(29) Bourke, M. E. (1967) 'A study of mating behaviour in Merino rams,' *Aust. J.*

*Expl. Anim. Husb.*, **7**, 203-205.

(30) Tomkins, T. and Bryant, M. J. (1972) 'Mating behaviour in a small flock of lowland sheep,' *Anim. Prod.*, **15**, 203-210.

(31) Mattner, P. E., Braden, A. W. H. and George, J. M. (1973) 'Studies on Mating Flock of Sheep. (5) Incidence, duration and effect on flock fertility of initial sexual inactivity in young rams,' *Aust. J. Expl. Agric. Anim. Husb.*, **60**, 35-41.

(32) Moule, G. R. (1970) 'Australian research into reproduction in the ram,' *A.B.A.* **38**, 185.

(33) Dott, H. M. (1964) 'Results of inseminations with ram spermatozoa in different media,' *J. Reprod. Fertil.*, **8**, 257-58.

(34) Colas, G., Dauzier, M., Ortavant, R. and Signoret, J. P. (1968) 'Resultats obtenus au cours de l'étude de quilques facteurs important de l'insemination artificielle ovine, *Ann. Zootech.*, **17**, 45-57.

(35) Salamon, S. (1962). 'Studies on the artificial insemination of Merino sheep (3) The effect of frequent ejaculation on semen characteristics and fertilisation capacity,' *Aust. J. Agric. Res.*, **13**, 1137.

(36) Salamon, S., and Robinson, T. J. (1962). 'Studies on the artificial insemination of Merino sheep. II. The effects of semen diluents and storage on lambing performance,' *Aust. J. Agric. Res.*, **13**, 271-81.

(37) Salamon, S. (1972) 'Fertility of deep-frozen ram spermatozoa for three years,' VIIth Int. Congr. Anim. Reprod. Art. Insem., Munich, *II*, 295.

(38) Colas, G. and Brice, G. (1970) 'Fertility of ewes treated with FCA and inseminated with frozen semen: Preliminary results,' *Anim. Zootech.*, **19**, 353-51.

(39) Gordon, I. (1975). 'Hormonal control of reproduction in sheep,' *Proc. Br. Soc. Anim. Prod.*, **4**, 79-93.

(40) Stamp, J. T. (1964). 'Report on visit to the Soviet Union.'

(41) Dutt and Simpson, 'Environmental temperature and fertility of Southdown varies.'

(42) Roberts, E. M., Clapham, B. M., McMahon, P. R., and Richard-Bell, L. (1960) 'A report on wool production in the Soviet Union.'

(43) Meat and Livestock Commission (1972) Sheep improvement. Scientific Study Group Report. MLC Bletchley, October, 1972.

# 4  Reproductive Physiology of the Ewe

## Introduction

The reproductive physiology of the ewe is a complex subject, difficult to discuss without becoming too deeply involved in endocrinology, the study of the ductless glands. Much scientific research has been devoted to this study; a great deal is still in progress. Since this book professes to deal with sheep husbandry, it would be inappropriate to attempt to delve too deeply into scientific profundities. In order to make the subject as easily understood as is possible, I have decided to deal with its various aspects in the following manner:

(1)  To give a brief description of the ewe's reproductive behaviour as the shepherd sees it.
(2)  To describe the essential anatomical and physiological factors which are known to determine that behaviour; and
(3)  To discuss the application of this scientific knowledge designed to control and, if desirable, alter the ewes' reproductive behaviour in the interests of more intensive production.

## Mating behaviour

In all British sheep breeds, with the important exception of the Dorset Horn, there is a definite mating season. Although the ram shows a certain sexual interest in the ewes throughout the year, the ewes are entirely unresponsive to his approaches until the autumn. They will then stand to him and accept service. In shepherds' language, the ewes have come in 'heat'. If the ewe becomes pregnant as a result of this first service she does not come in heat again until the following autumn. If, on the contrary, she fails to settle, she goes out of heat after an average heat duration of about 30 hours in the ewe, but considerably less in the ewe lamb. Heat will recur usually around 16–18 days later, but the interval can vary in length from one ewe to another. If the ewes are run with sterile rams, for example with vasectomised rams, it can be shown that ewes will continue to come in heat throughout the winter.

The Dorset Horn or the polled derivative, together with many foreign breeds, including the Merino, Finn and Rambouillet, have a rather different pattern of breeding behaviour. A Dorset ewe, unlike the ewes of other British sheep breeds, will continue to come in heat throughout the entire year unless pregnant. Even in this exceptional case, however, there is evidence of a greater regularity of heat periods in the autumn months, and she usually misses one or two cycles during the months of May and June.

## Reproductive organs

Before discussing the sexual physiology of the ewe and the factors that affect it, a clear understanding of the elementary anatomical facts concerning the ewes' reproductive system is essential.

The main reproductive organs of the ewe are, then:

(1) The *Ovaries* – two in number – lying posteriorly in the pelvic region of the abdomen.
(2) The tubes – called *fallopian tubes* – connecting the ovaries with the uterus or womb. The upper, dilated end of each fallopian tube is called the infundibulum.
(3) The *uterus* or womb – a small organ when the ewe is not in lamb, but grown to a large size by the end of pregnancy.
(4) The lower part of the uterus, called the *cervix*, or neck.
(5) The *vagina*, connecting the uterus with the exterior.

These organs and their spatial relationship one to another are illustrated in Fig. 4.1.

The ovaries, bean shaped and 12–25 mm in length in the ewe, contain numerous microscopic female germ cells called *ova*. Many thousands of undeveloped ova lie within the substance of the ovary. When a ewe is coming into heat one or more of these ova lying on the surface of the ovary enlarge within surface projections which look like tiny superficial blisters and are called the Graafian follicles. The term, as is also the case with the fallopian tubes, is taken from the name of the original discoverer. The Graafian follicles contain a fluid called the follicular fluid. Towards the end of heat, ovulation occurs.

The follicle ruptures, releasing the ripe ovum, which is drawn into the infundibulum of the fallopian tube. The sperm, meanwhile, following service by the ram or by artificial insemination (AI), have ascended up through the cervix, uterus, and finally the fallopian tube, where, in its upper third, fertilization of the ovum usually occurs. The fusion of one ovum and one spermatozoa forms a single cell – the zygote – and the zygote is the beginning of a lamb.

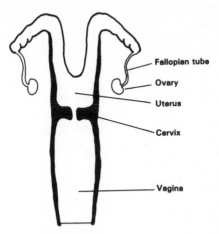

Fallopian tube

Ovary

Uterus

Cervix

Vagina

*Fig. 4.1* Reproductive organs of the ewe.

After the rupture of the Graafian follicle the minute scar left on the ovary's surface is filled by a blood clot, which quickly becomes transformed into an important organ of internal secretion – the corpus luteum, or yellow body. The corpus luteum in an unfertilised ewe remains only until shortly before the beginning of the next heat period. It is, in fact, the action of the corpus luteum's internal secretion, called progesterone which, by preventing further ovulation, accounts for the 16-day periodicity of heat recurrence in the ewe. If a ewe becomes pregnant, however, the corpus luteum persists as a more permanent organ of internal secretion, disappearing shortly before the end of pregnancy.

An important function of progesterone is to prepare the internal lining of the uterus for the reception and attachment of the zygote. The successful attachment of the zygote to the maternal uterine lining depends, apparently, upon the action of this hormone. The stage of implantation is a critical phase in the lamb's development.

Attachment of the embryo usually occurs around 15 days after fertilisation, but may be delayed or fail altogether. There is no doubt that the failure of implantation as well as that of ovulation or fertilisation is one cause of infertility in ewes.

The attachment between the zygote and the uterus develops into the placenta or afterbirth. The zygote, following a 5-month intra-uterine development, is born as a fully formed lamb.


## Control of reproduction

The natural tendency of sheep is towards an autumnal mating season. The breeding seasons appear to be more distinct at higher latitudes. There are distinct breed differences in both the commencement and duration of the breeding season. In Britain, hill breeds tend to breed later and to have a shorter breeding season, whereas the Finnish Landrace and the Dorset Horn have considerably longer breeding seasons. It is possible that in most breeds individuals vary considerably, as stated by Hafez,[1] giving rise to a wide range in both onset and cessation of breeding activity;[2] and as shown by Lees[2] with Clun Forest ewes, in Wales, some successful matings can occur in more or less any month of the year.

The question then arises as to what factors are responsible for this autumnal mating maximum or peak of sexual activity.

### Effect of light

British sheep breeds transported to the southern hemisphere alter their breeding season to correspond to the reversal of the seasons there, so that transported ewes will come in heat during the months corresponding to their homeland spring.

The mating season of sheep in both hemispheres occurs during a period of decreasing duration and intensity of daylight. It is natural to assume that decreasing daylight in autumn was the stimulus arousing sexual activity in the ewe.

Experiments on this subject at Cambridge from 1934 to 1949 confirmed quite definitely that light is the main factor, albeit not the only one, controlling both the onset and cessation of the mating season in sheep. A paper by Yeates,[3] gives a full account of these experiments with a wide list of references, while a more recent review by Williams[4] examines the phenomena of photoperiodicity in British ewes.

Manipulation of the patterns of the light/dark ratio can be used to induce oestrus in ewes. Various patterns of decreasing daylength both in the rate of daily reduction and also the absolute size of the reduction at any one time can affect the reaction interval, which is the time from the initiation of the treatment and the subsequent appearance of oestrus. The large variation in the time of oestrus onset is a practical disadvantage in the use of photostimulation.

For example, quite different treatments, described by Ducker *et al.*,[5] imposed on 1 July and involving either an absolute decrease in daylength of 11.75 hours or a daily reduction of 10.9 minutes gave a reaction interval of 33.6 days, while ewes in natural conditions did not come into heat until 66.2 days after the commencement of the experiment.

The reversal of the light/dark pattern can commence while ewes are still in lamb and successful results were obtained by Robinson[6] with North Country Cheviot and Finn × Dorset Horn ewes. Breeding success was achieved by abrupt increase in daylength to 18 hours, at the end of the second month of gestation, this being maintained for one month, then followed by a gradual reduction in the light period, until 2 months after the subsequent mating.

A similar method of an abrupt increase to 18 hours for 1 month, but followed by an abrupt decrease to 8 hours, was successful in inducing a fertile oestrus with a 91 per cent conception rate in March lambing Scotch Halfbred ewes about 3 months after parturition has been described by Newton and Betts.[7]

Only those procedures which do not depend on a light-proof building are realistic and, indeed, the cost of buildings and additional feed are major limitations in using light control to induce out-of-season oestrus. The method has not found favour largely because of the development of increasingly reliable methods using exogenous hormones.

It has been shown by Fraser and Lang[8] that a sudden decrease in daylength near to the normal breeding season will cause a concentration in the occurrence of oestrus. It is not possible to control oestrus directly by the manipulation of temperature. It has been suggested that low temperature tends to stimulate, and high temperature to inhibit, the manifestation of oestrus.

### Effect of suckling

Other factors, apart from light, influence the length of the rebreeding interval and the duration of the ewe's breeding season.

Although certain breeds, for example the Dorset, will mate while suckling lambs most breeds, even those with an extended mating season such as the Dorset Horn, will not always do so.

Thus, Anderson[9] found that high-grade Merinos in Kenya do not come in heat during lactation.

Hammond[10] using a Suffolk-cross ewe showed that suckling delayed the start of the breeding season.

On the other hand, several studies by Hunter *et al.*[11] have suggested that the length of lactation period does not necessarily affect the onset of oestrus and that nutrition could be an important factor. The explanation may lie in the rate and amount of body weight recovery during lactation, being of importance in relation to the rebreeding interval.

## *Effect of nutrition*

It is known that nutrition has a very definite effect on the number of lambs conceived at a mating. There is the further possibility that nutrition may also influence the sexual activity on which the mating of the ewe depends. There is, indeed, some evidence to support this view.

Engela and Bonsma[12] found that the anoestrous period – the period of sexual inactivity in the ewe occurring during spring and early summer – may be shortened by better nutrition. They further state that ewes are less sexually active when there is a shortage of vitamins in the diet than when kept on green grazing. On the contrary, Hafez[13] claimed that undernutrition of ewes was without effect on the date of onset and duration of the breeding season.

## Intensification

British sheep breeds and their crosses under traditional methods of British sheep husbandry produce annually on the average 1½ lambs and one fleece. Apart from the Dorset Horn, their breeding season is limited to the autumn months. It is the unanimous opinion of agricultural economists that under modern conditions this level of restricted production is too low to be profitable. How and by what means can this level be raised? As regards reproduction in the ewe, two main methods have been suggested and explored, namely:

(1)  To raise the prolificacy of the ewe so that what is commonly termed the lambing percentage – the number of lambs born per hundred ewes mated – may be increased substantially above its present level.
(2)  To extend and control the mating season so that lambs may be conceived and therefore born at any season of the year.

These are clearly distinctly different problems, and are therefore best dealt with separately.

**Increasing prolificacy**

There are three main methods of increasing the prolificacy of ewes (1) Genetic; (2) Nutritional; and (3) Hormonal, and before attempting to decide which of these holds most promise it is well to discuss each method in greater detail.

*Genetic methods of increasing prolificacy*

Among British sheep breeds and crosses there are some that are undoubtedly more prolific than others. The popularity of the Clun Forest breed, of the Scottish Halfbred and the Masham, Mule and Greyface crosses depends to some extent on their proven prolificacy. Comparisons of the relative prolificacy of the numerous breeds and crosses is perhaps somewhat invidious because of the variety of conditions under which the various types are kept. However, the Meat and Livestock Commission[14] data summarising performance of some common crossbred ewes under commercial conditions do suggest possible differences between breed types.

The potential fertility of our sheep breeds and crosses is certainly not exploited to the full under circumstances where better performance is possible. In lowland and upland recorded flocks the top one-third produce a weaned lambing percentage some 7–15 units higher than the average of 149.

The popular traditional crosses, as well as new types such as the Cambridge and the Damline, are capable of much higher performance levels and a lambing percentage of 180 and even above 200 is being achieved in the best commercial conditions.

When responsible for the Edinburgh School of Agriculture flock during the 1950s and 1960s we had, quite regularly, a lamb crop of 180–195 weaned and this level of performance continues to the present time.

There are, however, certain foreign sheep breeds far more prolific than any of our native breeds or their crosses. Cremer[15] gave the following figures for the East Friesian milch sheep. Of 5806 recorded births, only 10.28 per cent were singles, 53.84 per cent were twins, 31.06 per cent were triplets, and 2.47 per cent quadruplets. The Romanov is the most prolific breed in the Soviet Union. In one district the average number of lambs born of 6030 ewes was 2.38. Smirnov[16] recorded individual cases of very high prolificacy in this breed. Thus, one ewe produced 35 lambs in 6 years; another 64 in 12 lambings; a third 8 lambs in one lambing.

The Finnish Landrace is related to the Russian Romanov, being of the same short-tailed Northern type. The average number of lambs born per ewe in recorded flocks in Finland is 2.4. In the year 1962 the Animal Breeding Research Organization (ABRO) imported a small flock of Finnish Landrace sheep to Edinburgh. There were 5 shearling ewes, 5 ewe lambs, and 5 rams in this original importation. The fertility of the breed has been tested and confirmed under Scottish conditions. Read[17] states:

'The results show clearly enough that the Finnish sheep have confirmed their reputation as unusually fertile animals. In the main this is due to large litter sizes but is also a reflection of success at mating which is especially remarkable in the case of ewe lambs.'

Very carefully designed experiments by the ABRO have shown that part of the fertility (roughly 50 per cent) of Finnish sheep is inherited by crossbred ewes sired by Finnish rams.

Although there is this difference in fertility between breeds – a difference clearly genetical – selection within a breed to increase prolificacy does not, on present knowledge, promise much improvement.

In any flock of any breed there is considerable variation in this important character. In any one lambing some ewes bear single lambs, others twins, a few bear triplets. Ewes tend to follow the same individual pattern of prolificacy in subsequent lambings; a characteristic which has been given the rather clumsy technical title of 'repeatability'. Reeve and Robertson[18] defined as being 'a definite tendency for ewes producing multiple births at the first lambing to do so at subsequent lambings'. Smirnov[19] gave a very striking illustration of repeatability in Romanov ewes. In the stud flocks, ewes which bore a single lamb at first lambing subsequently produced on the average 1.89 lambs annually; those that bore twins at first lambing subsequently produced 2.15 lambs annually; those that bore triplets at first lambing subsequently produced on the average 2.68 lambs annually, and only these ever produced quintuplets. Heptner[20] again in Romanov sheep, stated that ewes bearing a single lamb at first lambing will subsequently produce about 186 lambs per 100 ewes, those bearing twins, 215 and those bearing triplets, 268.

A rapid and practical method of raising a flock's fertility is therefore to cull ewes bearing single lambs at their first lambing, retaining those that have borne twins or triplets.

Repeatability is one thing; inheritance is quite another. Individual prolificacy in a ewe is just as much a phenotypical character as the shape of her ears and as such, may not be inherited at all. Breeding only from ewes with multiple births need not necessarily increase the prolificacy of a breed or flock in subsequent generations. It is true that Marshall and Hammond[21] in an official bulletin published in 1945, stated categorically that 'the tendency to produce twins is transmitted not only through the ewe but through the ram', and, indeed, tended to blame the practice of selecting the better-grown single ram lamb in preference to less-well-grown twins for the relatively low level of fertility in English sheep breeds.

More recent authorities are much more cautious. Thus, Reeve and Robertson[23] in a comprehensive review of factors affecting multiple births in sheep published in 1953, wrote:

> 'There are some differences of opinion as to the importance of genetic variations, but the work on Swedish and American breeds suggests that they form only a small fraction of the total variation within a flock, even when the effect of age is eliminated, so that genetic progress, based on selection of ewes with the highest average litter size, is likely to be rather slow while selection on the basis of individual lambings would be practically ineffective.'

Personally, I am in entire agreement with the final sentence. A rapid increased in breed fertility by the simple procedure of culling the singles and breeding from the twins sounds just a little too easy to be altogether true.

*Nutritional methods of increasing prolificacy*

Within the genetic limitations of fertility considerable variation exists in many sheep breeds and crosses, and this can be influenced by nutritional manipulation.

Since certain breeds, such as the hill or mountain breeds, are maintained customarily on a low nutritional plane, error may arise in attributing a low average lambing percentage in a breed to genetical causes when, in fact, their genetic potential for fertility has never been attained because of ill-feeding. This is clearly demonstrated where hill ewes are transferred to lowland pasture, when their lambing percentage is frequently of a different order from that recorded previously under hill conditions. Fifty years ago, in 1927, White and Roberts[23] found that Welsh ewes mated on their native hills had a lambing percentage of only 90 per cent, whereas with the same ewes mated on good lowland pasture the percentage at once rose to 120 per cent. I obtained very similar results with Scottish Blackface ewes and, indeed, it is the universal experience of flockmasters. The low lambing percentage found in hill sheep under hill conditions is the fault, primarily, of the conditions and not of the sheep. It follows that improved nutrition should lead to greater prolificacy in ewes, and within limits it may be expected to do so.

When hill ewes are brought down to lowland pastures to mate in autumn they are still lean but are in rapidly improving condition when mated. Their response, by an increased conception rate, is one example of a very general practice in traditional sheep husbandry called 'flushing the ewes'. The effect is due mainly, if not entirely, to an increased proportion of multiple births. A massive weight of evidence has accumulated on the subject since the original observations of Heptner,[24] Heape[25] and Marshall[26] confirmed shepherds' practice.

Flushing has been defined by Thomson and Aitken[27] as the practice of giving ewes which are in fairly poor condition an improved diet for a few weeks before mating so that they are in rapidly rising condition when they meet the ram.

This practice consists in letting ewes down in condition after their lambs are weaned, usually by running them on inferior pasture or other bare keep; then, by improving their nutrition a few weeks before tupping, bringing them into rapidly improving condition when they meet the ram. The precise nutritional procedure employed in securing this rapidly improving condition of the ewes would seem to be of minor importance. A change on to fresh pasture, aftermath, and trough feed have all been used by different shepherds on various occasions, and all with reputed success.

From an analysis of the considerable amount of experimental work done on this subject, Reeve and Robertson[28] concluded that 'not only the level of nutrition at and before the time of mating but also changes in this level may have a considerable effect on the frequency of twins'.

From his classic work on the effect of flushing on reproductive performance Coop[29] suggested:

(1) A static effect – liveweight at mating has an effect on lambing percentages and ewes in good condition have a lower incidence of barrenness and a higher ovulation rate than those in poor condition.
(2) A dynamic effect – the change in liveweight occurring before mating and during the mating period many also have an influence on ovulation rate (OR), in some circumstances.

Since Coop's work, research has been concentrated more on examining independently the main components of lambing performance, namely, OR and embryonic mortality. The former has been investigated in relation to such factors as size, liveweight, body condition, rate of weight change, genotype and the amount, availability and type of food, before and during the mating season.

Widely different and, indeed, conflicting results, which appear difficult to reconcile, have been produced as shown in the study by Geisler and Fenlon.[30] Nonetheless, body condition at mating has been shown by Gunn et al.[31] to markedly influence the fertility of Scottish Blackface ewes. Doney and Gunn[32] have also demonstrated that the effect of body condition on OR response may vary quite markedly between hill breeds.

Morley and his colleagues[33] in Australia have examined the relationship between OR and liveweight in Corriedale, Merino and Romney Marsh ewes, and have concluded that in these types there will be an increase of approximately 2 per cent in OR for every kilogram increase in liveweight at mating.

Using data from a survey of commercial flocks, Geisler and Fenlon[34] suggested that even though body condition improved with increasing liveweight this did not necessarily lead to a discernible effect on lambing performance. This result may be accounted for by the fact that it has been postulated by Gunn et al.[35] that there may be threshold levels of body condition, which may vary between individuals within a breed, and between breeds, above which no ovulatory response is obtained. Also, there may be a level of condition below which the incidence of barrenness will increase.

Ducker and Boyd[36] and Bastiman[37] state that body condition, which can be assessed by condition scoring, is probably a much better practical method of assessing the breeding potential of ewes at mating time, than by using liveweight data which requires more effort to acquire, and can vary considerably between animals of the same breed type within a flock.

The effect of bodyweight change, that is the dynamic effect, before and during mating, has also been demonstrated by Allen and Lamming[38] and Killeen.[39] It could be that it is more readily apparent in ewes in a poor or relatively lean condition, and again there may be differences in the responses of individual ewes, and also breed differences in responsiveness to weight change.

For example, both the Scottish Blackface and Cheviot breeds were found by Gunn and Doney[40] and Gunn et al.[41] to respond to premating nutrition only at intermediate levels of condition.

In a study with Greyface ewes by Gunn and Maxwell,[42] those gaining, maintaining, and losing weight about mating time produced 196, 178 and 158 per cent lambs respectively. There was no difference in initial weight between those ewes gaining or losing weight, suggesting that weight loss probably should be avoided.

There would appear to be little merit in deliberately inducing a severe drop in condition of ewes postmating, unless there are sound reasons in relation to the availability of pasture or other feed. Indeed, a severe drop in body condition, even when followed by flushing, can be counterproductive by reducing lambing rate, as found by Botkin and Lang.[43]

An improvement in condition in most systems is likely to be based on the provision of adequate pasture of reasonable quality, and the customary practice is to save pasture for the period before and during mating, which can be a time of year when there is minimal, or little herbage growth.

When the allowance of pasture for ewes being flushed was increased from 2 to 8 kg dry matter (DM) per ewe daily a weight loss of 39 g changed to an increase of 110 g. This was associated with an OR of 1.2 at the lower allowance and 1.8 at the higher one in a study by Rattray et al.[44]

Gunn et al.[45] at the Hill Farming Research Organization (HFRO) have found that North Country Cheviot ewes in good condition (condition score (CS) > 3) consumed 1.5 kg pasture DM just before mating compared with 1.3 kg (CS > 2.5/2.75) and 1.4 kg (CS < 2.25) for moderate and lean ewes respectively when high amounts of pasture (1500–2000 kg DM/ha) were available. They had similar intakes on low amounts of pasture (500–1200 kg DM/ha).

Reproductive performance, particularly of ewes in moderate body condition when they are allowed to eat to appetite, may be linked to the voluntary intake of herbage, its quality and the amount available as suggested in Table 4.1.

**Table 4.1**    Lambing rate, litter size and the proportion of ewes lambing to first mating.

| Condition score (5 weeks before mating) | Pasture amount | Lambing rate | Litter size | Proportion of ewes lambing |
|---|---|---|---|---|
| 3 | High | 1.40 | 1.62 | 0.86 |
|   | Low | 1.10 | 1.50 | 0.73 |
| 2.5/2.75 | High | 1.54 | 1.70 | 0.91 |
|   | Low | 1.43 | 1.57 | 0.90 |
| 1.5 | High | 1.47 | 1.61 | 0.91 |
|   | Low | 0.93 | 1.33 | 0.70 |

The amount of pasture required will obviously depend on the condition of the ewes pre-mating and stocking rates should be carefully adjusted to ensure maximum intake. It is not always possible to provide and maintain liberal amounts of pasture before and during mating. If this is the case, a gradual improvement in the condition of the flock in late lactation and post-weaning may be a sounder management strategy than taking a lot of condition off ewes simply to replace it later.

Ova wastage is another potential source of loss and the two studies by Edey[46,47] show that it is increased with both under- and overnutrition in the post-mating period. The duration and severity of undernutrition are particularly important, as is also a sudden change in nutritional provision.

Most of this loss occurs between 13–30 days of pregnancy and some is unavoidable due to genetic defects and other factors. Extremes including high temperatures and cold combined with wind and rain have been implicated by Gunn and Doney[48] as causes of early embryo loss as well as other stresses, such as handling and transporting ewes during the period of implantation.

The aim in management should be to maintain body weight for about 30 days post-mating and to avoid unnecessary disturbance of the flock.

*Hormonal methods of increasing prolificacy*

The precise physiological mechanisms whereby the changes in body condition, in the short or longer term, are mediated in ovulatory response have not been worked out.

The occurrence of oestrus, and the number of eggs shed, are dependent on the response of the ovaries to the stimulus of the gonadotrophic hormones, mainly follicle stimulating hormone (FSH), and luteinising hormone (LH), both of which are secreted by the pituitary gland.

During the oestrus cycle, plasma progesterone levels rise slowly from around 3–4 of the cycle, reach a plateau about mid-cycle and fall rapidly 2 days before oestrus, remaining low throughout oestrus. There is a high release of LH 6–8 hours after the onset of oestrus, to around 50–100 times the levels during the other periods of the cycle, with a rapid fall after 8–10 hours, as studied by Land.[49]

The number of eggs normally shed can be artificially increased by super ovulation techniques.

Gonadotrophic hormones are present in high concentration in the blood serum of pregnant mares at a certain stage in their pregnancy, and pregnant mares serum gonadotrophin has been widely used in inducing a high ovulation rate in ewes.

The effect of such injection is to raise the level of gonadotrophic hormones in the ewe's blood (her own plus the amount injected) to an artificially high level. Multiple ovulation results. Instead of 1–3 ova being shed at each heat period, much greater numbers of follicles ripen simultaneously. The number is roughly proportional to the dose. Thus, Robinson[50] found the following relation between the strength of the injection measured in international units (IUs) of PMSG and the mean OR as assessed by counting the corpora lutea on the ovaries of ewes slaughtered after injection. Injection of 500 IU PMSG led to the shedding of 4.1 ova; 1000 IU to 10.6 ova; 2000 IU to 15.8 ova. To prove reliable, the injection of PMSG should be made on the twelfth day of the oestrus cycle – that is to say, about 4 days before the ewe is due to come into normal heat again.

Although PMSG injection into ewes during their breeding season leads to multiple ovulations reaching double figures, the increase in the number of lambs born is less dramatic. A high proportion of the ova shed are lost, owing to failure of fertilisation or of implantation, to early absorption or abortion. Thus, in Robinson's[51] experiments the lambing percentage (lambs born to ewes mated), in spite of a very much greater difference in OR, was only 165 per cent compared with 147 per cent in uninjected controls.

The response obtained from the use of PMSG is likely to vary between breeds, those having a higher inherent fertility being more responsive.[52] The dose required to stimulate successful super-ovulation may also be higher for breeds with a larger-than-average litter size.

Gordon[53] conducted an extensive series of field trials in this country. During four breeding seasons (1952–56) and using over 2500 ewes of nine breeds, he tested the effects of PMSG injection upon flock fertility, using dosage levels of between 250 and 1000 IU. In the majority of the trials the date of the oestrus preceding injection was ascertained by running the ewes with raddled vasectomised rams. Injection was performed on the twelfth or thirteenth day following the sterile service. Fertile service with normal rams took place 4–5 days later when oestrus had recurred. The results of these field trials were decisive and statistically significant PMSG increased fertility in commercial flocks, more markedly in those of initially low fertility. The lambing percentage of injected ewes was 1.81, compared with 1.51 in the uninjected controls, this increase being due to the greater number of multiple births. The relative proportions of singles, twins, triplets, quadruplets, and above, expressed as percentages, are shown in Table 4.2

Table 4.2

|  | PMSG Injected ewes | Uninjected ewes |
|---|---|---|
| Singles | 35.5 | 51.4 |
| Twins | 51.6 | 46.2 |
| Triplets | 9.7 | 2.4 |
| Quadruplets | 2.6 | 0.0 |
| Litters (5 or more lambs) | 0.6 | 0.0 |

Because of the considerable variability in response to PMSG the appropriate dose will depend on the breed used and the objectives of the system. Nonetheless, there is a lack of information on breed response since PMSG has been used only to a limited extent in sheep husbandry practice, as it is only in more intensive systems that larger litters are sought.

However, Gordon[54] states that the induction of a mild superovulatory effect with a breed such as the Galway, which has an average litter size of 1.28, may bring lambing performance up to a more acceptable level. The ovulation rate in this breed as a result of the use of PMSG is shown in Table 4.3.

**Table 4.3.**

|  | PMSG Nil | Dose level 375 | (IU) (44) 750 |
|---|---|---|---|
| No. laporotomised | 77 | 75 | 71 |
| No. ovulated | 56 | 65 | 66 |
| No. ovulations – Total | 69 | 99 | 138 |
| Mean | 1.23 | 1.52 | 2.09 |

It is possible that much of the advantage gained by increased fertility may be lost by a higher mortality amongst the new-born lambs, as shown by Gordon[55] and in Table 4.4.

**Table 4.4** Early mortality (expressed in percentages) of lambs born to PMSG injected and control ewes.

|  | PMSG injected ewes | Uninjected (control ewes) |
|---|---|---|
| Singles | 6.5 | 6.4 |
| Twins | 10.5 | 9.3 |
| Triplets | 20.9 | 13.9 |
| Quadruplets | 38.9 | — |
| Litters | 90.6 | — |

The high rate of mortality of triplets and quadruplets is by no means peculiar to flocks injected with PMSG. Under any system of sheep husbandry it is difficult – it may prove impossible – to secure a high rate of twinning, which, under conditions of lowland sheep farming is always desirable, without at the same time becoming burdened with triplets or quadruplets which may prove something of a liability. The difficulty may be dealt with by twinning on the odd triplets to ewes lambing singles.

There are several techniques available to deal with the problem of rearing surplus lambs. One of these has been described by Allan Fraser in his book *Sheep Farming*.

Alternatively, the problem can be met by the artificial rearing of surplus lambs by one of the several methods recently developed. It seems highly probable that with greater attention to the feeding of the pregnant ewes and a better-planned distribution of the surplus lambs born, the wastage of lambs following PMSG injection of ewes might, at least, be partially avoided.

An entirely new field of investigation has developed which has considerable potential for sheep husbandry practice in the future. Basic research has led to the discovery that ewes can be immunised against certain ovarian steroid hormones.

There are two basic immune techniques: 'active immunisation' against hormones is similar in principle to vaccination to increase antibodies. The other method is 'passive immunisation' using antiserum against the hormone and this is produced in another animal.

Immunisation against a number of hormones is possible and the development of suitable immunogens will be an active area of research. In the 'active' method the immunogen is injected into the ewe about 2 months before mating with a second booster injection 1 month after the first. In subsequent years a single booster 1–2 months before mating will be all that is required.

Results obtained in New Zealand by Smith et al.[56] show a consistent response with 14–28 per cent more lambs from immunised Coopworth and Romney ewes (Table 4.5).

**Table 4.5**  Effect of immunisation and nutrition on reproductive performance of ewes.

| Nutrition | Immunised | | | | Control | | | |
| | High | | Low | | High | | Low | |
| --- | --- | --- | --- | --- | --- | --- | --- | --- |
| Year | 1980 | 1981 | 1980 | 1981 | 1980 | 1981 | 1980 | 1981 |
| No ewes | 52 | 36 | 52 | 36 | 52 | 36 | 52 | 36 |
| Ovulation rate | 2.18 | 2.19 | 1.81 | 2.12 | 1.83 | 2.16 | 1.52 | 1.60 |

These data indicate that the effects of immunisation and improved nutrition are additive. Immunisation appears to be effective irrespective of level of nutrition at mating and works even in ewes actually losing weight.

The level of response in OR seems to be related to the level of antibody titre produced by immunisation. While there is some variation in titre levels between ewes it is not so great as to suggest that this method may offer a more reliable and predictable means of manipulating ovulation rate than those currently available.

This new discovery opens a vast field for investigation. Before commercial application can be recommended, the response of breed and the interaction with body condition and level and changes in nutrition will need to be investigated.

As our knowledge of the fundamental aspects of the physiology of reproduction improves so may other ideas emerge which lead to improved methods for controlling reproduction in practice.

It now remains for us to discuss which of the three methods – genetic, nutritional or hormonal methods of increasing prolificacy are best adapted to sheep husbandry.

At first sight the simplest solution would seem to be the use of a breed of sheep such as the Finnish Landrace which has an average litter size, according to Read,[51] of 3.4. However, other traits are also important and this breed has some limitations.

Provided a breed possesses this innate and heritable prolificacy, nothing more in the way of husbandry is required than to provide adequate feeding for the ewes and planned provision for the greater number of lambs born. There are, however, certain disadvantages equally evident.

Prolificacy is only one of the many desirable characteristics of a sheep breed, so that a breed possessing this quality in full measure might be lacking in others of equal importance, for example wool quality, mutton quality, and rapid growth rate. An obvious answer to this difficulty is crossing, thereby aiming to combine the prolificacy of one breed with the superior wool, mutton, or growth rate of others. Unfortunately, this does not work out quite so simply in practice. As regards prolificacy, the first cross would be roughly intermediate between the two breeds. This was demonstrated in the sheep-breeding trials conducted by ABRO. Whereas the average number of lambs born to Finnish Landrace shearling ewes was 3, that of Finnish Landrace × Blackface first-cross shearling ewes was 1.9 states Read.[58] With any further crossing, the prolificacy of the Finnish sheep would suffer further dilution.

An alternative breeding policy might be to preserve the exceptional prolificacy of a breed such as the Finnish Landrace by using the ewes as basic breeding stock and crossing them once, and once only, with rams of the highest mutton quality, such as Southdown, to produce lambs which, both male and female, would be used not for further breeding but for early slaughter.

It would be an advantage were there a rather clearer appreciation of the desirable level of prolificacy of ewes in relation to the system of husbandry under which they are customarily maintained. Increased prolificacy, whether secured by breeding, feeding or hormone manipulation, is not necessarily advisable in every situation.

In hill sheep flocks, as generally managed in Britain and in extensive systems elsewhere, even twin lambs may be undesirable. The lambs born are greatly dependent on the ewe's milk and while the pasture may be adequate to support the milk yield of a ewe suckling a single lamb it may not suffice for the production of milk for twins. Where the provision of higher quality pasture is possible and smaller lambs can be finished then twins may be desirable.

Again, in flocks kept primarily for wool production a degree of prolificacy greater than that required to maintain stock numbers is in no way essential. Indeed, it is recorded by Heape[59] that in the ancient sheep husbandry of Spain, Merino lambs surplus to that purpose were slaughtered.

Under lowland conditions of sheep husbandry, where the seasonal lamb crop is marketed off-pasture, anything beyond a 200 per cent lambing – 2 strong lambs to each ewe – may be an embarrassment. Indeed, even on the best-cultivated pastures, whether in Britain or in New Zealand, it has been proved that the milk yield of ewes of the breeds customarily kept is not sufficiently high to allow the potential growth rate of even twin lambs to be fully expressed. In this situation an endeavour to increase milk yield rather than prolificacy would seem sounder policy.

Furthermore, the need for greater labour efficiency, by increasing the flock size per man, militates against seeking high litter numbers in some flocks. It is in systems aimed at more frequent breeding, such as the Rowett indoor frequent breeding system which aims at a 205-day reproductive cycle, described by Robinson,[60] or a more simplified version producing 3 lamb crops in 2 years, that hormone treatment could have an important role.

No doubt it is with the possibility of such intensification on a commercial scale that so much experimental work is being devoted to the subject at the present time. In such a husbandry situation with the production of 'broiler lamb' as the eventual objective, prolificacy much beyond the presently accepted level becomes a first essential. Whether the best approach is through the avenue of breeding by the importation and multiplication of naturally prolific breeds such as the Finnish or Romanov, or by that of PMSG injection, may well be a matter of opinion and one for discussion. Among research workers the answer would most probably depend upon whether they happened to be trained in the discipline of genetics on the one hand or of endocrinology on the other. Considered more objectively from the implication angle of sheep husbandry itself, it might be argued that PMSG injection or immunisation applicable to those sheep breeds best designed for maximum meat production might prove the more effective and direct solution.

## Control of the mating season

As already mentioned, it is characteristic of the majority of British sheep breeds to mate in autumn and lamb in the spring.

The Dorset Horn breed is perhaps the most striking exception, since 10–40 per cent of ewes will show oestrus activity in the April–May period. As Lees[61] states, continuous access to the ram outwith the breeding season may stimulate some ewes in other breeds to a fertile oestrus.

For centuries, the mating of sheep has been controlled through human agency, by which the mating period is arranged to permit lambs being born at the most desirable period, which naturally varies according to conditions of husbandry. In some primitive societies the ram may be run with the flock throughout the year. Even in these circumstances peaks of sexual activity, and consequently of lambing, tend to occur.

Since traditionally sheep systems in Britain are primarily based on pasture, the majority of ewe flocks are mated to lamb from January to May depending on circumstances.

Thus, under Scottish lowland conditions the rams may be put out to the ewes in October with a view to lambing on early spring pasture in March. Under hill conditions, on the contrary, a March lambing would be disastrous, since hill pasture remains dormant until late April. Consequently, under hill-farming practice in Scotland the rams are put out towards the end of November with a view to lambing in the last week of April and in early May. Nevertheless, left to Nature, our hill breeds would mate very much earlier, as many hill shepherds have learnt to their cost should the rams break loose before the appointed time.

Under upland conditions, in a study by Maxwell[62] lambing in March gave rise to a 14 per cent mortality in new-born lambs whereas in the same experiment an April lambing flock had a loss of only 6–8 per cent of lambs born.

Control over the mating season may have several aims. Within the context of normal lambing times it may shorten the lambing season thus allowing more effective deployment of labour and also lead to the production of more uniform lots of lambs for sale.

The successful use of AI is virtually dependent on efficient oestrus synchronisation which is also an important component of intensive systems in which stimulation of out-of-season breeding is also essential.

## Synchronisation of oestrus

Much basic research on the endocrinology of the oestrus cycle has contributed to the development of techniques for the synchronisation of oestrus within a ewe flock. After a period of relative failure, probably due to a too-rapid introduction into practice of inadequately understood techniques, modern methods are now much more reliable.

When the egg or ovum is shed from the follicle in the ovary it may or may not be fertilised and, further, it may not develop into an embryo. In the absence of an embryo the follicle develops into a corpus luteum, a gland which then secretes progesterone for around two-thirds of the oestrus cycle, when synthesis of this hormone begins to decline. This period is usually referred to as the 'critical phase' when changes occur in the reproductive tract to adapt it for the nourishment of an embryo. Once the uterus is in a suitable state, a uterine secretion, the hormone prostaglandin $F_2$, causes the corpus luteum to regress (luteolysis) and under the influence of pituitary gonadotrophic hormones follicular development and oestrus follow soon thereafter.

Because follicular growth is prevented by progesterone the principle of oestrus synchronisation is based on the use of this hormone, or more frequently a synthetic analogue, administered at a relatively high level, to artificially prolong the luteal phase. If this is coupled with a sudden withdrawal, for all ewes in a flock, the majority can be expected to show oestrus within 2–3 days.

Two progestogens have been widely and successfully used: Cronolone (G. D. Searle) at 30 mg per ewe and MAP (Upjohn Ltd) 60 mg per ewe according to Smith et al.[62]

The development of a polyurethane pessary (plastic sponge) impregnated with the progestogen was a major advance. They are usually left in the vagina for a 14-day period.

Early experience in commercial practice occasionally produced low levels of fertility and best results are achieved either by service at the second heat following pessary withdrawal or by the administration of PMSG at sponge withdrawal.

Two active areas of investigation are referred to by Gordon[63] in his review of the hormonal control of reproduction in sheep. One is the use of subcutaneous implants with a range of progestogens, but he suggests that this technique is not yet superior to vaginal sponges. Another is the injection of $PGF_2$ but this method is as yet only of research interest.

The control of reproduction leading to more or less continuous breeding throughout the year, or at any time chosen, is the aim in intensive systems. Both oestrus synchronisation and progestogens have important roles.

A standard technique has been developed by Gordon[64]. This involves the use of vaginal sponges inserted for a 14-day period. These are dusted with an antibiotic and impregnated with a suitable progestogen. On removal, a single intramuscular injection of 370–750 PMSG is given. Forty-eight hours after sponge withdrawal, rams of known breeding ability are introduced in the ratio of 1 ram per 10 ewes.

This technique has been used in Eire at all seasons of the year and with pre-pubertal, dry anoestrus, cyclic dry and anoestrus milking sheep. Although a high oestrus response to the treatment has been reported, conception rates vary from normal levels of 78 per cent in autumn to 37 per cent in spring. These results have been obtained under outdoor conditions and housing of ewes should introduce greater scope for control. Robinson's technique[65] involves hormonal control of oestrus and abrupt weaning when lambs are about 7 weeks of age. Synchronisation of oestrus is achieved using vaginal sponges impregnated with Cronolone SA-9880, (G. D. Searle) and 400 mg progesterone and withdrawn after 12 days. Over three consecutive cycles of 205 days, 74 per cent of ewes completed all cycles to average 3.68 lambs per year.

Synchronisation of oestrus, apart from its application in more frequent breeding systems has many advantages in other systems. However, the poor results obtained initially by farmers led to a decline of its use in practice. Research has produced improved products and these combined with mating at the second oestrus after synchronisation offer a fairly reliable method which merits wider use in many husbandry conditions and systems.

*Stimulation by the ram*

The stimulation of early oestrus by introduction of the ram has been shown to advance the onset of the breeding season. This technique has been known to farmers, and indeed practised by them, for a considerable time.

Thus, Underwood *et al.*[66] made the observation that when the ram was turned into a flock of Border Leicester and Meino cross-ewes a peak number of ewes came into heat some 18–20 days later. These authors suggested that the presence of the ram had stimulated the occurrence of oestrus in the ewes.

It has been suggested that within 6 days of the ram being introduced a first silent heat may occur. It certainly has been shown that if rams are introduced no earlier than 3 weeks after lambing some stimulation of oestrus may occur.

The mechanisms which are triggered with the introduction of the ram are not well understood. Robinson's[67] suggestion that, 'it is not unlikely that in those breeds in which the anoestrum is not very deep, introduction of the ram towards the end of the anoestrum may stimulate the pituitary to exceed the threshold for ovulation and advance the onset of the season' seems possible.

This simple technique, although effective, is limited in scope. It seems unlikely that it can advance the breeding season more than a couple of weeks, but in aiming towards early fat-lamb production that may be a matter of practical importance. There is the further advantage in that an advance in the onset of the breeding season appears to be correlated with a degree of synchronisation in that the mating season, and consequently the lambing season, is somewhat condensed, as stated by Robinson.[68]

There is some evidence to suggest that rams stimulate oestrus in non-cyclic ewes through olfactory receptors in the ewe. In other words, smell is important.

In attempting to summarise the methods available for controlling the mating season of ewes, some obviously offer greater possibilities of practical application to husbandry than do others. Thus, again making the broad distinction between genetic and environmental factors, breeds of sheep that tend to cycle naturally throughout the year are at an evident advantage. The Dorset Horn, among British sheep breeds, has this particular distinction, and it has in fact and for many years now been used extensively throughout the country when early lambing or out-of-season lambing is the flockmaster's aim. Fortunately, the breed possesses other characters of commercial value, being prolific, milky, docile, and of good mutton and wool quality. The answer to the problem in this case appears so simple that it might seem unnecessary to explore alternative solutions.

However, the Dorset Horn or its crosses which are perhaps now more widely used may not suit every husbandry situation, and it is certainly valuable to have at command methods less dependent upon environmental factors, which provide control of the mating season in other sheep breeds and crosses.

Certain of these environmental control methods are much simpler than are others.

The introduction of rams, whether vasectomised or entire, previous to the onset of the normal autumnal breeding season in view of the experimental evidence now available, can be regarded as a method of proved utility in advancing mating and, consequently, permitting an earlier lambing. The technique has the further advantage of securing a greater synchronisation of oestrus.

Light is the primary environmental factor controlling the natural onset of oestrus in seasonal breeding sheep. Artificial alteration of illumination can alter the breeding season. The technique requires housing, but not necessarily the darkened sheds used originally by research workers.

Associated with photostimulation, or on its own, the use of progestogens in conjunction with gonadotrophins, is an effective means for synchronisation of oestrus and of achieving out-of-season lambing.

## Effects of age

The age of the female influences two aspects of reproduction – onset of breeding activity and level of performance during the reproductive lifetime.

Sexual maturity or the onset of puberty is reached considerably before the attainment of adult mature size and can occur at 40–80 per cent of adult weight. There are undoubtedly breed and even strain differences which determine the age when puberty is attained. However, nutrition is also of considerable significance since, in general, faster growth during rearing will favour an earlier onset of oestrus. Behavioural signs of oestrus in ewe lambs is usually weak and the first oestrus in the breeding season tends to be later than in adults. The conception rate of ewe lambs is extremely variable and 80 per cent lambing can be considered good.

Although the mothering ability of ewe lambs may be comparable with adult ewes, there is a tendency for high perinatal mortality in their offspring. This may be partly attributable to litters having low birthweights and to an increased incidence of dystokias.

Breeding from ewe lambs is not detrimental to subsequent reproductive performance since any retardation in growth and development which may occur can be overcome, by 2–3 years of age, if nutritional requirements are satisfied.

Dyrmundsson[69] has comprehensively reviewed the work on ewe lambs relating to puberty and early reproductive performance and states that nutrition during the early life of the female may have a profound influence on its subsequent reproductive performance. Poor nutrition during the juvenile phase, that is up to first oestrus, particularly if severe and of long duration, can lead to long-term persistent effects on adult reproductive performance (Table 4.6).

**Table 4.6** Mean liveweights at 12 and 54 months of age and lambing percentages in five successive years of Scottish Blackface ewes in different nutritional environments during rearing and adult life.

| Nutrition Level | | Liveweight (kg) | | Lambing % in 5 successive years | | | | | |
|---|---|---|---|---|---|---|---|---|---|
| Rearing (0–12 months) | Adult life (12–78 months) | 12 Months | 54 Months | 2 Years | 3 Years | 4 Years | 5 Years | 6 Years | Mean |
| H | H | 38.5 | 65.7 | 169 | 176 | 167 | 161 | 177 | 170 |
| L | H | 27.5 | 60.1 | 114 | 143 | 129 | 167 | 147 | 140 |
| H | L | 38.7 | 53.1 | 96 | 129 | 139 | 138 | 122 | 124 |
| L | L | 27.2 | 46.3 | 84 | 124 | 163 | 127 | 92 | 118 |

H = high; L = low plane of nutrition

Such effects are unlikely to occur if compensatory growth is possible during the rearing phase. However, the effects of extreme treatment, with no compensatory growth in early life, cannot be overcome by means of a superior nutritional environment in adult life. This finding suggests that maturity and size of ewe has in some way an influence on reproductive potential.

The length of the period of nutritional deprivation which will affect reproduction is, however, uncertain. In Australia work with Merinos, where lambs had an unrestricted plane of nutrition from birth to 8 weeks of age, followed by severe restriction to 14 months, subsequent performance was as good as those well-reared throughout. There may therefore be a critical development period, but existing knowledge is insufficient to indicate the extent to which the rate of development, either at the foetal stage or later, may be of importance.

When high reproductive performance is sought, then replacement female stock should be well-grown at weaning and also when first mated. Farmers are prepared to pay more for well-grown replacement stock and this undoubtedly reflects on investment in potential.

Lambing performance increases with age to reach a peak at 3–5 years and begins to decline rapidly after 8–10 years. However, very few ewes are retained for breeding to such an advanced age.

### References

(1)  Hafez, E. S. E. (1952) 'Studies on the breeding season and reproduction of the ewe,' *J. Agric Sci*, Camb. **42**, 241.
(2)  Lees, J. L. (1969) 'The reproductive pattern and performance of sheep.' *Outlook on Agric.* **6**, 82–88.
(3)  Yeates, N. T. M. (1949) 'The breeding season of the sheep with particular reference to its modification by artificial means using light,' *J. Agric Sci.* Camb. **39**, 1.
(4)  Williams, H. W. (1970) 'The photo-periodicity of British ewes,' *Span*, **13**, 54–56.
(5)  Ducker, M. J., Thwaites, C. J. and Bowman, J. C. (1970) 'Photo-periodism in the ewe. (2) The effects of varying patterns of decreasing daylength on the onset of oestrus in Clun Forest Ewes,' *Anim. Prod.*, **12**, 115–23.
(6)  Robinson, J. J., Gill, J. C. (1971) 'A note on observations on the effect of photo-stimulation and hormone treatment on oestrus activity and lambing performance of North Country Cheviot ewes,' *J. Agric Sci.*, Camb. **77**, 343–45.
(7)  Newton, J. E. and Betts, J. E. (1972) 'A comparison between the effect of various photoperiods on the reproductive performance of Scottish Halfbred ewes,' *J. Agric Sci.*, Camb. **78**, 425.

(8)  Fraser, A. F. and Laing, A. H. (1969) 'Oestrus induction in ewes with standard treatments of reduced natural light,' *Vet. Rec.* **84**, 427–30.

(9)  Anderson, J. (1938) 'The induction of oestrus in the ewe,' *Vet. J.* **94**, 328.

(10)  Hammond J. (Jnr) (1944) 'On the breeding season in sheep,' *J. Agric. Sci.*, Camb. **34**, 165.

(11)  Hunter, G. L. and Aarde, I. M. V. Van (1973) 'Influence of season of lambing and post-partum intervals to oestrus and ovulation in lactating and dry ewes at different nutritional levels,' *J. Reprod. Fert.* **32**, 1–8.

(12)  Engela, D. J. and Bonsma, H. C. (1938) 'One reason why rams sometimes fail to work,' *Fmg S. Africa*, **13**, 178.

(13)  Hafez, 'Studies on the breeding season.'

(14)  Meat and Livestock Commission U.K. (1981) *Feeding the Ewe*, Sheep Improvement Services.

(15)  Cremer, E. (1935) 'Studies on the Fecundity of the East Friesian Milch Sheep,' *Z. Schafz.*, **24**, 120; A.B.A., 1935, **3**, 147.

(16)  Smirnov, H. (1935) 'Prolificacy of the Romanov sheep,' *Probl. Zivotn.* No. 8, **7**; A.B.A., 1936, **4**, 195.

(17)  Read, J. L. (1967) 'Sheep and Lambs,' Animal Breeding Research Organization Report, p.7.

(18)  Reeve, E. C. R., and Robertson, F. W. (1953) *Factors Affecting Multiple Births in Sheep*, Anim. Breed, Abst, No. 21, p. 211.

(19)  Smirnov, 'Prolificacy of the Romanov sheep.'

(20)  Heptner, R. A. (1940) 'Current problems of increasing productivity of Romanov Sheep,' *Sovbok, N.* 2/3, A. Burk, 1941, 9, 317.

(21)  Marshall, F. H. A. and Hammond, J. (1945) *Fertility in Animal Breeding*. Min. Agric. and Fish, Eng. Bull. No. 39, 6th edn.

(22)  Reeve and Robertson, 'Factors affecting multiple births in sheep.'

(23)  White, R. G. and Roberts, J. A. F. (1927) 'Fertility and sex ratio in Welsh Mountain sheep,' *Welsh J. Agric.* **3**, 170.

(24)  Heptner, 'Current problems of increasing productivity of Romanov sheep.'

(25)  Heape, W. (1899) 'Abortion, barrenness and fertility in sheep,' *J. Roy. Agric Soc. Eng.* (Ser. 3), **10**, 217.

(26)  Marshall and Hammond, 'Fertility in animal breeding.'

(27)  Thomson, W., and Aitken, F. C. (1959) 'Diet in relation to reproduction and viability of the young. (ii) Sheep: World survey of reproduction and review of feeding experiments,' *Tech. Commun. Bur. Anim. Nutr.*, **20**, 67–76.

(28)  Reeve and Robertson, 'Factors affecting multiple births in sheep.'

(29)  Coop, I. E. (1966) 'The effect of flushing on the reproductive performance of ewes,' *J. Agric. Sci.* Camb., **67**, 305–23.

(30)  Geisler, P. A. and Fenlon, J. S. (1979) 'The effects of body weight and its components on lambing performance in some commercial flocks in Britain,' *Anim. Prod.* **28**, 245–69.

(31)  Gunn, R. G., Doney, J. M. and Russell, A. J. F. (1969) 'Fertility of Scottish Blackface ewes as influenced by nutrition and body condition at mating,' *J. Agric. Sci.*, Camb., **73**, 289–94.

(32)  Doney, J. M. and Gunn, R. G. (1973) 'Progress in studies on the reproductive performance of hill sheep,' *Rep. Hill Farming Res. Org.* **6** 69–73.

(33)  Morley, F. W. H., White, P. H., Kenney, P. A. and Davis, I. F. (1978)

'Predicting ovulation from liveweight in ewes.'

(34) Geisler and Fenlon, 'The effects of body weight on lambing performance.'

(35) Gunn, Doney and Russell, 'Fertility of Scottish Blackface ewes.'

(36) Ducker, M. J. and Boyd, J. S. (1977) 'The effect of body size and condition on the ovulation rate of ewes,' *Anim. Prod.*, **24**, 377–85.

(37) Bastiman, B. (1972) 'The effect of ewe condition at tupping on lambing performances,' *Expl. Husb.* **22**, 1, 22–24.

(38) Allen, D. M. and Lamming, G. E. (1961) 'Nutrition and reproduction in the ewe,' *J. Agric. Sci.*, Camb., **56**, 69–79.

(39) Killeen, I. D. (1976) 'The effects of body weight and level of nutrition before, during and after joining on ewe fertility,' *Aust. J. Exp. Agric., Anim. Husb.* **7**, 126–36.

(40) Gunn, R. G. Doney, J. M. (1975) 'The interaction of nutrition and body condition at mating on ovulation rate and early embryo mortality in Scottish Blackface ewes,' *J. Agric. Sci.*, Camb. **85**, 465–70.

(41) Gunn, R. G., Doney, J. M. and Smith, W. F. (1979) 'Fertility in Cheviot ewes,' *Anim. Prod.* **29**, 17–23.

(42) Gunn, R. G. and Maxwell, T. G. (1978) 'The effects of direction of liveweight change about mating on lamb production in Greyface ewes,' *Anim. Prod.* **26**, 392 (Abstr).

(43) Botkin, M. P. and Lang, R. L. (1978) 'Influence of severe dietary nutrition during the dry period. In subsequent ewe production,' *J. Anim. Sci.*, **46**, 1147–50.

(44) Rattray, R.Wq., Jagusch, K. T., Smith, J. R. and Tervit, H. R. (1978) 'Flushing ewes on pasture and pasture silage,' *Proc. N.Z. Soc. Anim. Prod.* **38**, 101–104.

(45) Gunn, R. G., Smith, W. F., Senior, A. J., Barthram, E. and Sim, D. A. (1983) *Pre-mating Pasture Intake and Reproductive Responses in both Country Cheviot ewes in Different Body Conditions*, Brit. Soc. Anim. Prod. Winter Meeting.

(46) Edey, T. N. (1969) 'Pre-natal mortality in sheep – a review,' *A.B.A.* **37**, 73–190.

(47) Edey, T. N. (1976) 'Embryo mortality,' in: eds G. J. Tomes, D. E. Robertson and R. J. Lightfoot *Sheep Breeding*. Proc. 1976 Int. Congr. Murest. W. Aust. Inst. Tech. pp. 400–410.

(48) Gunn, R. G. and Doney, J. M. (1979) 'Ewe management for control of reproductive performance,' *Adas Q.Rev.*, **35**, 231–45.

(49) Land, R. B. (1970) 'Physiological studies and genetic selection for sheep fertility,' A.B.A., **42**, 155–8.

(50) Robinson, J. J. (1951) 'The control of fertility in sheep,' *J. Agric. Sci.* Camb, **41**, 6.

(51) Robinson, 'The control of fertility in sheep.'

(52) Newton, J. E., Betts J. E. and Large, R. V. (1979) 'Increasing litter size in three breeds of sheep by superovulation,' *J. Agric. Sci.* Camb. **75**, 353–60.

(53) Gordon, I. (1958) 'The use of progesterone and serum gonadotrophin (PMSG) in the control of fertility in sheep,' *J. Agric. Sci*, Camb., **50**, 123–51.

(54) Gordon, I. (1975) 'Hormonal control of reproduction in sheep,' *Proc. Brit. Soc. Anim. Prod.* **4**, 79–93.

(55) Gordon, 'The use of progesterone.'

(56) Smith, J. F., Cos, R., McGowan, L. T. and Wilson, P. A. (1982) 'Increasing

lambing percentage through immunisation,' *Proc. Ruakura Frms Conf.* (34th) Hamilton, New Zealand.

(57) Read, 'Sheep and lambs.'

(58) Read, 'Sheep and lambs.'

(59) Heape, 'Abortion,' barrenness and fertility in sheep.'

(60) Robinson, J. J. (1974). 'Intensifying ewe productivity,' *Proc. Brit. Sci. Anim. Prod.* **3**, 31–40.

(61) Lees, T. L. (1968). 'The reproductive pattern and performance of sheep.' *Outlook on Agric.*, **6**, 82–88.

(62) Smith, Cos, McGowan and Wilson, 'Increasing lambing percentage.'

(63) Gordon, 'Hormonal control.'

(64) Gordon, 'Hormonal control.'

(65) Smith, Cos, McGowan and Wilson, 'Increasing lambing percentage.'

(66) Underwood, E. J., Stien, F. L., and Davenport, N. (1944). Studies in sheep husbandry in Western Australia: (V) The breeding season of Merino, crossbred and British breed ewes in the agricultural districts, I. *Dept. Agric. West. Austra.*, **21**.

(67) Robinson, T. J. (1954). 'Reproduction in the ewe', *Biol. Review*, **26**, 11.

(68) Robinson, 'Reproduction in the ewe'.

(69) Dyrmundsson, O. R. (1973). Puberty and early reproductive performance in sheep. (i) Ewe lambs,' A.B.A. **41**, 6, 173–289.

# 5 Production: Sheep Meat

A study of meat production follows quite logically that of the sexual physiology of sheep, for the biological purpose of the sexual physiology is the conception of a viable lamb or lambs, and the qualities of a butcher's lamb develop continuously from the date of its conception.

A considerable period of the 'life' of a lamb is spent *in utero*, since gestation is about 21 weeks, while the time from birth to slaughter is about 12–16 weeks in the more intensive systems and from 16–24 weeks in many others.

### Factors affecting birthweight and its importance

Quite commonly, the growth of a lamb is assumed to begin from its date of birth, but this is obviously false reasoning. A convenient stage for measurement and weighing of a lamb, however, is at its birth. The birthweight of lambs has, in fact, been used extensively in research for the accurate assessment of a variety of factors affecting the lamb while still unborn. It is a check-up, as it were, of the results of the first 5 (intra-uterine) months of a lamb's life.

Birthweight is an important factor in lambs' survival, particularly in a severe environment. It may also have some influence on subsequent growth rate of the lambs, and Cunningham and Maxwell[1] have traced the general relationship between birthweight and survival illustrated in Fig. 5.1.

With single lambs, survival tends to be greatest at just above the mean birthweight for the breed type. Multiples do not always reach the breed average and survival rate tends to improve with increasing birthweight. At heavier birthweights, mortality increases due to an increase in the incidence of dystokia and the consequences which follow. In contrast, at less than average birthweights losses are more frequently attributable to starvation, and in hill conditions to a combination of starvation and exposure as shown in the study by Dalton *et al.*[2]

In an examination of data involving growth-retarded foetuses, and consequently below-average birthweight, Robinson[3] has suggested that the

96

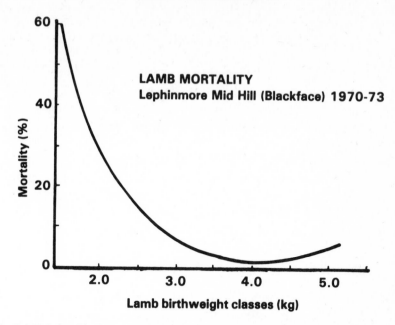

*Fig. 5.1* Relationship between lamb mortality and birth weight in a flock of Scottish Blackface ewes (after Cunningham and Maxwell[1]).

skeletons of small lambs at birth are more developed than younger foetuses of the same weight, and argues that neonatal mortality of small lambs is unlikely to arise from inadequate functional competence at birth. His data indicate that large lambs have more fat in their bodies than small lambs. Of the total lipid present in the new-born lamb, about 50 per cent in the adipose tissue is in the form of brown fat, which is the primary source of energy during the 24 hours following birth. It has a vital role in survival and clearly illustrates the reasons for the survival advantage of heavier lambs. An effective practical means of avoiding starvation and reducing losses of small lambs is to administer colostrum (30 ml/kg birthweight) directly into the lamb's stomach by using a syringe and catheter within a few minutes of birth.

Various factors, although intimately related in practice, are for convenience of discussion considered separately.

### Breed, size and age of ewe

This has an easily understood effect on the birthweight of a lamb, the larger breeds such as the Lincoln having heavier lambs (5.4 kg) than a small breed such as the Welsh (4 kg). Within breeds the larger ewes tend to give birth to larger lambs. The age of the ewe also affects the birthweight of the lamb, although it is frequently difficult to isolate this influence from that of size of ewe, since the younger ages of ewe, not being fully grown, tend to be smaller.

The average birthweight of lambs from ewe hoggs (bred from in their first winter) is always below that of lambs from mature ewes, but then of course, the ewe hoggs are themselves smaller in size.

Allan Fraser suggests that with gimmers, provided they are fully grown, the birthweight of their lambs will be average for the breed and it is only when they are under-grown that birthweight will be below that of mature ewes. Under adverse conditions 2-year-old ewes rarely reach mature size, and in a survey of three Scottish Blackface flocks and one Cheviot flock by Gunn and Robinson,[4] lamb mortality for progeny from 2-year-old ewes was from 100 to 150 per cent higher than amongst lambs from older ewes, with lamb birthweight being 0.31 and 0.54 kg lighter, respectively.

## Breed and size of sire

In Hammond's[5] classical experiments in using artificial insemination (AI) to produce crossbred foals between shire horses and Shetland ponies, the size of the sire appeared to have little or no effect on the size of the foal.

'In reciprocal crosses between large and small breeds, such, for example, as the large Shire horse and the small Shetland pony, the size of the off-spring and weight of the placenta are directly proportional to the size of the mother and are not just intermediate between the sizes in the parent breeds.'

This generalisation, if true, would contradict the general belief of practical stockbreeders, who are extremely cautious in using a large sire on a small dam because of the presumed danger of a difficult birth. It is a generalisation which cannot be safely applied to sheep-breeding, where the size of the lamb at birth is not by any means necessarily directly proportional to the size of the dam. Interesting evidence on this point has been published by Hunter *et al.*[6] In experiments where fertilised ova from one breed of sheep were successfully gestated in the bodies of ewes of another breed, it was found that the birthweight of the lambs, when born, might be quite disproportionate to the size of the ewe. Thus, when ova from Border Leicester ewes were implanted in Welsh Mountain ewes, the birthweight of the Border Leicester lambs, thus artificially gestated, was much beyond the average of Welsh Mountain lambs, although still below that of normally gestated Border Leicester lambs.

In these experiments the average birthweight of pure-bred Border Leicester lambs was 6–6.8 kg; that of pure-bred Welsh Mountain lambs

3.6–3.9 kg; of three Border Leicester lambs implanted in, gestated in, and born out of, Welsh ewes the mean birthweight was 5±0.17 kg, and all three lambs required assistance at birth.

In a less dramatic but equally conclusive experiment, Kincaid[7] also showed that the size of the sire affects the birthweight of the lamb. He mated a flock of 100–150 ewes of mixed breeding with Hampshire rams (a large breed) and Southdown rams (a small breed). The flock was divided into two groups and the breed sires were alternated between groups each year, for a period of 5 years. Lambs sired by Hampshire rams were 0.5 kg heavier at birth than those sired by Southdowns, the difference being highly significant as judged by statistical standards.

It is clear, therefore, that the ram is *not* without effect on the birthweight of its lambs, and that this factor requires careful consideration in practical sheep husbandry.

*Number of lambs born*

Birthweight is considerably influenced by whether lambs are born single, one of twins, or one of triplets. There is an increase in litter size up to around 5 years of age, when it either plateaus, or begins to decline.

A reduction of around 20 per cent in individual lamb birthweight for each additional litter member has been suggested by Donald and Russell,[8] and the conclusion which can be reached from most of the estimates from research indicates that the birthweight of twin lambs invariably lies between 80 and 85 per cent of the equivalent single lamb. Since the birthweight of twin and triplet lambs is likely to be limited by maternal and foetal nutrition, it follows that better nutrition of the in-lamb ewe could reduce differences between singles, twins and triplets, under many circumstances. The precise differences can vary with flock, farm, season and particularly the level of feeding of ewes during late pregnancy.

*Sex of lamb*

The sex of the lamb also affects its birthweight, that for males being, on the average, about 0.22–0.25 kg greater than that of females. The difference between a male born co-twin to a female can be twice that of male co-twins, while a female co-twin to a male will be invariably lighter than those co-twin to another female.

*Other factors*

There is a tendency in some systems for lamb birthweight to increase with time of lambing within the normal lambing season. Perhaps this can be attributed mainly to indirect nutritional effects.

A depression of birthweights from housing of ewes has been reported in Britain by Bastiman and Williams,[9] but this has been overcome by shearing ewes at, or soon after, housing (see the study by Rutter et al.[10]).

## Feeding of the pregnant ewe

The effects of feeding the ewe during pregnancy, and particularly in the later stages, has been the subject of a great deal of scientific investigation. Indeed, nothing was known about this topic when in 1932, Hammond,[11] in dealing with the birthweight ratios between single, twin and triplet lambs wrote: 'Not sufficient evidence has been obtained to determine whether the ratio varies between ewes in good and those in a poor condition before lambing.'

Since then, numerous studies have been undertaken on the growth and development of the gravid uterus, including its components; the uterine wall and membranes, fluid and the foetus itself, at different phases throughout gestation.

The study by Joulbert[12] shows that the increase in foetal weight in early pregnancy is quite small, the foetus being approximately 0.3 g at 25 days and 5 g at 40 days.

Growth of the placenta is substantially completed by around the ninetieth day of pregnancy, while the differential growth of the uterine components is such that the fluids increase over the first 3 months, and again in the final month, whereas the foetus makes rapid gains after day 90, as shown by Robinson et al.[13] and in Table 5.1. At this stage, the foetus is around 15 per cent of birthweight, so 85 per cent of foetal growth takes place over the last 2 months of pregnancy (Fig. 5.2).

**Table 5.1**

Growth of products of conception of twin lambs

| Days gestation | Percentage gravid uterus | | | |
|---|---|---|---|---|
| | Empty uterus | Placenta | Fluids | Foetus |
| 62 | 30.2 | 30.3 | 32.9 | 6.7 |
| 90 | 17.8 | 23.3 | 31.4 | 27.5 |
| 112 | 13.1 | 13.5 | 18.5 | 54.9 |
| 140 | 9.4 | 7.6 | 20.1 | 62.8 |

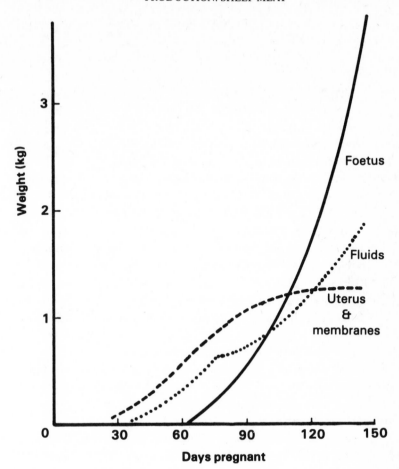

*Fig. 5.2* Foetal and uterine development.

There has been remarkably little investigation of·nutritional effects on foetal development in early pregnancy, but this is not surprising since the foetus is so small. Robinson[14] suggests that embryo loss, which largely occurs over the first month of pregnancy, will be increased only by severe undernutrition, requiring dietary energy levels at low as 15 per cent of maintenance for up to 7 days or so. There are suggestions, however, that overfeeding at this early stage may also significantly increase embryo loss.

No effect on foetal weight was found by Hulet *et al.*[15] when ewes were offered either 75 or 150 per cent of their estimated energy requirements from mating to 21 or 30 days of pregnancy.

Ever since Wallace[16] suggested that a loss of 7 per cent in gross body weight over the first 90 days of pregnancy did not affect foetal growth it has been widely assumed that nutrition over this period was relatively unimportant. Robinson[17] has suggested that foetal weight at 93 days gestation can be affected by previous nutrition and this aspect has been reviewed by Everitt.[18]

When a low plane of nutrition (weight loss 7–9 per cent) and a high plane (weight gain 6–16 per cent) were imposed, both before mating, and for 90 days of pregnancy, the cotyledons, a component of the placenta, were reduced in weight, by poor nutrition, to 90 per cent of those in the well-fed group. A slight reduction in foetal weight was also noticed by Clark et al.[19] Other studies, involving a more severe nutritional restriction and leading to mid-pregnancy weight loss of 11–12 per cent, have shown substantial reductions in foetal weight, these being around 87–89 per cent of the foetal weight of ewes improving in condition.

The important practical question is whether compensation may be possible in late pregnancy, to counteract the effects of poor mid-pregnancy nutrition. Robinson[20] states that young sheep which have not reached mature weight may not be capable of responding. Indeed, it has been suggested that even mature ewes, after a period of mid-pregnancy undernutrition, may partition additional nutrients, provided in late pregnancy, to replenishing body reserves and less to foetal growth.

The age of the ewe, litter size, her initial body condition, possibly the rate of weight loss, could all influence the extent to which undernutrition in mid-pregnancy may have an effect, and also the potential of the ewe to compensate at a later phase.

Until more precise information becomes available it would seem to be unwise to allow a mid-pregnancy weight loss of more than 7 per cent to occur because of possible effects on foetal development and birthweight. This would be represented by a change of 0.5 of a condition score (CS) grade.

In contrast, it has been suggested by Robinson[21] that gross overnutrition in mid-pregnancy can lead to a 40 per cent reduction in lamb birthweight. It is unlikely that this will be a problem frequently encountered in practice, through it may be a consideration in some circumstances.

It is well recognised that ewes improve in body condition in late lactation and post-weaning, reaching a peak condition in late autumn. Most systems rely on the capacity of the sheep to utilize these reserves during periods of inadequate nutrition. This is well illustrated by Russel et al.[22] in the case of hill ewes, which during pregnancy may lose 85 per cent of subcutaneous fatty tissue and up to 20 per cent of their empty bodyweight, or 12–14 per cent of their liveweight, yet produce a viable lamb at birth. Lodge and Heaney[23] found that the growth of the products of conception, including the mammary gland, occur at the expense of maternal tissues, with twin-bearing ewes losing 20 per cent more body weight than those carrying singles.

Lodge and Heaney[24] also discovered that catabolism of body fat continues even when ewes are in positive energy balance. Thus liveweight changes can readily mask changes in body composition, which tend to be greater in terms of energy loss in late pregnancy.

There may indeed be differences between breeds in their abilities to utilize body tissues, and it has been suggested by Robinson[25] that the effects of undernutrition on birthweight may be greater in the hill types, such as the Blackface and Welsh Mountain, than in the Finn-cross Dorset Horn. It seems reasonable to assume that a fatter ewe may mobilise tissue more readily than a thinner one, but this is an aspect which is inadequately quantified.

The relevance of nutrition depends not only on its adequacy, but also on its timing during pregnancy. The most tangible effects have generally been found consequent on changes in late-pregnancy feeding.

Early studies by Wallace[26] at Cambridge, where ewes were fed so as to gain 20 kg or lose 5 kg over the last 6 weeks of pregnancy, produced lambs at birth weighing 4.7 and 3.1 kg respectively, a difference of 53 per cent.

As a result of extensive field trials conducted in New Zealand, Coop[27] found that, although a high level of nutrition led to higher lamb birthweights it had little, if any, effect on their subsequent rate of growth and weaning weight.

Some of the apparently contradictory results which have appeared on the effects of late pregnancy nutrition can be explained on the basis of the nutritional status of the ewe. This can be determined by the use of biochemical parameters, such as blood plasma, free fatty acids (FFA) or ketone bodies, as shown by Russel.[28] By this means it can be shown that ewes on a relatively low dietary plane may be less undernourished than a group on a high plane bearing heavy foetal burdens.

Using criteria of this type Russel et al.[29] compared three groups of ewes in late pregnancy: (1) Control: adequately nourished; (2) Slightly undernourished: equivalent to Blackface hill ewes with single lambs; and (3) Severely undernourished: equivalent to Blackface ewes with twins.

Both single and twin lambs were affected to the same extent, in that singles in groups (2) and (3) had birthweights reduced to 91 and 75 per cent and twins to 90 and 70 per cent respectively of the controls. In contrast to previous research, where different but uniform planes of nutrition were imposed, ewes in this work were offered different amounts of food so as to sustain similar levels of nutritional status.

In the adequately nourished ewe in late pregnancy, as determined by studies on foetal oxidative metabolism by Girard et al.[30], energy is derived from glucose, lactate, amino acids, acetate and glycerol, in the order of 50, 25, 20, 5 and 1 per cent respectively. The corresponding figures for the severely undernourished ewe are 30, 15, 60, 0 and 1.

Many researchers are in agreement that the energy requirements of the foetus are in the order of 1.4–1.6MJ/kg foetus per day with a suggestion of a decline per unit weight in late pregnancy. Expressed in terms of multiples of maternal maintenance (0.42MJ/kg–0.75) requirements at day 100, 120 and 140 are:

| Day | 100 | 120 | 140 |
|---------|-----|-----|-----|
| Singles | 1.2 | 1.4 | 2.0 |
| Twins | 1.5 | 1.9 | 2.6 |
| Triplets | 1.6 | 2.1 | 2.8 |

This means that immediately before lambing, a 50 kg ewe with a 4.5 kg single lamb needs theoretically twice the energy intake of a similar dry ewe to maintain weight, (Fig 5.3).

Russel et al.[31] state that such absolute requirements need not be met in practice, since even a reduction of 25 per cent of these requirements will lead to only a 10 per cent fall in birthweight, at which viability is generally satisfactory. In any event, the buffer whereby the ewe can mobilise her body reserves without detriment to foetal weight is an important factor and is probably one reason why birthweight is an imprecise index of the adequacy of energy intake in late pregnancy.

Further, as mentioned earlier, there may be breed differences in their tolerance of undernutrition, before birthweight is affected, but this is a phenomenon which remains to be clarified.

It is now thought for example, by Robinson et al.,[32] that efficiency of energy retention in the foetus is improved as dietary energy concentration increases.

Because of the contribution of amino acids to foetal energy requirements, the detrimental effects of undernutrition may be more evident on low-protein diets, and this has been demonstrated in a number of experiments by Robinson,[33] where a reduction in lamb birthweight has occurred. This has led to the suggestion that the inclusion of protein foods of low degradability (for example, 10 per cent white fishmeal in a concentrate mixture) may have a beneficial effect on lamb viability and in ensuring an adequate supply of colostrum. The provision of an optimum diet for the pregnant ewe remains a complex and baffling problem. Computer-based models by Giesler et al.[34,35] have been developed to calculate energy requirements of pregnant

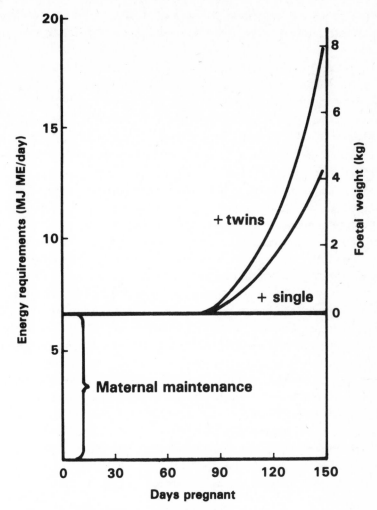

*Fig. 5.3* Energy requirements of the ewe during pregnancy.

ewes carrying any predicted numbers of foetuses in relation to response to energy in the diet. More than anything these new techniques highlight areas of deficient knowledge.

It has long been known to experienced shepherds that the important period of the pregnant ewe's feeding is the last 4–6 weeks, when extra feed, if available, puts size on the lamb and milk in the ewe. These facts, although until quite recently unsupported by exact figures, have for long been acted upon in the best Scottish sheep-husbandry practice. On the Scottish Borders, as far back at least as 1915, and probably much earlier, it was routine practice to put out the feeding troughs to the lambing flock 4 weeks before they were due to lamb and to give them concentrates, starting with 0.25 kg a head and working them up to 0.5–1 kg by the time they came to lamb.

The application of such management techniques is further improved where ewes of similar gestational date are grouped together, having been identified by the use of marker crayons at mating. Further preferential treatment according to body condition and age, for example, segregation of gimmers, can lead to more effective use of expensive food. A cheap and reliable means of identifying foetal numbers would make a significant contribution to feeding practice.

But ewes are normally fed in groups which have their 'pecking order' so that aggressive ewes tend to push out the lighter ones while differences in rate of eating have also been shown to occur (Foot et al.[36]).

Uniformity of intake is more likely to be achieved with roughages, suggesting that the best results will be obtained with high-quality conserved forages requiring a minimum of supplementation.

A level pattern of feeding concentrate supplements, as opposed to the usually recommended ascending pattern, has been found to be satisfactory with housed ewes, but the precise conditions under which this might be applicable have not been worked out, to indicate under which circumstances this procedure might be used in practice.

From a scientific view, the important influence of the feeding of the ewe, at different phases of pregnancy, on the development and birthweight of her lambs, has advanced rapidly in recent years. The practical flockmaster is well aware that well-fed pregnant ewes drop big and heavy lambs and that lambs from ill-fed ewes are both small and light. The effect of poor feeding is much greater in reducing the weight than the length and height of the lambs, unless restriction on nutrition is severe and prolonged.

### Growth rate of lambs

In the first month of postnatal life, when lambs are almost entirely dependent on their mothers' milk, the growth of single lambs is almost double that of twin lambs of the same flock. This difference between the growth rate of singles and twins narrows as the lambs grow older and are grazing on their own account, but the twin continues to lag somewhat behind. Observations on Scottish sheep (in this case Down-cross lambs out of Greyface ewes) showed that – within the same flock – single lambs gained at the rate of 454 g a day and reached 36.5 kg in 68 days. Twin lambs from the same flock averaged 4.1 at birth, gained 320 g a day and reached a similar weight at 101 days of age.

This difference in growth rate is unlikely to be due to a lamb being born single or twin, since a good twin lamb reared as a single can have a similar growth rate to a single, and a single raised as a twin will have the growth rate of a twin. These facts emphasise the importance of the correlation between milk yield of a ewe and the growth rate of her lamb.

Nonetheless, there can be a relationship between birth and weaning weights with many estimates suggesting an increase in the latter of 2–3 kg for each additional kilogram of weight at birth.

Age of ewe may also have some influence on growth rate, experience showing that lambs from ewe lambs and even 2-year-old ewes may grow less rapidly than those from older ewes, which in some types will not produce their heaviest lambs at weaning until they are 5–6 years old.

In a study of the daily weight gain of lambs from birth to 135 days of age, Cole[37] obtained the following figures for various breeds.

| | |
|---|---|
| Hampshire | 0.20–0.28 kg |
| Oxford | 0.20–0.34 kg |
| Rambouillets | 0.20–0.30 kg |
| Shropshires | 0.12–0.30 kg |

He concluded that the daily gain in weight of lambs varied much more between individuals than between breeds. Most probably, this finding of Cole's can be explained as being due to a corresponding variation in the milking capacity of the individual ewe concerned.

It would seem, therefore, that the primary factor governing the growth rate of a young lamb is the milk yield of its dam. All other factors, including the breed of sire, are subsidiary to that. Nevertheless, the practical preference for rams of certain breeds to cross with ewes of other breeds suggests that breeding as distinct from nutrition, is of considerable importance in the growth of lambs.

Within the British meat sires – mainly the Down breeds, growth rate potential and mature size tend to parallel each other. In a comparison by Donald et al.[38] of the extreme types, the Southdown and the Oxford, the latter produced slaughter lambs 5.9 kg heavier in liveweight, at very similar ages.

Likewise, heavier ewes tend also to produce faster-growing lambs but they may mature at a heavier weight, as also occurs with the heavier sire breeds.

Although the increase in liveweight of the animal is of direct interest to the farmer, the butcher, and ultimately the consumer, we are also concerned with the composition of this increase, and the value of the product and quality of the meat produced.

## The carcass

The carcass is what remains when all the offal is removed from a slaughtered sheep, and consists of lean tissue or meat, fat and bone.

The carcass weight may constitute between 40–70 per cent of the weight of the live sheep. This percentage is the 'killing percentage' and varies with the age and fatness of the sheep and whether it may have been shorn previous to slaughter, or has a full fleece. Young fat lambs will not kill out at more than 50 per cent and usually less.

The liveweight of a sheep taken on the farm before slaughter may not always be a good guide to carcass weight, unless an allowance is made for weight loss – shrinkage is the technical expression – during transit. This loss in weight can vary from 2–14 per cent of the original liveweight, being greater when sheep have been fed on succulent fodder or are starved for a prolonged period.

### Offal

After the sheep is killed the offal is separated from the carcass of mutton. Almost every constituent of the offal has its uses. The skin is the most valuable one, especially in a fine-woolled sheep slaughtered when in full fleece. The wool off slaughtered sheep – commonly called 'slipe' or skin wool – forms abour 30 per cent of the home-produced wool in Britain.

The skin is by far the most valuable part of a sheep's offal, its value being between 10–20 per cent of that of the whole sheep. The rest of the offal – head and feet, gut, and other internal organs – is of much less value, accounting for about 5 per cent of the value of a slaughtered sheep. Liver and suet are the most valuable of the offals, other than the skin.

## Growth and development

Hammond[39] was the first scientist to make a comprehensive and systematic study of growth development in the sheep, and details were published in his monumental work, *Growth and Development of Mutton Qualities in the Sheep.*

It was one of Hammond's objectives in studying the mutton qualities of sheep to discover to what extent and in what way the so-called 'improved mutton' breeds of sheep were actually superior to other breeds from the butcher's and mutton consumer's point of view. His method consisted in the detailed dissection, measurement, and weighing of carcasses of sheep of types as diverse as the half-wild Soay and the specialised Southdown, together with those of other breeds of intermediate type. The laborious nature of the research, together with the difficulty and expense of securing a sufficient number of carcasses suitable for dissection, made adequate statistical treatment of his results impossible. The number of carcasses dissected was too few, the variation between carcasses of the same breed too great, for the significance of results to be statistically assessed. Nevertheless, Hammond was able to draw several important conclusions which subsequent work on the subject has fully confirmed.

He found that the improved mutton breeds, for example, the Southdown and Suffolk, did, in fact, show to definite advantage over the breeds in conforming to the butcher's requirements.

Thus, in the improved mutton breeds, certain inedible portions of the body – the head, horns, and internal organs – are proportionately less developed. On the other hand, the development of the lumbar and sacral vertebrae – forming the hinder end of the backbone and the solid bony basis for the most valuable cuts – has been greatly increased. The points looked for in judging a mutton sheep – the rather lengthy yet blocky body set close to the ground, the finely-shaped head set well into the shoulders, the broad and even back, the well-developed buttocks and well-fleshed thighs – these are not fancy showyard points, but are truly based on the economic usefulness of the sheep when killed.

In the improved mutton breeds the distribution of fat is more economically useful. The more primitive, semi-wild types of sheep tend to store their fat internally around the kidneys and gut, forming those internal fat deposits which the butcher calls suet. It follows that although a primitive type of sheep has been storing fat and gaining weight over a considerable period, its carcass after slaughter may prove relatively lean. In the improved mutton breeds, however, the deposition of internal fat, following a fattening period, is relatively less, that of carcass fat relatively greater. The fat is laid down beneath the skin, forming the subcutaneous fat, and between the muscles or natural division of the lean meat. It is also deposited *within* the muscles, between the individual muscle fibres, causing the 'marbling' appearance so desirable in choice meat. There is, indeed, no way in which an improved mutton sheep differs more from a wild sheep than in the distribution of its fat.

Growth to maturity in the sheep follows a growth curve which is typically sigmoidal. That is to say, growth acceleration over the first few months begins to slow down by puberty, and from that stage declines progressively towards maturity.

It is the relative growth of bone, muscle and fat which are of importance. Genotypes vary considerably in the growth of bone relative to muscle and this is exemplified in types selected for 'strong' bone.

The potential for muscle growth also varies between breed types. For example, the late T. H. Jackson in a study comparing the Cheviot and Suffolk-cross lambs, over the same range of maturity, 40–80 per cent, obtained muscle gains of 6 and 7.6 kg, respectively. The same study also clearly illustrated the effect of nutrition on rate of muscle growth, with the Suffolk having a higher potential.

On a skeletal basis, muscle tends to represent a relative constant proportion of the fleece-free empty-body (FFEB) weight. This comprises the carcass, with skin and gut contents removed but otherwise intact and including gut tissues.

In a study by McClelland et al.[40] including four breeds ranging from the primitive Soay to the large Oxford Down, and also the Southdown and Finnish Landrace, total muscle as a percentage of FFEB was around 28.6 in all breeds.

It is the deposition of fat which is the most variable factor as it is influenced by breed, nutrition and sex. In practice, the aim is to finish lambs at broadly similar slaughter condition, which essentially is at the same amount of subcutaneous fat.

As the animal increases in size and weight with age, compositional changes occur due to differential growth of bone, muscle and fat (Table 5.3).

**Table 5.3** Development changes in dissected tissue content with increasing age, bodyweight, and stage of maturity

| | | | | |
|---|---|---|---|---|
| Age (days) | 100 | 172 | 208 | 236 |
| Liveweight (kg) | 25 | 32 | 40 | 47 |
| % Mature weight | 40 | 50 | 62 | 73 |
| **Weights (kg)** | | | | |
| Carcass | 9.5 | 13.3 | 17.5 | 21.2 |
| Muscle | 5.4 | 7.2 | 8.9 | 10.4 |
| Fat | 2.1 | 3.9 | 5.7 | 7.8 |
| Bone | 2.0 | 2.1 | 2.8 | 3.0 |
| **Percentage of Carcass** | | | | |
| Muscle | 57.9 | 56.8 | 52.4 | 51.2 |
| Fat | 20.0 | 25.6 | 30.2 | 34.0 |
| Bone | 21.7 | 17.0 | 16.3 | 14.7 |

Considerable controversy has occurred about the interpretation of growth data, but this will not be examined here. As data has accumulated, notably that referred to above and the comprehensive studies of Kirkton[41] and his colleagues in New Zealand, a quantitative description of the growth of various parts and tissues of the body has become possible.

The growth of bone and muscle increase at a rate which can be reasonably predicted in proportion to structural body size. Fourie et al.[42] indicate that for every 1 per cent change in starved liveweight there is a similar change in total muscle, but bone grows more slowly (0.77 per cent). The implication of this is that lean/bone (L/B) ratio is closely related to the size of the animal.

The rate of increase of fat deposition is relatively greater in terms of growth than the other important tissues and this is reflected in an increase in the percentage of fat in the carcass. At constant carcass weight the amount of fat is the most variable feature and can be markedly influenced by both nutrition and the genotype of the sheep. Nonetheless, as found by Wood et al.[43] there appears to be a distinguishable order in the relative growth of the tissues and fat deposits and this was found to be: subcutaneous fat > caul fat > kidney knob and channel fat (KKCF)> intermuscular fat > lean > bone. Thus, internal fat deposits are later maturing than intermuscular fat, which explains why subcutaneous fat is a good guide to finish.

The so-called early-maturing breeds are those which achieve the successive growth phases and the deposition of fat at relatively light weights, whereas the so-called late-maturing breeds, although faster growing, will achieve a comparable 'finish', which is essentially similar subcutaneous fat cover, at a heavier weight and at a later age. It should be added that to enable lambs to express any early-maturing qualities that they possess then, a high level of nutrition is essential. On a low plane of nutrition, the development of an early-maturing type reverts to a late-maturing pattern and the deposition of fat in any type can be reduced, allowing the animal to be slaughtered at a heavier weight. This is attained, however, at an increased cost in food use and a decline in the overall efficiency in the system.

## Carcass composition

From the practical and utilitarian point of view, the composition and conformation of the carcass are generally regarded throughout the meat trade us the two main factors of importance.

The relative percentages of bone, lean meat, and fat in a carcass vary with the age, breed, and fatness of the sheep. Of these three main constituents of a sheep's carcass, fat is by far the most variable. Naturally, when there is an increased percentage of fat in a carcass, there is a corresponding relative decrease in the percentage of lean meat and bone.

The distribution of these three constituents in different parts of the carcass is also of economic importance.

Bone, of course, is inedible, so that from the butcher's and consumer's point of view, the less the relative proportion of bone there is in a carcass, so much much more valuable will that carcass be.

So far as lean meat is concerned, it is true to say that from the nutritional point of view there can hardly be too much of it, since it is the protein of the lean that gives meat of any kind its special nutritional value. As concerns the economic value of the carcass, however, the distribution of the lean meat in different parts of the carcass is also of considerable importance. It is customary to divide up a sheep's carcass into various cuts, and these cuts are of very different value. The precise methods of cutting-up the carcass depend a good deal on local custom and trade tradition, but in general it may be said that cuts from the hind end of the sheep are of greater value than those from the fore end. To the consumer, the legs of mutton and the loin chops are the best parts of the sheep.

In addition to inedible bone and the valuable muscle or lean meat, a sheep's carcass also contains a variable proportion of fat, and fat is the constituent that both in amount and in distribution varies most widely between individual sheep and between sheep breeds. Excessive fat in a lamb or mutton carcass is no longer wanted either by butcher or consumer, but a certain amount of superficial (subcutaneous) fat is still required.

It is preferable that there should be sufficient subcutaneous fat without there being excessive amounts in the other main fat deposits. These include intermuscular fat, which is important in cooking, and the KKCF.

The carcasses of the main types of British lambs have been surveyed by Kempster and Cuthbertson[44] and Table 5.4 shows the means of the main characteristics.

**Table 5.4** Carcass characteristics of main types of British lamb (overall means)

|  | Mean |
|---|---|
| Carcass weight (kg) | 16.6 |
| Composition of carcass (%) |  |
|     Lean | 55.5 |
|     Bone | 15.5 |
|     Subcutaneous fat | 12.6 |
|     Total fat | 27.6 |
|     Total lean in high-priced cuts* | 56.3 |
|     Eye muscle area (cm$^2$) | 16.8 |

*Best joints are leg, chump, loin and best-end neck

There are several methods used for determining carcass composition. Physical separation, which involves dissecting out the main tissue components, bone, lean, fat and connective tissue, is both time consuming and expensive. An alternative is to grind the carcass, mix until uniform, and chemically analyse a sample. Another less laborious method involves separation of fat and the remainder of the carcass into its main anatomical features, such as the more important and larger muscles. For some purposes the yield of the main commercial cuts – gigot, loin back, shoulder, flank, etc. – can be of value in assessing carcass quality.

Simple measurements which could be used to predict the lean and fat content of the carcass would be invaluable and especially so, if they could also be made in the live animal. Several measurements have been investigated by Kempster *et al.*[45] including fore cannon bone weight, femur length, and humerus weight. The most convenient and valuable of these has proved to be subcutaneous fat depth which can be made in the region of the last rib, above the greatest depth of the 'eye' (in — *longissimus thoracis*) C, or the rib fat thickness, J, as illustrated in Fig. 5.4.

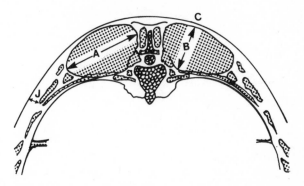

*Fig. 5.4* Eye muscle (A and B) and fat thickness (C and J) measurements taken on the side to be dissected (from *Anim. Prod.* (1980) **31**, 316).
  A – the greatest width of the 'eye' muscle (*m. longissimus thoracis*);
  B – the greatest depth of the 'eye' muscle (*m. longissimus thoracis*) taken at right angles to A;
  C – the depth of subcutaneous fat directly above B; and
  J – the depth of subcutaneous fat above the ventral edge of *m. serratus ventralis*.

It has been suggested that the accuracy of prediction of the lean and fat content of the carcass will be improved even further if carcass weight is included in the prediction equation, with either the C or J measurements, or both.

The widespread application of this measurement is dependent on the fact that subcutaneous fat should be distributed similarly between breeds, and further, that its relationship to lean, intermuscular and KKCF fat is broadly the same in all breeds. While some studies indicate that this may be so, there are nonetheless good reasons, as evaluated by Murray[46], to suggest that prediction equations have to be used with great caution.

Attempts to measure backfat depths in live sheep have proved to be generally disappointing in the past. Improved ultrasonic equipment is now available and, using this technique, an accuracy of prediction of 90 per cent has been claimed by Gooden et al.[47]

## Conformation

In practice this is usually assessed visually, with preference being given to a 'blocky' type of carcass, which implies a tendency to short bones. Differences in conformation between breeds is well recognised, with hill breeds tending to have poorer carcasses than those from lowground or Down sires as shown by Kempster and Cuthbertson.[48]

However, many studies particularly in those of Kempster[49] and Jackson,[50] indicate quite clearly that conformation is a poor index of the size of higher-priced joints. Good conformation may reflect a tendency in a carcass to increased fatness, better L/B ratio and thicker muscles. This latter characteristic may be of considerable relevance to different cutting methods, as related to culinary purpose and eye appeal. It has been suggested by Kirkton[51] that longer carcasses contain more muscle and bone and less fat than blockier carcasses of similar weight.

## Breeds

Breeds can be compared at similar ages, stage of maturity, carcass weight or backfat thickness, with the problem that each method can lead to different interpretations.

It has been suggested that the most biologically relevant basis for comparisons is equal stages of maturity, related to mature body weight. On this basis, a comparison of the Southdown, Oxford Down, Finnish Landrace and Soay (an unimproved breed), at 40, 50, 60 and 70 per cent of estimated mature body weight respectively, showed few differences between breeds. Apart from the Soay, which had a lower percentage of fat, all other breeds had similar carcass composition in terms of fat, muscle and bone. McClelland et al.[52] state that total muscle as a percentage of body weight was around 28.5 per cent for all breeds at all four stages of maturity. A broadly similar conclusion was reached in a comparison of the Clun Forest and Colbred (classified as dam breeds) and the Suffolk and Hampshire (termed ram breeds). Wood et al.[53] found that, when compared at similar carcass weights the Colbreds were bigger and leaner than anticipated, but as butcher's animals, had a lower percentage of prime cuts. The Clun was the fattest breed, the Colbred the leanest, with the others intermediate. It was found that subcutaneous fat was influenced much more by carcass weight than breed and this illustrates the importance of selecting the appropriate weight and condition for a particular breed when choosing animals for slaughter.

Breeds also appear to differ in the way in which they store fat, the ewe breeds mentioned above, having heavier weights of both KKCF and caul fat (omental) than the ram breeds.

Many breeders deliberately select sheep, particularly the meat-type ram breeds, for bone development, with emphasis on the thickness of the fore cannon bone. Certainly there is evidence to show that L/B ratio varies considerably between breeds, the Southdown in New Zealand, having been shown to be consistently superior to the Romney at comparable carcass weights. It is questionable if selection based on cannon bone development is likely to contribute to improvement in muscle weight or L/B ratio and its only value may be as an indicator of maturity.

Even though conformation may have little influence on carcass composition, important breed differences in carcasses do occur. 'Eye muscle' area was found by Wood et al.[54] to be 15.9, 16.1, 17.4 and 17.3 cm$^2$ in the Clun, Colbred, Suffolk and Hampshire breeds respectively.

A larger chop will inevitably have a greater appeal in retail display. Indeed, this suggests that benefits could accrue were it possible to take breeds to heavier weights without producing excessive fat or sacrificing growth rate.

In several experiments the claim that the Texel breed produces high-quality carcasses appears to have been substantiated. The results suggest that, in comparison to Suffolk crosses, Texel-cross carcasses will have an advantage of up to 7 per cent in lean meat content, with 5 per cent larger eye muscle area, and a L/B ratio 5 per cent higher, at similar weights. However, Suffolk crosses are considered to be markedly superior – 6–7 per cent in daily liveweight gain – but differences within breeds and their crosses may be as great as between them.

## Carcass classification

Carcass classification has an important role to play in identifying those carcasses which will yield meat most suitable for retail sale and for the consumer. The scheme developed by the Meat and Livestock Commission (MLC) in Britain has two main categories which are of relevance, fat class and conformation class.

The former involves a subjective assessment of the degree of subcutaneous fat cover on a scale 1–5, where 1 = very lean and 5 = very fat, the preference being for carcasses in grades 2 and 3 for home market requirements, with 1 being acceptable for export purposes. Although carried out by eye appraisal, the grades reflect important differences in the fat content of the carcasses with subsequent waste at the retail stage and during consumption (Table 5.5).

**Table 5.5** Estimate of waste fat expressed as a percentage of carcass weight

| | MLC fat class | | | | |
|---|---|---|---|---|---|
| | 1 | 2 | 3 | 4 | 5 |
| Total dissectable fat[+] | 14.3 | 20.3 | 26.6 | 32.7 | 38.9 |
| Fat trim in trade | 2.0 | 3.7 | 6.7 | 11.2 | 15.0 |
| Plate waste* | 5.8 | 10.7 | 14.3 | 16.3 | 19.1 |

+Include KK and CF but not depots in the body cavity
*Assumes consumer eats 10 parts lean to 1 part fat
After Cuthbertson and Kempster[55]

In Britain, carcasses are classified by the MLC on the basis of weight, age/sex group, fatness and conformation.

The classification system introduced in 1986 is shown in Fig. 5.5.

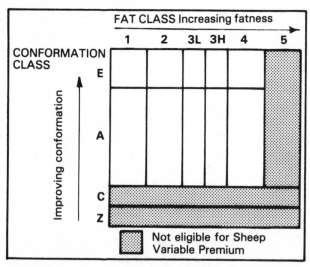

*Fig. 5.5* Sheep carcass classification grid (Meat and Livestock Commission, UK).

Carcasses which are of excessively poor conformation are not scored for fat. There are basically 5 main classes for fat score but classes 3 and 4 are subdivided and those in 4H are not now eligible for the EEC variable premium. Compared with the previous classification system there are more conformation classes – E being exceptionally good – and the previous average A class is now subdivided into three H, R and O in descending order of merit. The purpose of the system is to allow meat traders to specify their preferred carcass types and to encourage pricing according to market requirements.

In selecting lambs for slaughter, the farmer has to base his assessment on condition scoring of the live lamb. This involves handling along the back to estimate the depth of muscle and fat over the backbone and feeling the foot of the tail, since fat is deposited in this region late in the finishing process, while the breastbone may also be checked. Those skilled in the art of handling can achieve remarkable agreement with results obtained in subsequent carcass classification.

The whole subject of mutton production is bound up with question of social change. Consumer preference may not be – in fact, is usually not – merely a fashionable fad. The work men do and the way men live have a profound effect on the types of food they prefer to eat. It has been the task of the sheep breeder and sheep feeder – and the task has often been a difficult one – to adapt his mutton production to a changing demand.

In Britain, modern consumer preference is for lean meat, particularly amongst younger age-groups. Where there is a positive aversion to fat visual appearance is of importance in presentation at the retail stage and white fat is preferred to yellow fat. Eating quality is primarily determined by tenderness, juiciness and flavour. It is fat which imparts to lamb its characteristic flavour and this may vary depending on the feeds used. Meat produced from lambs, or hoggets up to a year old, is generally tender provided appropriate cooking methods are used.

When the temperature of any part of a carcass falls to below 10°C within 10 hours after slaughter, 'cold shortening' will occur, leading to the meat becoming tougher. It would appear that this problem has now been overcome by electrical stimulation, which involves applying a high voltage current to the carcass, immediately after slaughter, or soon after dressing. This allows freezing post-slaughter, as is customary in New Zealand, without detriment to meat quality.

Because of the high percentage of women with outside employment and the demand for convenience foods, new methods of cutting lamb are being developed, and the incorporation of lamb in new products is likely to be an important feature in market presentation in the future.

The role of the farmer is to match the breed or breed types he has chosen to finish, and considering their size and growth potential, to relate these to feeding behaviour and available food supplies, so as to produce a carcass best-suited to market demands. Thus, for early fat lamb production, fast growth coupled with early finishing at moderate weights, say 17–19 kg carcass weight (CW) will be the aim and will limit the choice of breed.

In recent years the Suffolk breed has become one of the most popular terminal sires, since it combines good growth potential with flexibility in finishing. In early or mid season it can produce an acceptable carcass at lighter weights, 19–20 kg CW yet it has the potential to be carried through, if given suitable management, to produce an acceptable carcass at heavier weights.

In general, it is desirable to market at the optimum slaughter weight for different breeds, it overfat carcasses are to be avoided.

## References

(1)  Cunningham, J. M. and Maxwell, T. J. (1979) 'Improved sheep production on hill farms', *Vet. Ann.* 19 Issue. (eds) C. S. G. Grundsell and F. W. G. Hill, Scientechnica.

(2)  Dalton, D. C., Knight, T. W. and Johnson, D. L. (1980) 'Lamb survival in sheep breeds on New Zealand hill country,' *N. Z. J. Agric. Res.*, 23, 167–73.

(3)  Robinson, J. T. (1981) 'Prenatal growth and development in the sheep and its implications for the viability of the newborn lamb,' *Livest. Prod. Sci.*, 8.

(4)  Gunn, R. G. and Robinson, J. F. (1963) 'Lamb mortality in Scottish hill flocks,' *Anim. Prod.* 5, 67–76.

(5)  Hammond, J. (1943) 'Physiological factors affecting birthweight,' *Proc. Nutrition Soc.* 2.

(6)  Hunter, G. L., Adams, C. E. and Ronson, L. E. (1954) 'Successful inter-breed transfer of ova in sheep,' *Nature*, 174, 890.

(7)  Kincaid, G. M. (1943) 'Influence of the sire on the birthweight of the lamb,' *J. Anim. Sci.*, 3, 152.

(8)  Donald, H. P. and Russell, W. S. (1970) 'The relationship between liveweight of ewes at mating and weight of newborn lamb,' *Anim. Prod.*, 12, 273–80.

(9)  Bastiman, B. and Williams, D. O. (1973) 'Inwintering of ewes. The effect of housing,' *Exper. Husb.* 24, 1–6.

(10) Rutter, W., Laird, T. R. and Broadbent, P. J. (1972) 'A note of the effects of clipping pregnant ewes at housing,' *Anim. Prod.*, 14, 127–30.

(11) Hammond, J. (1932) *Growth and Development of Mutton Qualities in the Sheep* (London).

(12) Joulbert, D. M. (1956) 'A study of pre-natal growth and development in the sheep,' *J. Agric. Sci.* Camb, 47.

(13) Robinson, J. J., McDonald, I., Fraser, C. and Crofts, J. M. J. (1977) 'Studies on reproduction in prolific ewes. I. Growth of the products of conception,' *J.*

*Agric Sci.*, Camb., **88**, 539–82.

(14) Robinson, J. J. (1977) 'The influence of maternal nutrition on ovine foetal growth,' *Proc. Nutrition Soc.*, **36**, 9–16.

(15) Hulet, C. V., Foot, W. C. and Price, D. A. (1969) 'Factors affecting growth of ovine foetuses during early gestation,' *Anim. Prod.* **11**, 219–23.

(16) Wallace, L. R. (1948) 'The growth of lambs before and after birth in relation to the level of nutrition,' *J. Agric. Sci.*, Camb., **38**, 93–153.

(17) Robinson, Foot and Price, 'Studies on reproduction in prolific ewes.'

(18) Everitt, G. C. (1968) 'Pre-natal development of uniporous animals with particular reference to the influence of maternal nutrition in sheep,' *in: Growth and Development in Animals*: Butterworth.

(19) Clark, C. Fiona, S. and Speed, A. W. (1980) 'The effect of pre-mating and early pregnancy nutrition on foetal growth and body reserves in Scottish Halfbred ewes,' *Proc. Brit. Soc. Anim. Prod.* Abstr. 83; *Anim. Prod.*, **30**, 485.

(20) Robinson, 'The influence of maternal nutrition'.

(21) Robinson, 'The influence of maternal nutrition'.

(22) Russel, A. J. F., Gunn, R. G. and Doney, J. M. (1968) 'Components of weight loss in pregnant hill ewes during winter,' *Anim. Prod.*, **10**, 43–51.

(23) Lodge, G. A., and Heaney, D. P. (1978) 'Composition of weight change in the pregnant ewe,' *Can. J. Anim Sci.*, **53**, 95–105.

(24) Lodge and Heaney, 'Composition of weight change.'

(25) Robinson, J. J. (1980) 'Energy requirements of ewes during late pregnancy and early lactation,' *Vet. Rec.*, **106**, 282–84.

(26) Wallace, 'The growth of lambs before and after birth.'

(27) Coop, I. E. (1950) 'The effect of level of nutrition during pregnancy and during lactation on lamb and wool production of grazing sheep', *J. Agric. Sci.*, **30**, 311.

(28) Russel, A. J. F. (1978) The use of measurement of energy status in pregnant ewes. In: *The Use of Blood Metabolites in Animal Production*, B.S.A.P. Occ. Pub. **1**; 31–40.

(29) Russel, A. J. F., Doney, J. M. and Reid, R. L. (1967) 'The use of biochemical parameters in controlling nutritional state in pregnant ewes and the effect of undernourishment during pregnancy on lamb birthweight,' *J. Agric. Sci.*, Camb., **68**; 351–58.

(30) Girard. J., Pintado, E. and Ferre, P. (1979) 'Fuel metabolism in the mamalian foetus,' *Annales de Biologie Quinale, Biochemie Biophysique*, **19**, 181–91.

(31) Russel, A. J. F., Maxwell, T. J. and Foot, J. Z. (1973) 'Nutrition of the hill ewe during late pregnancy,' 6 Report IMU Farming Res. Org.

(32) Robinson, J. J., McDonald, I., Fraser, R. C. and Gordon, I. G. (1980) 'Studies on reproduction in prolific ewes. 6. The efficiency of energy utilisation for conceptus growth,' *J. Agric. Sci.*, Camb **94**, 331–38.

(33) Robinson, 'The influence of maternal nutrition'.

(34) Giesler, Pamela and Heal, Heather (1979) 'A model for the effects of energy nutrition on the pregnant ewe,' *Anim. Prod.*, **29**; 357–69.

(35) Giesler, Pamela A. and Jones, Meryl C. (1979) 'A model for the calculation of the energy requirements of the pregnant ewe,' *Anim. Prod.*, **29**, 339–55.

(36) Foot, Janet Z. and Russel, A. J. F. (1973) 'Some nutritional implications of group-feeding hill sheep,' *Anim. Prod.*, **16**, 293–302.

(37) Cole, C. L. (1940) 'Record of performance in sheep,' *Quart. Bull. Michigan Agric. Exp. Stat.*, **23**, 6.

(38) Donald, H. P., Read, J. L. and Russell, W. S. (1970) 'Influence of litter size and breed of sire on carcass weight and quality of lambs,' *Anim. Prod.*, **12**, 281–90.

(39) Hammond, *Growth and Development of Mutton Qualities*.

(40) McClelland, T. H., Bonaiti, B. and Taylor, St C. S. (1976) 'Breed differences in body composition of equally mature sheep,' *Anim. Prod.*, **23**, 281–93.

(41) Kirkton, A. H. (1976) 'Growth, carcass composition and palatability of sheep' *Proc. Symp. Carcass Classif., Adelaide*, Aust. Meat Board, Sydney, Paper SI.

(42) Fourie, P. D., Kirkton, A. H. and Jury, K. E. (1970) 'Growth and development of sheep. II. The effect of breed and sex on the growth and carcass composition of the Southdown and Romney and their crosses,' *N.Z. Inst. Agric. Res.*, **13**, 753–70.

(43) Wood, J. D., McFie, H. J. H., Pomeroy, R. W. and Twinn, D. J. (1980) 'Carcass composition in four sheep breeds, the importance of type of breed and stage of maturity,' *Anim. Prod.*, **30**, 135–52.

(44) Kempster, A. J. and Cuthbertson, A. (1977) 'A survey of the carcass characteristics of the main types of British lamb', *Anim. Prod.*, **25**; 165–79.

(45) Kempster, A. J., Avis, P. R. D., Cuthbertson, A. and Harrington, G. (1976) 'Prediction of lean content of lamb carcasses of different breeds,' *J. Agric. Sci.*, Camb., **86**; 23.

(46) Murray, D. M. (1978) 'An evaluation of some methods of predicting carcass composition of sheep,' *Aust. J. Exp. Agric. and Anim. Husb.*, **18**; 196–201.

(47) Gooden, J. M., Beach, A. D. and Purchas, R. W. (1980) 'Measurement of subcutaneous backfat depth in live lambs with an ultrasonic probe,' *N.Z. J. Agric, Res*, **23**, 161–65.

(48) Kempster and Cuthbertson, 'A survey of carcass characteristics'.

(49) Kempster, A. J., Croston, D. and Jones, D. W. (1981) 'Value of conformation as an indicator of sheep carcass composition within and between breeds' *Anim. Prod.*, **33**, 39–49.

(50) Jackson, T. H. and Mansour, J. A. (1974) 'Differences between lamb carcasses chosen for good and poor conformation,' *Anim. Prod.*, **19**, 93–105.

(51) Kirkton, 'Growth, carcass composition and palatibility of sheep.'

(52) McClelland, Bonaiti and Taylor, 'Breed differences in body composition.'

(53) Wood, McFie, Pomeroy and Twinn, 'Carcass composition in four sheep breeds.'

(54) Wood, McFie, Pomeroy and Twinn, 'Carcass composition in four sheep breeds.'

(55) Cuthbertson A. and Kempster, A. J. (1978) 'Sheep carcass and eating quality.' in: Management and Diseases of Sheep: CAB: British Council course.

# 6 Production: Wool

Largely because of its importance to the textile industry, an enormous literature has accumulated on all aspects of wool production and its industrial uses. It would require a large volume devoted to wool alone even to review the available knowledge. One of the most useful of such volumes is that by Onions.[1] In this chapter only those questions of wool production of most direct interest and importance to the sheep industry are discussed.

Not every sheep produces wool. Indeed, it has been estimated that almost half the world's sheep population are either without wool altogether or carry very light fleeces. As might be anticipated, these non-woolled breeds are native to the hotter, semi-tropical regions of the world.

The sheep's fleece, when present, is a peculiar derivative of the protective skin covering of mammals in general, in which there is an outer coat of coarse hair and an under-coat of finer fibres which may be fur, but in the sheep is wool. The mammalian skin covering, called technically the 'pelage', might therefore be compared to a jacket and vest, the main purpose of the jacket being protection from weather and that of the vest to conserve body heat. In wild sheep and in certain domesticated breeds, particularly hill or mountain breeds, this distinction is clearly evident. In other breeds, such as the Merino, it is entirely lost due to the hypertrophied development of the 'vest' and the atrophy, as it were, of the 'jacket'.

Within recent years, the pattern and development of the mammalian coat, and notably that of the sheep's fleece, have been greatly clarified by investigations of the skin structure both before and after birth. Research work on wool has, so to speak, gone under the sheep's skin to discover more about the wool growing out of it. This work, largely initiated and subsequently reviewed by Carter[2], has shown that in the prenatal life of the lamb the skin follicles from which the hairs of the outer, hairy coat are to develop appear first. These are called the primary follicles (P for short), and are characteristically arranged in groups of three, forming a follicle group. Each primary follicle has three accessory structures associated with it, namely a strand of muscle called the arector pili, a simple sweat-gland and a branched yolk gland. In the unborn Merino lamb these primary follicles are laid down in the skin between 50–100 days after the lamb's conception.

It is from these primary follicles that the outer hair coat of the wild sheep and the kemps and hairlike fibres of domestic sheep breeds are subsequently formed.

From the ninetieth day until birth, and indeed for some considerable time thereafter, secondary follicles are laid down in a somewhat irregular pattern around the group of three primaries. These secondary follicles (S for short) have only one accessory structure associated with them; a simple unbranched yolk gland, or they may have none at all. It is from these secondary follicles that the under-coat of the wild sheep and the true wool fibres of domesticated sheep arise.

The number of secondary follicles surrounding the group of three primary follicles is expressed as the S/P ratio and varies much between breeds, being widest where the number of secondary follicles is large, 15–25:1 as in the Merino, and narrowest where the number of secondary follicles is few, 3:1 as in the Scottish Blackface.

The distinction between hair and wool fibres is not a clearcut one, since fibres intermediate between the two occur in the birth-coat of all sheep breeds and in the adult fleece of all breeds apart from the Merino. Nevertheless, there are important differences between a hair and a typical wool fibre.

In a hair fibre (Fig. 6.1) there are three main parts to be observed microscopically either on transverse or longitudinal section – an external cuticle, an intermediate solid cortex, and an internal hollow medulla. The cuticle is composed of smooth scales, the cortex is made up of closely packed horny scales; the medulla contains hollow spaces and cells loosely arranged.

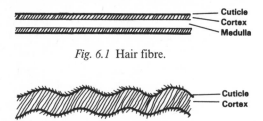

*Fig. 6.1* Hair fibre.

*Fig. 6.2* Wool fibre.

In a wool fibre (Fig. 6.2) the upper edge of each cuticular scale projects beyond the surface of the fibre. Owing to this overlapping arrangement for the cuticular scales, their general appearance is similar to that of tiles on a roof. This peculiar arrangement of the wool fibres' cuticular scales has a bearing upon the felting properties of wool in textile manufacture.

Again, in a true wool fibre the cortex is solidly packed throughout the fibres' substance and the medulla is entirely absent. This solid cortex is associated with the elasticity and other important physical qualities of wool which give it especial importance in the textile trade.

In those fibres of the sheeps' fleece which grow from the primary follicles the cortex is not solid and a medulla is present either throughout the longitudinal axis of the fibre or in sections interrupted by stretches of solid cortex. These aberrant wool fibres, the kemps and heterotypes as they are called, are shed or moulted periodically, whereas in true wool fibres, unless interrupted in growth by malnutrition or disease, growth is continuous.

All skin fibres, whether hairs or wool fibres, develop in much the same way from pits or invaginations of the skin called the follicles (Fig. 6.3). At the base of the pit the papilla or hair root is formed, and this is the only part of the fibre where active cell division occurs. The cells of the fibre are first formed there and are pushed up the follicle by those behind. As the cells are pressed up the follicle they become dead, horny scales. The emergence of the wool fibre on the skin's surface is therefore analogous to the pressing out of toothpaste from a tube. The papilla is the only living portion of the fibre, the remainder is its dead secretion.

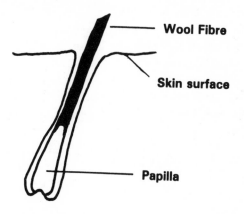

*Fig. 6.3* Wool follicle.

Fibres developed from the primary follicles are medullated or partially medullated, and therefore hairy in character. There are the kemps and heterotypes which are found in the birth-coat of lambs in all sheep breeds and in the adult fleeces of all breeds except the Merino. It was the German zoologist Haeckel who first formulated the biological law which states that ontogeny (the development of the individual) tends to recapitulate phylogeny (the evolution of the species). This law is exemplified in the development of the sheep's fleece, the birth-coat of the lamb bearing a closer resemblance to the double coat of the wild sheep than does that of the fleece of the adult sheep of most domesticated breeds.

The kemps and heterotype fibres of the birth-coat of the lamb are shed, and it depends upon the breed as to what type of fibre the primary follicles form to replace them. In most sheep breeds these fibres are again kemps or heterotypes which are shed periodically and replaced throughout the life of the adult sheep. In the Merino, however, once the kemps and heterotypes of the lamb's birth-coat are shed, the primary follicles, like the secondary follicles, proceed to produce unmedullated fibres, which, besides being true wool fibres, show continuous growth and are not periodically shed.

This character of continuous growth in the wool fibre is one of the most interesting features in the biology of sheep. It is true that in certain of the more primitive breeds of sheep, such as those of the Faroe and Shetland Islands, there is a tendency to an annual moult, and the older method of wool-gathering was to pluck sheep rather than to shear them. In some breeds of Scottish sheep – the North Country Cheviot is an example – should ewes suffer a check during wintering, there may be more wool on the fences and hedges than on the sheep's backs when they come to be shorn. The Scottish Blackface, on the other hand, tends to hold its wool much more firmly, and on the wilder grazings in the old days of wether hirsels, a wether missed at previous gatherings might be brought in for clipping with three years of wool-growth burdening its back.

Annual cyclic changes in wool growth have been observed in many breeds and it may be affected by nutrition, physiological state such as lactation, climatic variation and genetically determined inherent rhythms (see, for example, the study by Corbett[3]). Large differences in winter nutrition have only minimal effect on wool growth in the Scottish Blackface sheep since about 80 per cent of growth occurs in the period June–November. Doney[4] has shown that improved nutrition during lactation will increase fleece weight.

The Merino, in this respect, is the most remarkable sheep breed of all. Both in theory and in practice, it is a salutary exercise of thought to consider what a monstrous abnormality, a pathological development, the fleece of a specialised woolbearing sheep, such as the Merino, really is. Because the sheep lives under conditions most nearly approaching those of Nature, we have become inclined to regard the sheep as the most natural of all domestic animals. That, of course, in relation to wool growth, is quite untrue. As Marston[5] has written:

'The highly evolved wool sheep is an outstanding example of artificial selection. Its fleece, unlike the pelage of its primitive ancestors which is shed periodically, grows continuously, and would become an intolerable burden if left unshorn. Its integument is densely populated with follicles that produce wool at a rate which is grossly in excess of the animal's need for heat conservation and which imposes nutritional demands that are in many ways unique.'

The sheep's fleece, particularly in its highest development in the Merino, in reality approaches closer to a pathological uncontrolled growth such as cancer than to an organ of biological utility to the sheep itself. The Merino sheep, starving, still keeps on growing wool, although at a reduced rate so that, as Marston[6] says, 'the fasting sheep continues to grow wool virtually at the expense of its other tissues'.

The importance of these theoretical considerations in relation to practical sheep husbandry is that there may be an unwarranted tendency to exaggerate the protective value of the fleece to the sheep itself, in the sense that what may be in fact a pathological burden has come to be regarded as a physiological protection.

The assumed biological value of the fleece is to keep the sheep dry and warm. The coat of the wild sheep is, presumably, well adapted to this purpose, the coarse hairs of the outer coat shedding the rain like slates on a house top, the soft and fine under-coat acting as the insulating material of the house walls. Incidentally, wool itself is not a particularly good heat insulator; it is the air entangled among the fine wool fibres that most effectively serves this purpose. It might therefore be presumed that the further the character of the fleece deviates from that of its primitive pattern, the less its protective value is likely to be. Actually, there is not very much reliable information concerning the relative value of different types of fleece as protection against weather.

Differences in 'hardiness' between breeds have been attributed in practice to fleece type. The property of the fleece which would seem to be important in this context is its insulation value. The Cheviot and Scottish Blackface breeds have very different fleece types yet as Doney[7] points out, they have very similar thermal insulation values. This is, Armstrong et al.[8] state, because depth of wool, and to some extent weight or volume of wool per unit area are the critical factors in relation to insulation properties. Exposure to even moderate winds can increase heat flow by 3–5 times, so resistance to changes in fleece structure might also be an important component in adverse conditions.

In Britain, hill farmers generally prefer a coarse type of birth-coat in lambs. That resistance to cold is strongly influenced by coat morphology has been confirmed. An analysis by Purser and Karam[9] of hill farm records of the Welsh Mountain breed showed that in bad weather mortality in short-coated lambs was 42 per cent, and only 10 per cent in long-coated types.

The presence of even a moderate covering of coarse hair appears to be adequate. There are three main kinds of fibre which can be present in the lamb birth-coat – the pre-curly tips, the curly tips and the post-curly tips (Fig. 6.4). Ryder[10] has illustrated that there are two types of pre-curly tip – the halo and the sickle-shaped. The latter is long and hairy but is shed from the lamb and replaced by wool fibres in the adult.

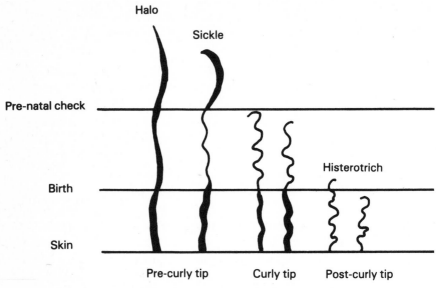

*Fig. 6.4* Fibre types in the lamb birth-coat (after Ryder[10]).

Resistance to cold and wind has been shown to be 4 times greater in Welsh Mountain lambs with long as opposed to short birth-coats. Slee[11] states that Merino lambs are considerably less resistant to cold than either Scottish Blackface or Welsh Mountain lambs, and Merinos would be much more subject to hypothermia than the hill breeds in the British climate.

Of course, differences of this nature may not be entirely attributable to birth-coat morphology and it is highly probable that physiological factors influence resistance to cold and contribute to breed differences. Nonetheless, despite its fine wool, the Merino appears to survive and thrive on the high Southern Alps in New Zealand.

Is it that we have tended to exaggerate the biological value of the sheep's fleece, especially in the more specialised woolbearing breeds? Coop[12] describes a custom fairly recently introduced in the South Island of New Zealand, and said to have several practical advantages, of shearing the ewes shortly before lambing. The ewes are shorn in August, the coldest month of the New Zealand year, so that the ewe loses the protection of her fleece – if the fleece has, in fact, any important protective function – at the very time she might be imagined to need it most! Contrary to theoretical expectation, however, results in practice are reported to be good. Fewer ewes are lost from 'backing' or 'casting', and young lambs find the ewes' teats with greater ease. The custom is spreading, so that the benefits apparently outweigh the disadvantages, should any such exist. There is evidence, moreover, that the practice results in wool of better textile quality. Tests carried out by the New Zealand Woollen Mills Research Association (WMRA) showed that while wool shorn before lambing was practically free from tenderness, breaking, and cotting, many of the fleeces shorn after lambing displayed these faults. Pre-lambing shorn wool also gave better spinning performance and yielded appreciably stronger yarns.

Shearing in early March, some 4–6 weeks before lambing, has also been advocated in Norway and has been described by Nedkvitine[13] and for housed sheep in Britain, where it has been suggested that birthweights are increased by this practice.

Whatever the biological value of the fleece may be to the sheep itself, undoubtedly the main use of sheep's wool has been to provide apparel for mankind to wear. Probably the first wool used for this purpose was the wool attached to the skins of slaughtered sheep. Then the wool that sheep shed naturally may have been used for the same purpose. In primitive breeds and in many breeds not so primitive under conditions of hard wintering, the wool becomes loose and is wholly or partially shed in spring or early summer.

In most cases, however, a sheep is shorn annually either by hand shears or, nowadays, more commonly by machine, so that the shorn fleece represents approximately 12 months' growth.

Sometimes sheep are clipped twice within a year, and it has been claimed that, by this means, it is possible to secure a greater weight of wool, about 10 per cent more within a given period.

Lambs, particularly Merino lambs, are sometimes shorn when about 6 months of age. This delicate and very beautiful lambs' wool is highly valued in the wool trade.

Shearing remains a labour-intensive operation. The use of chemical defleecing agents, which cause a partial or complete cessation of wool growth so that the fleece might be removed mechanically, is an active area of research. Unfortunately, there is considerable individual variation in the animal response to the chemicals used and to the dose rates. There is a risk of sheep being comparatively nude with a danger of exposure, or sunburn, as might occur in Australia, where mechanical removal of wool could otherwise be used. In Britain, it is the new growth of wool in early summer which causes the 'rise' and allows shearing to proceed.

Once a sheep is shorn, from a wool point of view, the fleece is the important consideration. Fleeces, first of all, vary according to the type of sheep from which they have been clipped. Broadly, the textile trade division of fleeces is into three groups: (1) Merino wools; (2) Crossbred wools; and (3) Carpet wools.

The designation of the first group is largely self-explanatory. In this class primary and secondary fibres are mostly equal in structure and growth rate. Fibre population density is usually above $5000/cm^2$ and fibre diameter is less than 25 microns so that quality number is 60s or more.

That of the second group is less straightforward, for crossbred wool, in the trade sense, does not mean necessarily wool from crossbred sheep, since wool from many pure sheep breeds falls into this group. Fleeces from longwools, Downs, Corriedales, Cheviot, etc. all come under the very broad classification of crossbred wool. An approximate definition of crossbred wool might be any wool, other than Merino wool, used in the manufacture of clothing such as fine tweeds, and knitting yarns.

The range in fineness, 25–35 microns or more and consequently in quality, is immense. Many of the longwool wools have a high lustre and are relatively coarse, that is, above 35 microns in fibre diameter, whereas the wools derived from the shortwoolled types are finer, 30–35 microns, and tend to be denser.

Carpet wools include all coarse and hairy fleeces not suited for ordinary clothing manufacture, but used in making carpets, rugs, coarse blankets, mattresses, and other articles of a similar nature. The majority of Eastern sheep breeds produce wool of carpet class, and the best wool for this purpose is, in fact, of Eastern origin.

Of these three classes, Merino wool is the finest and most valuable and constitutes about 40 per cent of the world's total wool clip. Australia produces nearly half of the world's Merino wool.

Crossbred wool forms rather more than 40 per cent of the world's wool clip, the most important producing countries being New Zealand and Argentina. Most British wools are of this type.

Carpet wools are, in general, the least valuable of all wools, although there are exceptions. They constitute about one-fifth of the world's wool clip. These wools come mainly from the Soviet Union and the East.

Merino fleeces provide the wool from which the finest and most valuable woollen clothing is manufactured, but even within the Merino class there are wide ranges of quality and of textile utilization. The precise use to which wool is put in textile manufacture depends primarily upon its: (1) Fineness; (2) Length and strength; and (3) Colour.

The physical properties of wool have been described in detail by Onions.[14]

(1) Fineness. Fineness is the quality upon which the classification not only of Merino wools but of all wools other than carpet wools is customarily based. It depends upon variation in the average diameter of the individual wool fibres in one fleece as compared with another. Fine wool is wool in which the individual fibres are of narrow diameter. Coarse wool is that in which the individual wool fibres are of wider diameter on cross-section.

Weight-for-weight, fine wool will spin out to a greater length than coarse wool, and that fact forms the basis for the conventional classification of clothing wools, a classification based on fineness, and known as the Bradford spinning count. While the Bradford count is based on a definite technical process, it is essentially a rule-of-thumb method of estimating fineness in wool, although, like many other rule-of-thumb methods, sufficiently accurate for all practical purposes in experienced hands.

What, then, is the real meaning of the Bradford spinning count?

Hawkesworth's original definition has been metricated so that it is the number of hanks, 510 m long, which can be spun using the worsted system, from 450 g of the particular wool. A 60s count really means that every 450 g (approximately 1 lb) can be spun into yarn or thread 60 times 510 m or 30 600 m long.

Traditionally, an expert wool-buyer would pick up a fleece, examine it both by sight and touch, and then classify it according to his impression of its fineness by giving it a number in reference to its presumed count. Thus he would call a fine Merino wool 70/74s; a crossbred wool of medium fineness 56/58s; and coarse long wool 34/36s. It will be noted that all the figures named are even numbers, since a difference of two counts is regarded as the closest possible limit of differentiation.

A representative selection of various breeds and the average spinning count of their wool is shown below.

| | |
|---|---|
| Merinos | 64s–80s or above up to 100s |
| Rambouillet | 60s–80s |
| Southdown | 56s–60s |
| Suffolk | 50s–56s |
| Cheviot | 50s–56s |
| Romney Marsh | 40s–48s |
| Lincoln | 36s–40s |
| Scottish Blackface – long | 32s–36s |
| – short | 46s–50s |

Additional to breed differences, hoggs have a slightly higher count – that is to say, a rather finer-fibred wool – than older sheep. Rams tend to have a rather lower count – that is to say, rather thicker-fibred wool – than wethers or ewes. Sheep on poor pasture incline to grow a finer wool than do sheep of similar breeding on rich pasture.

Using modern electronic techniques based on ultrasonic or laser technology, the measurement of fibre diameter has become accurate and relatively fast. Apart from sophisticated laboratory equipment, which will provide not only a mean diameter, but also information on the distribution of this measurement in the sample, there is also equipment of adequate accuracy available for on-farm use.

Intensive investigation in Australia by Whiteley and Charlton[15] has shown that in a fine-woolled flock, differences between sheep account for only about 16 per cent of the total variation in fibre diameter, while 80 per cent of this variation occurs within individual staples, so that between-staple differences are quite small – 4 per cent. This finding has had a profound effect on clip preparation and the need for wool grading prior to sale from the farm. In general, it is the finer wools which tend to be more valuable.

(2) **Length and strength.** Second only to fineness, length of staple is of greatest importance in wool classification. That is because of the separation of the wool manufacturing industry into two broad divisions: (1) The combing industry producing worsteds; and (2) The clothing industry producing woollens.

The combing industry has the more elaborate technique producing on the whole a more valuable finished product and tending to use better wool and fewer substitutes. The principle of wool-combing is to secure a parallel arrangement of the wool fibres, and to do this efficiently and without undue wastage, length and soundness in the raw wool are both essential.

On the contrary, in the initial stages of woollen manufacture no attempt is made to secure parallel arrangement of the individual wool fibres, in fact, rather the reverse.

The important consideration here is that length of staple is not necessary for woollen as opposed to worsted manufacture. In point of fact, the shorter fibres thrown out by combing, and known to the trade as 'noil', form a by-product of the wool-combing industry that is sold to woollen manufacturers.

The minimum length of Merino wool staple for the English system of combing is usually put at 6.2 cm long, and for the French or continental combing system 3.1 cm long. The figures for crossbred wools are higher, being 8.75 and 5 cm respectively.

Finally, in dealing with this question of fibre length, it should be emphasised that the point of real importance to the manufacturer is the effective length. Many wools, otherwise excellent, have a line of weakness running transversely across the fibres due to ill-health or malnutrition, thinning the wool fibres at some period of their growth. Such wools, when processed, break at this line of weakness, so that effective length of the fibre may be considerably shortened, even halved.

The amount of fibre breakage is strongly related to the strength of the wool fibres. Traditionally this has been assessed by repeated pulling of the staple between the buyer's fingers until it breaks. Better techniques have been developed including machines for measuring the tensile strength of wool.

As with fineness, a similar degree of variation in length occurs within the individual staple, on a flock basis, with only 10 per cent variation between fleeces and the same between staples from the one fleece.

(3) Colour. This term as applied to wool has no reference to the spectrum. It means the degree of brightness and whiteness of the wool.

Dingy, stained, or discoloured wool has the important disadvantage to the manufacturer that it will not take a light dye, and must therefore be dyed with the darker shades of brown or black. In some cases the natural brilliance of wool is ruined by certain colouring materials such as 'bloom' dips used by the sheep producer with the object of making his sheep appear more attractive before sale.

Commercial instruments are available to measure the degree of whiteness of wool. The presence of individual coloured fibres in a fleece which may arise because of bacterial or urine staining can reduce the value of a fleece. Individual dark or black fibres which may be produced naturally during fleece growth are, particularly in fine wools, almost impossible to detect. Freedom from pigmented fibres is essential in the production of pastel-dyed fabrics. Coloured wools are in demand, however, from the craft industry.

There are a number of other properties which influence the value of a fleece in the eyes of the wool merchant. There is the question of yield or condition. When a fleece newly shorn off a sheep is weighed all the weight is not due to wool. In every fleece there is a certain amount of dirt. There is also a certain amount of dried sweat, called 'suint'. There is also a variable, sometimes a great deal of grease, often called 'yolk'. Fine wools such as the Merino contain a great quantity of grease; crossbred wools much less, and carpet wools least of all. In the wrinkle-skinned 'Vermont' type of Merino the grease may, on occasion, weigh as much or more than the actual wool.

When, therefore, wool is cleaned and scoured there is always a loss in weight. The proportion by weight of the scoured wool to the unscoured fleece is called the 'yield'. Thus, for example, if half the weight of the fleece is due to actual wool the fleece is said to scour out at 50 per cent; the yield is 50 per cent and the shrinkage is also 50 per cent. Naturally, it is an important duty of the wool-buyer to estimate shrinkage, since it is wool, not grease, he wishes to buy. The grease has some value as a by-product, one derivative being lanoline.

Grease, suint, sand, and earth will scour out by the usual processes employed by the manufacturer. Vegetable material contaminating wool is not so easily removed. Adhesive seeds called burrs may, if present in large numbers, seriously lower the value of wool, since they require to be picked out by hand, an expensive procedure; or destroyed by heating with weak acid, a process called 'carbonising', which damages the elasticity and other important textile qualities of the wool fibre.

In Australian wool sales the catalogue data, in addition to mean fibre diameter, also contains information on yield and vegetable matter content. The development of equipment for the rapid and accurate measurement of these characteristics is an active area of research.

The requirements of the wool trade have been succinctly summarised by Carter[16] as 'High yielding clean fleeces with no pigmented fibres, uniform in the specified diameter and length, free from all vegetable matter and other contaminants from the environment.'

## Wool classing

Fleeces shorn from the sheep's back vary therefore, even within the same breed and flocks, in regard to many important technical qualities which include fineness, length of staple, soundness, yield, and colour. It is the business of the wool-classer to sort them out into their various classes or grades. Wool classing may be done either by the wool merchant or – as on the large Australian sheep stations – by specialised wool-classers employed by the flockmaster and who usually take complete charge of the organisation of the shearing shed. It should be emphasised that the business of wool classing is the classification of the complete fleeces, or rather of the individual fleeces after they have been skirted, that is to say, after the least valuable parts of the fleece – belly pieces, soiled wool, or dags – have been clipped off.

The traditional system was based on estimating the number of crimps on fibres as an indication of fineness and using this to divide the clip into 'fine' and 'coarse' lines. But crimp has been shown not to be a good indicator of fibre diameter of wool and therefore for classing fleeces within a flock. Today, a great deal of the Australian clip is prepared by a system of objective preparation. This involves putting the fleeces into one main line, with smaller lines separated out for exceptionally tender, coarse, short, long or cast fleeces. This, of course, allows more time for supervision of other shed operations.

Much of the wool sold in Australian auction is marketed under a sale-by-sample system. This is a combination of traditional and modern techniques. Samples on display are examined by the time-honoured practice of eye and hand appraisal of staple length and strength, colour and occurrence of staining, to assess processing quality. This is supplemented by objective information which includes mean fibre diameter, clean fibre (yield) and vegetable matter content, derived from an analysis of some samples from bales.

In Britain a system of wool grading to determine price is practised by the buyer – the British Wool Marketing Board (BWMB).

An extensive and complex list of grades based on breed type, fineness, length and cleanliness is used to determine the price to the farmer. The grading system is entirely subjective and is used to bulk fleeces, of comparable quality, into lots of 5 t for sale by auction. Penalties are imposed if fleeces are tinted, badly stained, branded with pitch, tar or paint or tied with twine. Fleeces which are cloggy and undocked or have inferior wool wrapped inside also suffer a price reduction.

The uses of wool are manifold. The same type of wool may be used in the manufacture of a variety of finished goods. Moreover, a single piece of cloth may contain many kinds of wool. The wools from several continents may be combined in the one suiting. To go any further into the intricacies and technicalities of woollen textile manufacture is beyond the scope of this work – but it is important to realise that there is no 'best wool'. Wool is required for such a wide variety of purposes – not exclusively those of textile manufacture – that there is a market, at a price, for each and every kind of wool. Enough has been said already, also to make it clear that the requirements of one wool manufacturer may be entirely different from those of another.

In general, industrial competition, encouraging the use of machinery with increased output, and a need to minimise supervision, requires the production of uniform yarns for wool-spinners. Increasingly combers are requiring more precise specification of the basic raw material – greasy wool – and this will inevitably lead to an increase in the use of objective measurements.

Wool in many respects would seem as though especially designed by

Providence for the health and comfort of mankind. People have become so accustomed to the use of woollen clothing in cold climates that they are apt to overlook the very wonderful qualities – particularly as regards personal hygiene – that wool possesses. For forming durable fabrics wool has the two great qualities of elasticity and felting power. The elasticity of the fibre is of first importance in textile manufacture, and the way in which wool fibres felt together results in a firm fabric with important heat-retaining powers due to air enmeshed among the felted fibres.

The following extract from *Wool – Study Course*, an Army Education Welfare Service document published in 1945 in New Zealand, summarises very clearly the main health benefits which wool provides.

'Wool has another important property which enhances its value for underclothing. Dry wool absorbs as much as 35 per cent of water into its substance before it becomes wet, and even when wet, wool garments never become "clammy" because elasticity is not very much affected. This is not all, however. During the process of absorbing water, an appreciable amount of heat is generated. For athletes and persons performing strenuous work, putting on a wool garment during rest periods prevents danger from chills, because the body is actually warmed at the same time as perspiration is being absorbed. Again, when one passes from a warm dry room to cold outer air in winter, a wool suit takes up moisture from the air and generates a great deal of heat. A suit of clothes weighing $3\frac{1}{2}$ lbs will, in this way, give off as much heat as a human being gives off in one hour.'

Nor should it be forgotten, particularly with regard to children's clothing, that wool has the safety value of being relatively non-flammable, a characteristic which nowadays can be significantly improved by chemical treatment during the manufacturing process.

### Synthetic fibre

Despite these manifold advantages of natural wool, the wool-growing industry is in serious competition with synthetic and other fibres. In cheaper fabrics, cotton has been mixed with wool for a very long time. Artificial silk broke the textile market for lustre wools many years ago.

It is the newer types of artificial fibre such as rayon, nylon, etc. which compete with wool. The sources of these fibres is extremely variable. For example, nylon is made from coal or petroleum products, alginate rayons from seaweed.

Despite the wide publicity, lavish advertisements, and the transatlantic type of boosting that heralds each fresh development in synthetic fibre, it may be doubted whether any synthetic fibres so far invented has all the peculiar and valuable qualities of sheep's wool. Yet the synthetic chemist has not said his last word. Having achieved so much, he can be expected to achieve still more.

For certain purposes, synthetic fibre is undoubtedly superior, particularly in feminine apparel, where comfort is so willingly sacrificed to transparency. The greatest advantage that synthetic fibre possesses over wool lies in its uniformity. One sample of synthetic fibre is the exact replica of another. The same cannot be said of wool – an extraordinarily variable product. The classified grades of wool from a single country may number over a thousand.

During the last decade the most important end-use for synthetic fibre has been for carpets, where the demand has increased at a phenomenal rate. Between 1970 and 1978 the use of synthetic fibres for carpet manufacture, in 8 major manufacturing countries, increased from 488 to $1012 \times 10^3$ tonnes.

Because of competition from synthetic fibres the International Wool Secretariat (IWS) was formed in 1937. It is financed by a small levy on wool and includes amongst its most prominent supporters Australia, New Zealand and Uruguay. The main activities of the IWS are publicity, propaganda and research. Its reasearch activities have been successful in developing new products, as well as processes, to enhance the qualities of wool. It provides a wide range of data on wool and its uses.

## References

(1)  Onions, W. J. (1962) *Wool, an Introduction to its Properties, Varieties, Uses and Production*. Benn.
(2)  Carter, H. B. (1955) 'The hair follicle group in sheep,' *Anim. Breed. Abstr.*, **23**, 107.
(3)  Corbett, J. L. (1979) 'Variation in wool growth with physiological state,' *In:* (eds), J. L. Black and P. J. Reis, Physiological and Environmental Limitations to Wool Growth. Univ. New England Publ. Unit, Armidale, Aust.
(4)  Doney, J. M. (1964) 'The fleece of the Scottish Blackface sheep,' *J. Agric. Sci. Camb.*, **62**, 59–66.
(5)  Marston, H. R. (1955) 'Wool growth,' *In:* (ed.) J. Hammond, *Progress in the Physiology of Farm Animals.* V6, p. 543. Butterworth.
(6)  Marston, 'Wool growth.'
(7)  Doney, J. M. (1963) 'The effects of exposure in Blackface sheep with particular relevance to the role of the fleece,' *J. Agric. Sci.*, Camb., **60**, 267–73.
(8)  Armstrong, D. G., Blaxter, K. L., Graham, M. C. C. and Wainman, F. W.

(1959) 'The effect of environmental conditions on food utilisation by sheep,' *Anim. Prod.*, **1**, 1–12.
(9) Purser, A. F. and Karam, H. A. (1967) 'Lamb survival, growth and fleece production in relation to birth-coat type among Welsh Mountain sheep,' *Anim. Prod.* **19**, 201–210.
(10) Ryder, M. L. (1974) The Birth-coat of Lambs, *Ann. Rep. Anim.* Breed. Res. Org. p.33.
(11) Slee, J. (1978) 'The effects of breed, birth-coat and bodyweight on the cold resistance of new born lambs,' *Anim. Prod.* **27**, 43–49.
(12) Coop, I. E. (1950) *Shearing Ewes before Lambing.* Tech. Publ. Lincoln Coll. N.Z., No.4.
(13) Nedkvitine, J. J. (1972) 'Effect of shearing before and after lambing,' *Acta Agric. Scand.* **23**, 96–102.
(14) Onions, *Wool, an introduction to its properties.*
(15) Whiteley, K. J. and Charlton, D. (1975) 'The appraisal of fineness in greasy wool sale lots,' *J. Agric. Sci.*, Camb., **85**, 45–52.
(16) Carter, H. B. (1971) *Economic and Biological Perspectives in Wool Production.* Europ. Assoc. Anim. Prod. Sheep Comm. No.22.

# 7  Production: Milk

'I saw Libya, too, where the lambs are born with sprouting horns and their dams yean three times in the course of the year; where nobody from king to shepherd need go without cheese or milk, or fresh milk either, since all the year ewes have their udders full.'[1]

'In the first of the morning I drive my sheep to their pasture and stand over them, in heat and in cold, with my dogs, lest the wolves swallow them up. And I lead them back to their folds and milk them twice a day; and their folds I move; and I make cheese and butter, and I am true to my lord.'[2]

These quotations, the first from *The Odyssey*, the second from an Anglo-Saxon dialogue of date AD 1000, clearly show the importance of the third product of the sheep – namely, milk – both in classical times and in the shepherding of ancient England. The custom of milking the ewes persisted to much later dates. The old Scottish songs are full of it:

'I've heard them lilting at our ewe milking.'

In Britain the custom died out, but once again milk-producing flocks are being established, based on high-yielding milk breeds. In the newer sheep countries – Australia, South America and South Africa – it has never arisen, although there are now some milk-production flocks in New Zealand.

The sheep is still an important source of dairy products, notably in the countries around the Mediterranean, in Bulgaria, Portugal and the Middle East. Mavrogenis and Louca[3] state that in the poorer rural areas it is likely that sheep's milk will continue to provide a substantial part of the population needs of milk and milk products.

The native breeds used in most of these areas are regarded as dual-purpose for milk and meat and are usually not very productive. However, milk production potential has been developed to a high level in a number of breeds and the East Friesland, the Awassi and Sardinian are examples.

An improvement programme with the Awassi, continued over a 30-year period, raised typical lactation milk yields from 131 to 381 kg, with the best flocks achieving more than 500 kg. The best East Friesland flocks average over 600 litres. It is highly unlikely that many milk-type breeds yield more than 200 litres per lactation.

There is a great variation in milk yield between the individuals within a breed, a variation often greater than that between the average milk yield of one breed as compared with that of another. This result might be anticipated, since when sheep were being selected for wool or for mutton, there was not necessarily a simultaneous selection for high milk yield in the ewe. There is evidence via Parry[4] that in the Merino, high milk yield was actually discouraged, since heavy lactation was believed by the old Spanish shepherds to be prejudicial to the development of the ewe's fleece. Fleece weight is, in fact, reduced by lactation.

In mutton breeds the explanation is probably not so simple. The early growth of a lamb is governed by the milk yield of its dam, so that it might be thought that in selection for early maturity, strains of deeply-milking ewes would have been selected. In this connection, however, it must be remembered that when the English mutton breeds were first formed, it was the rare exception for a lamb to be killed before it was weaned or for that matter before it was at least 1 year old, so that rapid growth during the suckling period had not the same importance that it has today.

Moreover, in the rotational folding system that English mutton breeds were evolved to suit lambs, except at a very early age, were not altogether dependent on their mother's milk. The folding system, in its perfected form one of the most intensive systems of sheep husbandry ever invented, provided lambs with a constant succession of fresh and succulent forage crops, and as a rule concentrated trough feed as well. In this way it was possible for a mutton lamb to grow rapidly and to mature early without its dam being in any way outstanding in her milk yield.

Now that the fat lamb trade has become the most profitable branch of sheep husbandry, the relative milking capacity of the different breeds of mutton-producing sheep is obviously of far greater importance, for the rapidity with which fat lambs come to slaughter weight and condition depends, beyond all other things, on the quantity of milk their dams provide.

Since the dairy types of sheep are accustomed to hand-milking and letting down their milk, the milk can be weighed directly, so reliable figures of yield are more readily available.

A number of methods have been used for estimating the milk yield of mutton-type breeds, which normally suckle their lambs. These include weighing lambs, before and after suckling, to establish the amount of milk consumed, with the test period, which is usually 12–24 hours, divided into a number of intervals of time. Another popular method is the use of oxytocin to induce milk let-down, followed by hand- or machine-milking. These and other methods are subject to error, and possible bias, and their limitations have been discussed by Doney *et al.*[5] with the suggestion that results must be interpreted with due reserve.

The typical shape of the lactation curve follows a pattern of an increase in yield to around 3–4 weeks, followed by a short plateau, then a slow decline to 12–14 weeks, as illustrated in Fig. 7.1.

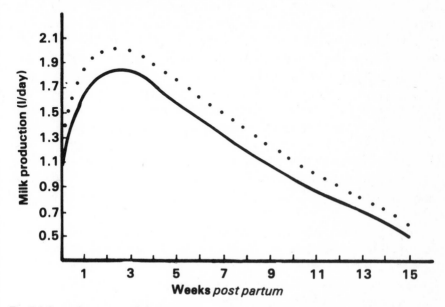

*Fig. 7.1* Lactation curves of single-rearing v. twin-rearing ewes estimated from the formula $Y(x) = ax^b e^{-cx}$. (.... twins, —— singles).[27]

Over a 16-week period Wallace[6] obtained a monthly yield breakdown for his Suffolk ewes of 37.9, 30.5, 20.4 and 11.6 per cent. Peak yields are higher and may be reached earlier in ewes suckling twins but a lower persistence in yield can lead to lactation yields similar to ewes suckling single lambs, as described by Treacher.[7]

Differences in total lactation yield arise mainly from the differences which occur over the first 3–4 week period of lactation. The main factors which affect milk yield in the mutton breeds are breed type, lamb birthweight, number of lambs, age of ewe, ewe size and nutrition in both pregnancy and lactation. There is no doubt that contemporary nutrition under many conditions of management is the most significant single factor, with the effect of demand for milk exercising an important regulatory influence.

In the annual production cycle of the ewe, lactation is the period of highest nutrient requirement. Fortunately, the appetite of the ewe expands substantially at this stage, increasing by 20–50 per cent over the first 2–3 weeks of lactation, after which it begins to decline. Indeed, it is rarely possible for ewes to consume sufficient nutrients in early lactation to meet the requirements of milk production, with the consequence that there is a mobilisation of body reserves and a loss of weight. Nonetheless, the utilisation of body reserves without an adequate diet is incapable in itself of sustaining high milk yields, in the early phase of lactation. It is improbable that body reserves can contribute more than 25–30 per cent of the energy requirements for milk production during the early weeks of lactation as a study by Peart[8] shows.

In general, states Robinson,[9] total milk energy yield increases in direct relationship to an increase in energy by the ewe and certainly up to levels of 20 MJ metabolisable energy (ME) per day above maintenance requirements.

Ewes producing around 2–3 kg milk daily and receiving a diet adequate in energy require supplementary protein. Indeed, for any particular level of energy intake there is a minimum level of protein intake, which if not met, will result in a decrease in milk yield. If ewes are receiving insufficient energy to achieve their potential yield then an increase in crude protein above the minimum requirement will lead to an increase in milk yield. When ewes yielding above 2 kg milk daily are fed below their energy requirement the efficient use of body reserves is dependent on an adequate supply of protein in the diet, since only small amounts are contributed from the amino acids in the body of the ewe, as the study by Robinson shows.[10] The precise protein requirements depend to a large extent on the type of protein and particularly the amount of undegradable protein in the diet. Thus, source of protein is also an important consideration, as has been demonstrated by Gonzalez *et al*.[11] Their Finnish Landrace-cross Dorset Horn ewes were offered a complete diet, based mainly on hay and barley and vitamin/mineral supplementation, to provide sufficient energy to meet the requirements of a daily milk yield of 1.5 kg. The complete diet was supplemented so that daily crude protein intake increased from 170 g to between 340–355 g, depending on the protein source used. The mean daily milk yields recorded were: basal diet, 1.9; groundnut, 2.3; soya bean meal, 2.5; meat and bone meal, 2.5; linseed meal, 2.7; fishmeal, 2.8; and blood meal 2.9.

These improvements in yield could be attributed to an increase in the amount of body tissue used or an increase in the efficiency of its use, or perhaps both factors. This emphasises the need for ewes to be in reasonable body condition at lambing, if they are to be stimulated to produce high yields.

Relatively high inclusion rates of foods of low protein degradability may have a role when energy intake is likely to limit milk yield. This could occur with ewes having a high milk yield potential in the early phases of lactation.

This must be regarded as an expensive and impractical means of meeting energy requirements.

The precise effects of nutrition during late pregnancy on the level of subsequent milk production cannot be defined with any exactitude. Since 95 per cent of the secretory tissue in the ewe's udder is laid down over the last 8 weeks of pregnancy and nutrition affects both lamb birthweight as well as the ewe's body reserves, then some indirect effects might be expected.

It has been suggested that good feeding during lactation can effectively eliminate any effects which arise from inadequate late-pregnancy nutrition unless the latter has led to a drop of 20–25 per cent, or more, in foetal weight. Nonetheless, acute undernutrition before lambing can reduce ewe milk yield and Peart[12] has recorded a reduction of up to 50 per cent.

After lambing a delay in the onset of lactation, poor maternal instinct and a continuing lack of milk due to a poor secretion rate are all familiar consequences of inadequate nutrition throughout late pregnancy and early lactation. Indeed, before the use of supplementary feeding of in-lamb ewes became more widespread in the Scottish hill areas, this picture represented a relatively common experience.

There is some uncertainty about the effects of prolongation of a period of undernutrition post-lambing. Obviously, ewes in good body condition may be capable of sustaining lactation better than thin ewes. The provision of an adequate diet at any time within the first 2–3 weeks of lactation will usually lead to the attainment of normal peak yields. On the other hand, the continuation of poor levels of nutrition for the first 4 weeks has been shown by Peart[13] to result in no response subsequently, no matter how good nutrition then may be. An improvement in the nutrition of the milking ewe, before peak yield is usually reached, would seem to be essential for the achievement of high milk yields.

There are relatively few experiments which compare the milk yield of breeds of sheep under identical conditions. Because of the variety of techniques used, the range of nutritional conditions, lamb numbers and other factors, it is doubtful if much of the data recorded on ewe milk yields can be compared in any meaningful way.

Under strictly comparable conditions the performance of crossbred ewes out of either Columbia-type ewes or Suffolks, by a range of sires, was shown by Torres-Hernandez and Hoikenboken[14] to give small, but nonetheless adequate differences so as to influence lamb growth (see Table 7.1).

The apparently small difference in daily yield over a 15-week period, gives an additional production of 27 kg milk and 2 kg fat from the higher yielding Dorset ewes compared with the lower yielding Finn-crosses. In accord with practical experience the profound effects of mastitis were demonstrated in

**Table 7.1** Milk production (g/24h) and milk composition. (Least square means for genetic effects over a 15-week period.)

| | Yield | Milk composition (per cent) | |
|---|---|---|---|
| | | Protein | Fat |
| Breed ewes sire | | | |
| Cheviot | 1164 | 4.67 | 6.66 |
| Dorset | 1198 | 5.04 | 6.73 |
| Finn | 937 | 4.83 | 6.57 |
| Romney | 1061 | 5.06 | 6.64 |
| Breed ewes dam | | | |
| Columbia type | 1095 | 4.73 | 6.92 |
| Suffolk | 1085 | 5.07 | 6.38 |

this experiment. When one or both halves of the udder were infected, 12 and 58 per cent less milk, respectively, was produced. Breed differences undoubtedly exist, but variation within breeds is often considerable, offering scope for selection in this trait.

Dairy-type breeds have sustained lactation patterns with yields remaining at higher levels for a longer period of time than occurs with the mutton-type breeds. This characteristic is transferred, in some degree, to crossbred ewes of dairy-cross mutton-types.

The milk yield of East Friesland-cross Scottish Blackface ewes was found by Doney et al.[15] to be not only greater than suckling either singles or twins, but was sustained at a higher proportion of peak yield over a 14-week lactation period. This may explain the increasing role of the East Friesian in cross-breeding for the production of dam breed types in Britain.

Apart from supply, the demand for milk is also an important determinant of milk yield. Twin lambs nearly always suck the ewe dry and lactation yield of twin-suckling ewes has been found by Treader[16] to be, on average, 41 per cent, with a range from 6–68 per cent, higher than for singles. Further small increases have been reported with triplet lambs.

There is a need to examine the performance of ewes suckling triplets, against a background of extremely liberal feeding, since I have found, under practical conditions, that ewes so treated will rear triplets exceedingly well and with good growth rates.

It is uncertain however, whether the additional milk produced from multiple suckling is a response to a more frequent suckling stimulus by two or more lambs or whether the single lamb is often incapable of utilising all of the milk produced, so that the udder is rarely emptied and a feedback mechanism reduces the rate of milk secretion.

Practical experience has shown that ewes suckling crossbred lambs are invariably pulled down in body condition to a much greater extent than those suckling a purebred lamb. Lamb genotype has an effect on milk yield and this has been proved by Peart and his colleagues at the Hill Farming Research Organisation (HFRO). Both Texel-cross Blackface and Suffolk-cross East Friesland lambs obtained more milk from Scottish Blackface ewes than purebred lambs. This effect of lamb demand seems to disappear after 5–8 weeks of lactation.

When ewes have a low milk-producing potential this could well be the main limiting factor to the amount of milk consumed by lambs during the first 6 weeks of life. In contrast, demand might limit yield when ewes have a high milk yield potential. Over the first month of life the growth rate of the suckling lamb is largely dependent on milk consumption. Farmers seeking rapid growth rate in lambs rightly attach major importance to ensuring ewes are managed so that they will milk well.

In this early phase milk is utilised with a high level of efficiency with 1 unit of liveweight gain requiring 5.4–5.7 units of milk. This is nearly a unit of gain per unit of milk dry matter (DM) intake.

The correlation between milk yield and lamb growth is about 0.8–0.9 over the first month of lactation, declining as it progresses to 0.4–0.5 kg by 8–12 weeks. From about 3 weeks of age young lambs eat an increasing amount of solid food, be it concentrate or pasture. After 5 weeks of lactation, the energy derived from the milk fraction of the diet is about 88 per cent and it will decline to 34 per cent at 10 weeks. When milk yield is poor, lambs will compensate by increasing solid food consumption and, if it is highly nutritious, there may be little effect on daily gain.

As milk intake increases, herbage intake declines as is clearly illustrated in Table 7.2.

The metabolisable energy of milk is utilised for growth with an effciency of about 70 per cent, whereas similar energy in grass is about half as efficient – 33 per cent. This means that about 4–5 g herbage DM is required to replace 1 g of milk DM and lambs seems to be incapable of achieving an adequate level of pasture intake to compensate until they reach 11–12 weeks. In the later stages of lactation, additional milk will not wholly compensate for an insufficient or poor-quality food supply to the lamb. A supply of high nutritious food is essential if high growth rates are to be maintained, which are so important in early fat lamb systems.

The value of young fresh nutritious herbage for stimulating milk production in ewes has long been recognised by flockmasters. It is essential to allow an adequate supply of pasture to build up rapidly, since there is evidence to suggest that below a herbage mass of 750–1000 kg DM per

**Table 7.2**  Herbage organic matter intake (HOMI) and daily liveweight gain of grazing lambs.

| | Milk allowance (kg over 12 weeks) | | | |
| --- | --- | --- | --- | --- |
| | 50 | 70 | 90 | 110 |
| 5-week old lamb daily HOMI | | | | |
| g/kg liveweight | 17 | 13 | 10 | 8.5 |
| g/lamb | 180 | 170 | 160 | 140 |
| Daily growth rate (g) | 236 | 274 | 310 | 329 |
| 11-week old lamb daily HOMI | | | | |
| g/kg liveweight | 32 | 29 | 30 | 28 |
| g/lamb | 680 | 707 | 858 | 870 |
| Daily growth rate (g) | 193 | 208 | 282 | 331 |

After Penning and Gibb.[17]

hectare intake of herbage by the ewe will fall. This can affect milk yield and in turn lamb growth, even when up to 900 g per day of a concentrate are offered.

The use of a concentrate containing low-degradability protein was examined under grazing conditions where herbage mass increased from 450 kg DM/ha to 1200 at 8 weeks, and 1600 at 15 weeks of lactation. The supplement, containing 14 per cent herring meal and sugarbeet pulp, was superior for twin lambs, compared with one including 14 per cent soya bean meal and barley, which is a more conventional mixture. The effects noted by Milne et al.[18] were that the daily lamb growth rate over an 8-week period increased from 296 to 337 g as a consequence of including a low-degradability protein source.

Pasture quality is clearly also an important aspect of the grazing management of the milking ewe. Munro,[19] studying the milk yield of Scottish Blackface ewes and using the conventional technique of weighing lambs before and after suckling, found a pronounced difference in yield according to the type of pasture they were grazed upon. Ewes grazing what she classified as good pasture gave an average yield of 159 lb in a 6-week suckling period compared with 96 lb in ewes grazing poor pasture. There were wide individual variations in both groups, but her figures suggest that on good pasture the yield of this breed of sheep is somewhere between 90 and 135 litres in a lactation and 'is of the same order as that of other British sheep breeds previously investigated'.

Owen[20] obtained interesting comparative figures over a 10-week suckling period between ewes on hill pasture, ewes of hill breeding on lowland pasture, and ewes from a pedigree flock bred and reared on lowland pasture for many generations. Some of the groups used were of rather small numbers considered in relation to the wide variation of milk yield within groups and certain of the differences recorded – 109.3 lb. for hill ewes, 171.6 lb. for hill ewes on lowland pasture, and 157.9 lb. for ewes of the pedigreed flock – may not be statistically significant. The results, however, do show a notable difference in milk yield between hill ewes on hill pasture and hill ewes brought down to lowland grazing. Owen's figures suggest that the milk yield of Welsh ewes on good grazing is in the region of 90 litres.

Neither Munro's nor Owen's results give very much support to the commonly held opinion that ewes of our hill breeds of sheep are outstandingly good milkers. Their results do show, however, the overriding importance of nutrition upon the amount of milk a ewe will yield.

The new-born lamb does not have immune antibodies in its blood as a protection against disease. Early ingestion of colostrum which contains immunoglobulins is therefore desirable. Pre-suckling colostrum contains 20 per cent by weight of protein, half of which is immunoglobulin in the form of large molecules. The intestine of the lamb rapidly loses its capabilities to absorb immunoglobulin 6–10 hours after birth and the gut only remains permeable for about a day. Colostrum, by providing a rich source of energy, has also an important role in maintaining the energy status of the lambs.

If there is an insufficient supply of natural colostrum, or a lamb is weak and unable to suckle, a supplementary supply of 100–200 ml can be given by stomach tube. Substitutes for the dam's colostrum can include frozen ewe, or cow colostrum, with the first being preferred, since it contains the essential sheep antibodies. Because of the factors which affect the concentration of immunoglobulins in lambs, those at greatest risk are female multiple-birth lambs of the larger breeds born late in the lambing season to very young or older ewes.

Artificial rearing methods have been widely investigated and suitable ewe milk replacers are available (see the study by Treacher[21]).

Fat, lactose and, to an extent, casein are the main sources of energy in ewes' milk. Since fat contributes the greatest part of the energy supply, most milk replacers contain about 20–30 per cent fat in the DM. Research by Gibney and Walker[22] has sought to establish the best sources of energy from substances such as tallow, coconut, olive oil, butter and groundnut oils and hydrolysed starch.

In an attempt to reduce costs, substitutes for casein derived from milk products have also been investigated with some success especially for lambs over 3–4 weeks of age.

Liquid milk replacer can be offered cold and *ad libitum*, but consumption of solid food tends to remain at a low level. If early weaning at around 3–4 weeks is the intention, then it is desirable to impose some restriction to encourage adequate solid food intakes. Complete diets incorporating some roughage have been proven to be satisfactory. Rolled or coarsely milled material is acceptable, but young lambs from 5 weeks of age or so can cope adequately with whole barley.

The milk of all sheep breeds is more concentrated than that of the cow or goat. This is very evident in the butterfat percentage, which is, on the average, much higher than that of cow's milk. Variations in composition of sheep's milk as given in the literature of the subject are extreme as shown, for example, by Ramos and Juarez.[23] Some of this variation may be due to difficulties in obtaining a truly representative sample for analysis, since significant differences have been found to occur in milk samples taken at different stages of milking. Milk quality can probably be modified by plane of nutrition during lactation, but the evidence is conflicting. Certainly, Barnicoat's[21] ewes on 'low-plane' nutrition yielded milk richer in fat, but lower in protein and other solids content. Treacher[25] also found plane of nutrition to have an effect on milk composition.

Gross changes in milk composition occur as lactation progresses. After an initial fall over 3–4 weeks, the content of fat, solids non-fat, protein and ash generally increase, while levels of lactose, after a small increase, fall throughout lactation.

The figures quoted by Ling,[26] as representative, are shown in Table 7.3.

**Table 7.3** Composition of Ewe–Cow milk (g/100 g).

|  | Ewe | Cow |
|---|---|---|
| Total solids | 18.4 | 12.1 |
| Solids not fat | 10.9 | 8.6 |
| Fat | 7.5 | 3.5 |
| Protein | 5.6 | 3.2 |
| Lactose | 4.4 | 4.6 |
| Ash | 0.87 | 0.75 |
| Calcium | 0.19 | 0.12 |
| Phosphorus | 0.15 | 0.10 |

In most cases the content of fat is usually within the range of 5–10 per cent, but in dairy types it may be narrower. The figures for the East Friesian are probably the most accurate available, and these, as recorded, vary between 5 and 8 per cent.

Obviously, ewes' milk is much stronger than cows' milk. To water cows milk before feeding it to orphan lambs, as is occasionally done, on the assumption presumably that the cow being the larger animal, secretes a more potent milk, is therefore, a mistaken practice.

Ewes' milk is less pleasant than cows' milk both in taste and odour. Nor is the butter so agreeable a product as is cows' butter. The colour is paler, consistency less firm, and keeping qualities poorer. On the other hand, ewes' milk can be made into attractive cheeses and is very well suited to that purpose. Actually, it is only as cheese that sheep's milk enters into trade to any important extent. In this form ewes' milk can be manufactured into a luxury-class food, and forms quite an important article of commerce in the countries of South-West Europe.

According to the Food and Agriculture Organisation, the production of ewes' milk in 1979 was around 7.3 million tonnes and similar to the output from goats.

Most of the ewes'-milk cheese is still made by peasant labour. The peasant way of life, although still persisting in some parts of Europe, is giving way to more intensive systems. The quality of cheese obtained from the individual sheep is too small to continue to justify the labour involved in hand-milking. However, machine-milking of large flocks, which are being more intensively managed, will ensure the survival of this system of sheep production, which now is being introduced in Britain and New Zealand.

## References

(1)   Rieu, E. V. (trans.) *The Odyssey*. Penguin.
(2)   Trevelyan, G. M. (1945) *History of England*.
(3)   Mavrogenis, A. P. and Louca, A. (1980) 'Effects of different husbandry systems on milk production of purchased and crossbred sheep,' *Anim. Prod.*, **31**, 171–76.
(4)   Parry, Dr (1806) *Communications to the Board of Agriculture*, Vol. 4.
(5)   Doney, J. M., Peart, J. H., Smith, W. F. and Louda, F. (1979) 'A consideration of the techniques for estimation of milk yield by suckled sheep and a comparison of estimates obtained by two methods in relation to the effect of breed, level of production and stage of lactation,' *J. Agric. Sci.*, Camb., **92**, 123–32.
(6)   Wallace, L. R. (1948) 'Growth of lambs before and after birth in relation to the level of nutrition,' *J. Agric. Sci.*, Camb., **38**, 93–153.
(7)   Treacher, T. T. (1978) *The Effects on Milk Production of the Number of Lambs Suckled and Age, Parity and Size of Ewe*, Proc. Europ. Assoc. Anim. Prod., Sheep Comm. No. 29.
(8)   Peart, J. M. (1968) 'Lactation studies with Blackface ewes and their lambs,' *J. Agric. Sci.*, Camb., **70**, 87–94.

(9)   Robinson, J. J. (1977) *Response of the Lactating Ewe to Variation in Energy and Protein Intake*, Proc. Europ. Assoc. Anim. Prod., Sheep Comm. No. 28.

(10)  Robinson, 'Response of the lactating ewe.'

(11)  Gonzalez, I. S., Robinson, J. J., McHattie, I. and Fraser, C. (1982) 'The effect in ewes of source and level of dietary protein on milk yield, and the relationship between the intestinal supply of non-ammonia nitrogen and the production of milk protein,' *Anim. Prod.*, **34**, 31–40.

(12)  Peart, J. N. (1976) 'The effect of different levels of nutrition during late pregnancy on the subsequent milk production of Blackface ewes and the growth of their lambs,' *J. Agric. Sci.*, Camb. **68**, 365–71.

(13)  Peart, J. H. (1970) 'The influence of liveweight and body condition on the subsequent milk production of Blackface ewes following a period of under-nourishment in early lactation,' *J. Agric. Sci.*, Camb. **75**, 459–69.

(14)  Torres-Hernandez, G. and Hoikenboken, W. (1979) 'Genetic and environmental effects on milk production, milk composition and mastitits incidence in crossbred ewes,' *J. Anim. Sci.,* **49**, 410–417.

(15)  Doney, J. M., Peart, J. N. and Smith, W. F. (1981) 'The effect of interaction of ewe and lamb genotype on milk production of ewes and on growth of lambs to weaning,' *Anim. Prod.*, **33**, 137–42.

(16)  Treacher, 'The effects on milk production.'

(17)  Penning, P. D. and Gibb, M. J. (1979) 'The effect of milk intake on the intake of cut and grazed herbage by lambs,' *Anim. Prod.*, **29**, 53–67.

(18)  Milne, J. A., Maxwell, T. J., Agnew, R. D. M. and Sibbald, A. R. (1982) 'The effects of supplementary feeding in early lactation and herbage mass on the performance of ewes and lambs,' *Proc. Brit. Soc. Anim. Prod.*, Paper 76.

(19)  Munro, J. (1955) 'Studies on the milk yields of Scottish Blackface ewes,' *J. Agric. Sci.*, Camb., **46**, 131.

(20)  Owen, J. B. (1955) 'Milk production in sheep,' Agric. (J. Min. Agric. Eng.), **62**, 110.

(21)  Treacher, T. T. (1973) 'Artificial rearing of lambs – a review,' *Vet. Rec.*, **92**, 311–315.

(22)  Gibney, M. J. and Walker, D. M. (1977) 'Milk replacers for pre-ruminant lambs: protein and fat interactions,' *Aust. J. Agric. Res.*, **28**, 703–712.

(23)  Ramos, M. M. and Juarez, M. (1981) 'The composition of ewes' and goats' milk,' *Int. Dairy Fed. Bull.*, **140**.

(24)  Barnicoat, C. R., Murray, P. F., Roberts, E. M. and Wilson, G. S. (1951) 'Milk secretion studies with New Zealand Romney ewes,' *J. Agric. Sci.*, Camb., **48**, 9–35.

(25)  Treacher, T. T. (1971)'Effects of nutrition in pregnancy and in lactation on milk yield of ewes? *Anim. Prod.*, **13**, 493–501.

(26)  Ling, E. R. (1961) *Milk: The Mammary Gland and its Secretion*. Ed. S. K. Ken and A. T. Cowie, Vol. 2, Academic Press.

(27)  Torres-Hernandez, G. and Hoikenboken, W. D. (1980) Biometric properties of lactation in ewes raising single or twin lambs. *Anim. Prod.*, **30**, 431–436.

# 8 Nutrition

The sheep is a ruminant sharing that distinction amongst domesticated animals with cattle, goats, deer and other species. There are wide differences between the nutrition of ruminants and that of non-ruminants such as the pig and the rat.

The anatomical distinction between the complex stomach of the ruminant with its four compartments (rumen, reticulum, omasum, and abomasum) and the simple stomach of the non-ruminant has, of course been fully appreciated for a very long time. The true physiological distinction is of more recent date.

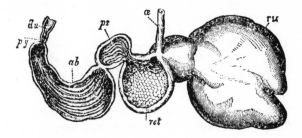

*Fig. 8.1* Stomach of a sheep, partially cut open to show the internal structure (from *The Sheep and its Cousins* by R. Lydekker, with kind permission of Messrs Allen and Unwin).

    oe, oesophagus, or gullet; ru, rumen, or paunch; ret, reticulum, or honeycomb; ps, psalterium, or manyplies; ab, abomasum; py, pylorus; du, duodenum, or commencement of the small intestine.

The onset of rumen development and function occurs during the first few weeks in the life of a lamb so that, as Large[1] states, by 6–8 weeks of age suckling lambs will have developed a good digestive ability, provided that pasture or solid food is available.

In the suckling lamb there is a reflex action which allows the oesophageal groove to function as a continuation of the oesophagus so that swallowed liquids are prevented from entering the rumen and reticulum, thereby passing directly to the abomasum, and leading to efficient use of the nutrients in milk. This phenomenon, examined by Orskov[2] can be exploited in feeding systems as a means of improving protein utilisation.

149

The rumen is the first stomach of the sheep and has a capacity of around 4–10 litres. It has been shown that voluntary intake in lambs is positively related to rumen size, but this is extremely difficult to measure accurately.

Partially chewed food is stored in the rumen and the regurgitation which occurs with herbage or coarse fodders at around 6- to 8-hour intervals leads to a second and more thorough chewing and mixing with saliva. 'Cudding' or rumination, as this process is called, occurs during periods of rest and it is important in management to avoid disturbance after feeding or grazing periods, as this may contribute to stomach disorders.

In addition to the swallowed food, the rumen contains micro-organisms, both bacteria and protozoa, in vast numbers. They have an important role in attacking the cell walls of plant material, permitting the cell contents to escape, then breaking down the cellulose of plant tissue into simpler forms of carboyhdrate on which digestive enzymes might act. In particular, the cellulose of forage such as hay and straw is broken down to that group of chemical compounds called the volatile fatty acids (VFAs) of which acetic acid and propionic acid are of greatest importance. The VFAs form, in fact, the main source of energy in ruminant metabolism and can pass directly through the rumen wall into the circulating blood. As the level of cereals in the diet is increased, the proportion of acetic acid decreases and that of propionic acid increases. When the ratio of the latter and glucose is high, the partition of energy in the lactating animal is so altered that more energy is retained in the bpdy as fat, and less is available for milk.

The type of fermentation in the rumen is dependent on so many factors that it is impossible to discuss them here, but it is nonetheless of significance to note that the range of species and varieties of micro-organisms present can be substantially modified by the type of foods in the diet. Thus, changes in diet have to take place slowly, especially when forages are to be replaced with cereals in the diet.

## Energy

The provision of adequate energy for both maintenance and production is the primary and most important function of food, the energy in all food, whether vegetable or animal, being derived ultimately from the energy of sunlight, trapped by the photosynthetic chloroplasts of the green plant. It is a peculiar property of the ruminant that, through the action of the symbiotic bacteria the rumen contains, it can utilise and set free the energy contained in the more complex forms of carbohydrate, such as cellulose, of which so much plant tissue is composed. That is one thing ruminants can do which non-ruminants cannot do, explaining the wide use and value of ruminants in animal production.

Carbohydrates form the bulk of the food commonly fed to ruminants, and carbohydrates are the main source of energy they require. Both protein and fat, can, however, be utilized along different metabolic routes for precisely the same purpose.

A rational approach to the feeding of livestock requires a knowledge of the energy value of feeds and assessment of the energy needs of livestock. Arising out of the work of Sir Kenneth Blaxter[3] and his associates, the metabolisable energy (ME) system has been adopted in Britain.

The use of a feeding system is necessarily restricted to periods when sheep are housed or when they are substantially dependent on hand-feeding. This occurs mainly with ewes in late pregnancy or early lactation, when lambs are on intensive systems, or ewe hoggs are in-wintered.

Under grazing conditions, the application of feeding standards, and in particular the use of supplementary feed, remains a complex problem across a wide range of pastoral conditions so that precise dietary provision is rarely possible.

**Protein metabolism**

Protein, together with carbohydrates and fats, constitute the three broad divisions into which foodstuffs are commonly classified. Proteins, properly speaking, are combinations of amino acids, and these amino acids, although usable for energy purposes, are also and more importantly required for the maintenance of the body's vital organs and for the formation of animal products, such as meat, milk, and wool, which are predominantly of protein composition.

Here, again, there are important differences between the metabolism of non-ruminants, such as the pig, and ruminants, such as sheep and cattle. The non-ruminant is capable of synthesising the majority of amino acids for itself. These are termed the non-essential amino acids. There are other amino acids, however, 10 in number, which the non-ruminant *cannot* synthesise in its own body, and which must therefore be provided in its food. These are called the essential amino acids.

In the ruminant, on the contrary, the distinction between essential and non-essential amino acids is of minor importance, because those amino acids which the ruminant body cannot synthesise are manufactured by the symbiotic bacteria its rumen contains.

A further distinction lies in the utilisation of nitrogenous compounds, other than amino acids, present in food. While the non-ruminant is incapable of making any use of such compounds, ruminants can and do, even of such relatively simple and cheaply synthesised compounds as urea. Again, the explanation is the ability of organisms such as bacteria to make use of simple nitrogenous compounds by building them up into bacterial protein, which can then be digested and utilised by the ruminant host.

In recent years significant advances have been made in the understanding of the complexities of nitrogen digestion in the ruminant (see, for example, Armstrong[4]). Briefly, there are two aspects which have a relevance to practical conditions. First, the extent to which proteins are broken down or degraded in the rumen, and second, the use of non-protein sources such as urea, biuret, etc. in diets.

Protein foods seem to vary widely in the extent to which dietary protein escapes microbial attack or proteolysis and the importance of this form of degradation has been reviewed by Ørskov[5], who has been associated with the important requirement of developing methods for assessing this characteristic of foods.

On diets high in nitrogen, considerable amounts of ammonia are absorbed from the rumen and eventually converted to urea, to be excreted in the urine. However, with poor-quality, low-nitrogen diets, a similar disadvantage does not apply and supplementation with non-protein nitrogen (NPN) can be considered. Herbages in semi-tropical areas invariably contain less than 4 per cent nitrogen and the provision of an NPN supplement can significantly improve the voluntary intake of the forage.

Low availability of protein in plant material can also occur. For example, Milne et al.[6] have shown that with heather (*Calluna vulgaris*), a low availability of protein occurs due to complexing with tannins, and a substantial improvement in intake can be obtained by the provision of a nitrogen source, such as urea, in supplementary feeds for grazing sheep, where this plant constitutes a large part of their winter diet.

Although the synthesis of microbial protein varies with the type, composition and amount of food in the diet, the relative proportion of amino acids which arrive at the duodenum can be remarkably similar. Clearly, protein nutrition is a complex matter and, apart from the correction of deficiencies and the provision of adequate amounts of protein during periods of high requirements, for example during lactation and early growth, dietary adjustments are rarely undertaken under pastoral conditions. Where controlled feeding is possible, the use of standards produced by the Agricultural Research Council (ARC)[7] are of value.

To summarise the main distinctions between the nutritional requirements of ruminants compared with non-ruminant mammals, without risking the loss of the sheep in the complexities of modern biochemistry, it may be said that these are, briefly, three in number.

(1) The ruminant can use complex forms of carbohydrate, so abundantly present in plant tissues, for energy purposes, while the non-ruminant cannot.

(2) The ruminant, besides being less exacting as to the precise amino acid constitution of the protein fraction of its food, can make a limited dietary use of much simpler nitrogenous compounds, such as urea, which the non-ruminant cannot do.

(3) The ruminant is relatively independent of the B group of vitamins preformed in its diet, whereas the non-ruminant, once again, is not.

It is plain, therefore, that under many conditions of animal husbandry – which is essentially the conversion of vegetable material into animal products useful to man – the ruminant has the advantage. It can do what the non-ruminant cannot do, although it may, if desirable, do everything the non-ruminant does. It is quite possible to feed sheep on pigs' rations, although the reverse procedure, besides being uneconomical, might prove injurious or even fatal to the pig.

It would be improper in a book dealing with sheep husbandry in particular, to probe any further into the subject of ruminant metabolism, a study which has made so many important advances in recent years, and in which research at the time of writing is extremely active. Those who desire further and authoritative information in the subject may be referred to Blaxter's[8] profound analysis of its energy aspects and, for more generalised information, to the appropriate section of the ARC's[9] Nutrition Requirements of Livestock.

## Teeth

Whatever the biochemical fate of food may be, the primary practical consideration is to get it into the mouth in the first place and, in mammals, to masticate it efficiently when once it arrives there. In this connection sheep have two important advantages over cattle. The split upper lip of both sheep and goats enables these species to graze short herbage and to graze it closer than cattle are able to do. Moreover, sheep masticate their food more efficiently than do cattle, so much so that while it is always advisable to crush or grind grain before feeding it to cattle as otherwise a proportion will pass through them undigested, this is unneccessary in the case of sheep.

Like other mammals, the sheep has two sets of teeth, a first or temporary set, often called milk teeth, and a second, permanent set. There are 20 teeth in the first set, 32 in the second. The permanent set consists of 8 incisors, 12 pre-molars, and 20 molars.

All the 8 incisor teeth, in both first and second sets, are situated in the lower jaw. There are no teeth at all in the front part of the upper jaw. In the sheep, in place of upper incisor teeth, there is a hard and fibrous pad against which the lower incisors cut. It is the lower incisor teeth that are of greatest practical importance for, just as in horses, the development and condition of the teeth are a guide in the estimation of a sheep's age. It is only a rough guide, since the rate of development and decay of teeth varies with the strain and breed of sheep, with the husbandry conditions to which they have been exposed, and between different individuals of the same flock. In judging the age of sheep by examining their teeth it is never wise to be too dogmatic, and certainly it is a shifting foundation to found a lawsuit on, although attempts to do so have been made.

Nevertheless, unless a sheep is branded, tattooed, or otherwise marked with its year of birth, an examination of the development and condition of the 8 lower incisor teeth is by far the best and most reliable guide to its age.

A lamb when born may show 1 or 2 temporary incisor teeth or none at all. Occasionally, as in piglets, the sharp edge of a newly-cut tooth may damage the ewe's teat while the lamb is suckling and require filing down on that account. By the time the lamb is about 2 months old it will have cut all 8 temporary or milk incisors. These milk teeth, compared with a sheep's permanent teeth, are much smaller, narrower, and more conical in form.

Between 10 and 14 months of age the two central incisors become loose and fall out. This is of practical importance when feeding hoggs, for it causes temporary interference with their powers of mastication. Roots, for example, must be sliced if fed to sheep of this age.

The two central milk incisors are replaced by two much larger and broader permanent incisor teeth. A yearling or once-shorn sheep has, therefore, in its lower jaw 2 broad central teeth with 3 much smaller milk teeth on either side of them (Fig. 8.2).

A 2-year-old or twice-shorn sheep usually has 2 further milk incisors replaced by permanent teeth. It shows, therefore, 4 broad permanent incisor teeth with 2 small temporary incisor teeth on either side of them.

A 3-year-old or 3-shear sheep has 6 central permanent incisors with only 1 temporary incisor left on either side of them.

A 4-year-old or 4-shear sheep has 8 permanent incisors. All the temporary incisors have been shed.

In most cases, therefore, it takes at least 4 years for a sheep to acquire a

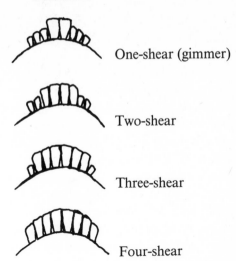

One-shear (gimmer)

Two-shear

Three-shear

Four-shear

*Fig. 8.2* Changes in the teeth of the lower jaw associated with the number of times the sheep has been shorn.

full and permanent set of incisor teeth – but there are exceptions. I have , on occasion, seen sheep with 8 permanent incisors that were yet, as my records showed, not more than 3 years of age. Kammlade,[10] in discussing this question, instances the results of mouth examination of 2300 sheep of known age at the Montana Agricultural Experimental Station. It is evident from the results of this extensive research that while a yearling sheep can be distinguished with certainty by its 2 broad teeth, in older sheep the teeth are a much less reliable index of age.

Once all 8 permanent incisors are present it is impossible to make more than an approximate estimate of a sheep's age. The age to which sheep retain a full and useful set of teeth varies as much in sheep as in people, and there is just as little known of the true underlying causes of this variability. Some sheep begin to lose their teeth at 5 or 6 years old, which seems unnaturally early, considering they may have been more than 4 years getting them. Others retain a full set of teeth until they are 8, or even 12 years old. Such a sheep, however, shows old age in its mouth, even although all the teeth have been retained, for the teeth are widely spaced and narrow, and they appear longer than normal because the gums have shrunk downwards, displaying the roots. The teeth, too, are usually discoloured and loose.

When a sheep has lost 1 or more of its permanent incisor teeth it is call 'broken-mouthed' and the condition may cause a considerable economic loss due to premature casting. This is shown in the much lower price obtained for 'broken-mouthed' sheep than for 'whole-mouthed' sheep of comparable age and size.

Although recognised to a much lesser extent, abnormalities of molar teeth may also occur, leading to an inability to chew or masticate and resulting in a loss of condition.

Simple physical wear may also be a problem, and Nolan and Black[11] show that it is significantly increased on bare pastures as stocking rates rise.

The basic causes of the 'broken-mouthed' condition remain unresolved. It is generally recognised that physical factors such as eating roots, especially when frozen, or badly designed hay racks or boxes predispose sheep to tooth loss.

Gunn[12] shows that factors influencing skeletal formation, such as calcium and phosphorous intake, protein deficiency, sub-clinical parasitism and deficiencies of cobalt or copper – the latter because of its association with bone abnormalities and collagenous tissue – have all been implicated.

The occlusion of the incisor teeth on the pad is important and may be inherited to some extent but Weiner and Purser[13] have shown that it is also influenced by nutrition between the ninth and fifteenth month of a sheep's life.

It has been suggested that an increasing incidence of broken-mouth condition may also be associated with land improvement.

Coop and Abrahamson[14] have shown that when grazing relatively long grass with high feed availability, broken-mouthed ewes appear to be able to harvest an adequate diet, but suffer a significant disadvantage on short pasture, relative to those with a full set of incisor teeth. A study by Sykes *et al.*[15] shows that on poor-quality hill pastures, where selection of herbage may be important, broken-mouthed Scottish Blackface ewes had a greater weight loss in late pregnancy, with a suggestion of poorer lactation performance, when compared with sound-mouthed ewes. It would be of considerable advantage in sheep farming if sheep could be relied upon to keep perfect teeth to an older age.

The replacement of incisor teeth with a set of false teeth has been experimented on and is now being developed but has achieved little success in practice.

## Hay

Another way of meeting the problem of securing a proper balance between stock and food throughout the year is to preserve a portion of summer's abundance for use during winter's scarcity, and the old-fashioned method of doing so was to make hay. There is no doubt that hay still remains the most popular conserved forage for sheep. Because of the greater rapidity in curing and ease of harvesting, grass tends to be fairly mature when hay is cut. Both the quality of the crop when cut, and the weather during the curing period, affect the nutritive value of hay. Maturity in fodder results in a high-fibre content, with a consequent diminution in other nutritional constituents, including protein. Improved methods of mechanical treatment to accelerate drying are being developed and the use of chemical preservatives is now an active field of investigation, particularly by Klinner and Sheppersen.[16] High-quality material can be made by barn-drying, either chopped or in conventional bales. The hay is removed from the field at 35–40 per cent moisture content and is cured by blowing air, which may be heated, through the mass.

Ferguson and Black,[17] at the Edinburgh School of Agriculture, fed pregnant Scottish Halfbred ewes, weighing on average 75 kg, hays varying in metabolisable energy (ME) from 6.3 to 10.25 MJ/kg dry matter (DM), with cereal-based concentrates varying from 0 to 30 kg per ewe. Hay intake was positively related to ME concentration and when hay of superior quality was offered *ad libitum*, concentrates were not required. But hay of such high quality is rarely available and the diet of in-lamb or milking ewes will require concentrate supplementation, the amount being adjusted according to the quality of the hay available. One of the advantages of hay is that batches of different nutritive value can be stored separately and used at the most advantageous time.

Hay of some kind is undoubtedly the best emergency ration for sheep in a storm, or a least that is true of Scottish conditions. Although in Australia they can feed grain during drought, concentrates of any kind are difficult to handle and to feed in snow. Haystacks, placed at strategic points on a hillside, perhaps held in reserve for years, may save the entire flock when a severe storm comes.

As a winter feed for sheep it is true to say of hay as Bottom said of it:

'Good hay, sweet hay, hath no fellow'

Some hay is neither good nor sweet, and is useless for sheep. Coarse or musty hay they will not eat. The ration of hay per ewe per day will vary, obviously with the quality of the hay and the size of the sheep. Something the order of 0.5–1 kg per day is usually recommended.

## Silage

Because of the uncertainties of haymaking and the high loss of nutrients that this may involve, alternative methods of preserving surplus summer pasture have been devised. The most readily used is silage.

Silage, just like hay, is best considered as being the product of a particular technical process rather than as a specific crop. Any vegetable material can be ensiled, which really implies the storage of cut but undried vegetation under conditions that exclude air. There are therefore some kinds of silage that are quite unsuitable for sheep. The beans, peas, and oats silage mixture popular on dairy farms some 40 years ago was well-nigh useless. They picked off the leaves and left the stems, so that cattle had to follow to clean up the wastage.

On the other hand, grass silage of good quality is highly valuable. It took some time before this fact was accepted in traditional sheep husbandry practice, and it is still probably true to say that grass silage, unless harvested at an early stage, is a less suitable feed for sheep than it is for cattle. The study by Wilkinson et al.[18] shows that sheep have low intakes when fed on silage, this being around 20 per cent less than for roughages, either fresh or dried, and from the same origin and of comparable ME value. However, reduction in intake can be as high as 50 per cent, according to Demarquilly,[17] particularly with a clostridial-type fermentation which causes the formation of butyric and acetic acids, imparting to silage a most obnoxious smell. A lactic acid fermentation is to be preferred, state Wilkins et al.,[20] and an increasing proportion of lactic acid in the total acids is associated with increased intake.

Consumption of silage tends to improve when pasture is wilted before ensilage and also when the particle size or, in practical terms the length of the cut material, is reduced, for example by using double chop harvesters during harvest. Indeed, intake of chopped was shown by Dulphy and Michalet[21] to be 90 per cent greater than for long material. Thomas et al.[22] show that comminution from particles of 10–20 to 2–3 mm long only increased intake by around 12 per cent and led to a depression in digestibility.

Nevertheless, sheep under commercial conditions have been successfully wintered on grass silage of high quality, using the silage as a substitute not only for hay but for roots as well. Thomson et al.[23] were pioneers of this practice. Mr Thomson's flock of 1300 Scottish Halfbred ewes were fed 5 kg per ewe/day before lambing and 6 kg after they had lambed, until spring pasture was fully available. The silage was spread on a fresh area of winter pasture each day. The appearance and condition of this flock was certainly a good advertisement for grass silage.

An essential supplement both to hay and silage, unless roots form part of the ration as well, is an adequate water supply. Thomson et al.[24] found that ewes on silage, whether before or after lambing, required access to water and that it was most essential in dry and windy spring weather.

Ewes can also be self-fed on silage provided the face is not too deep and the material is chopped. Its higher moisture content compared with hay leads to high levels of urine production and a need for more bedding, unless slats are used under housed conditions.

The use in France of maize silage for ewes has been reported by Tissier and Molenat[25] who obtained best results with finely chopped silage of 23–35 per cent DM content.

Because of its low intake characteristics, silage is not a particularly satisfactory feed for finishing lambs. Irish experiments by Nolan[26] and by Sheehan and Fitzgerald[27] have shown that lambs can lose weight, while more recent work by Reed[28] has shown that this can occur with silage having a dry matter digestibility (DMD) of 66, whereas wilted grass silage (DMD : 76), and grass/red clover silage (DMD : 71), produced carcass gains of 0.16 and 0.35 kg per week respectively.

Supplementation with concentrates is always possible, but the response will inevitably vary, since there is a trend for concentrates to substitute to some extent, rather than supplement forages as the quality of the latter improves. Large increases in the total intake of organic matter have been reported when dried grass pellets have been used to supplement silages, and Wernli and Wilkins[29] found that the intake of DM may be similar to that obtained with barley, but will be less if dried grass wafers are used.

A practical point in the feeding of both hay and silage to sheep is linked with the conservative temperament of the species. It takes some little time, longer in the case of silage, to accustom them to a novel diet. This period of initiation may extend to a week with hay and a fortnight with silage. Once the initiation is accomplished in any one year, it is unneccessary to repeat it in subsequent years, since sheep have long memories. It follows that it is a mistake to delay the first feeding of hay or silage to sheep until winter storm compels its use. Sheep should be accustomed to it before the necessity of its use arises. Otherwise, they may starve from timidity amidst nutritional plenty.

**Dried grass**

Dried grass, cut at the peak of its digestibility and nutritional value, is a feeding-stuff of acknowledged value excellent for sheep. Once in a late spring I fed dried grass to the ewe flock in an attempt to tide them over until the fresh grass came. I never remember having seen sheep more ravenous for any feed. They came racing from all corners of the field and practically mobbed me when I opened the sack. The cost, however, was prohibitive, and that question of cost will be difficult to overcome. In haymaking the drying agents, the beneficent sun, and the kindly wind, although unpredictable and uncontrollable, cost nothing, whereas the fuel required for grass-drying gets dearer every year.

This is inevitable because of the high temperatures used in the drying process – around 1000°C. Marsh and Murdoch[30] have shown that these can lead to a reduction in the digestibility of the organic matter of 5–8 units per cent. Packaging, in the form of cobs or pellets, can give rise to further reductions in digestibility.

Mechanical processing – grinding and pelleting, have been shown by Osbourne et al.[31] to increase intake by up to 40 per cent in sheep but have no effect on net energy values. Lambs fed on pelleted artificially dried grass will grow as fast (Greenhalgh et al.[32]) as with all commercial diets, performance being related to the quality of the herbage dried.

When dried grass in the wafered or pelleted form and silage prepared from the same crop were offered to sheep *ad libitum*, the intake of digestible organic matter (DOM) and the concentration of VFAs in the rumen were highest for pellets, intermediate for wafers and lowest for silage. The rumen fluid contained higher proportions of propionic and *n*-butyric acids with dried grass diets, whereas silage produced higher proportions of acetic, *iso*-butyric and other higher acids.

This explains the unsuitability of silage for fattening of lambs since propionic acid is important in the fattening process (Wernli and Wilkins[33]).

### Brassica forage crops

Roots are so familiar a crop with us that we are apt to forget what a revolution in stock feeding the introduction of turnips caused in Britain. It was really the introduction of the turnip into farm rotations that made winter production of meat possible. Before that, cattle and sheep were salted down for winter use at the end of the grazing season. A ration of turnips and hay, or in parts of Scotland of turnips and oat straw, made it possible to fatten mature stock for slaughter during the winter period.

The forage root acreage in this country has progressively decreased, although with improved mechanisation of their cultivation and techniques such as direct drilling of the leafy crops, there are signs of slight revival.

In composition these crops are low in DM ranging from 8–12 per cent but have relatively high digestibility (D) values of around 70–74 per cent. In rumen fermentation, swedes and turnip produce similar molar proportions of VFAs as other high energy crops.

Forage crops can be divided into those which usually occupy the land for a whole growing season such as swedes, turnips and amongst the leafy crops, cabbage and early sown kale, and those used as catch crops such as stubble turnips (Dutch) rape, and fodder radish.

A comparison of fodder crops by Jackson et al.[34] shows that both leafy crops, amongst which rape is the most important, and root crops, including turnips and swedes, are used widely for finishing of lambs.

In composition the turnip is mainly water, although sometimes one is inclined to agree with the old Scottish cattle-feeder, who, on being informed of this scientific fact, retorted that, 'the watter maun be by-ordinary watter'. Actually, the high water-content of roots has the important advantage that, when fed in quantity, no other source of water is required.

In assessing the potential profitability of production systems for finishing lambs or hoggets, there are a number of factors of economic importance. Amongst these are the buying and selling prices which can change seasonally, the variable cost for the production of a forage crop and the amounts of supplementary feeds that may be used. A prediction model has been developed by Geisler et al.[35] to evaluate a range of conditions and possible management policies for autumn catch crops. Folding of ewes on most crops is also a common practice in Scotland. Intake is controlled by limiting grazing time, usually from 2 to 4 hours each day, with the intention of limiting consumption to 5–6 kg per head daily.

A number of disorders may be associated with the leafy brassicas as they contain toxic substances, thiocyanates, which inhibit the uptake of iodine by animals and S-methylcysteine sulphoxide (SMCO). If excessively large quantities of brassicas are ingested, SMCO can induce haemolytic anaemia and jaundice. Levels of SMCO increase as crops mature and stock are at greatest risk when folded entirely on forage crops.

A run back on to a grass field over the introductory phase is desirable and will minimise scour and 'brassica poisoning' problems. In an early fat lamb production system involving January lambing, Keane[36] obtained a better lamb performance from rape, at a low level of utilisation, in comparison with swedes, but the carrying capacity was only 25 per cent of that of swedes.

The early maturing, rapidly growing crops, such as forage rape and stubble turnips, are more widely used for finishing 5- to 6-month-old lambs over a relatively short grazing period of 6–12 weeks.

With the leafy crops there may be a considerable amount of wastage if initial stocking rates are too high. There also appears to be a relationship between utilisation rate and animal performance. With stubble turnips, Jackson[37] obtained a daily lamb growth rate of 250 g at a 45 per cent utilisation rate, dropping to 140 g liveweight gain at a 60 per cent level of utilisation. In practice, liveweight gains are usually around 60–150 g per day.

With all forage crops, improved animal performance can frequently be obtained if other foods are used, such as concentrates or hay. As an example (see Drew[38]) a diet of swedes plus hay improved weight gain in hoggets by 20 per cent compared with swedes alone.

The finishing of large flocks of hoggs by folding on turnips or swedes, which are sliced in late winter as the hoggs shed their incisor teeth, is a traditional industry in the Borders of Scotland.

## Concentrates

Concentrates of any kind play less of a part in the feeding of sheep than in any other class of farm animal. In Britain, sheep obtain 90 per cent of their nutrients from fresh or conserved grass and similarly in the United States, where only 11 per cent of energy requirements are obtained from concentrates (see Wedin et al.[39]). Nevertheless, at certain times and in certain husbandry situations, they have a definite economic value, for example, in the artificial rearing of lambs under controlled environment conditions; in the creep-feeding of lambs under intensive grazing conditions; in the breeding-ewe flock both before and after lambing, and under hill conditions, they are of proven value.

The cereals – wheat, maize, barley, oats and so on – have a definite place in sheep-feeding. Maize, per weight unit, has the greatest fattening value of all cereals, both wheat and barley running it fairly close. Oats, particularly for breeding ewes, is the safest of all the cereals when fed to sheep. Sheep of all ages and breeds do well on it, and losses due to digestive disturbance are much less frequent than with any other grain.

In early weaning systems, concentrate diets are offered *ad libitum* to lambs of 15–20 kg liveweight. Animals at this stage will consume less oats than barley, because of the particle size and higher fibre content of oats, and oats are therefore not recommended in systems of this type. At 40 kg liveweight they can consume more oats, so achieving comparable intakes of digestible organic matter DOM. In his work at the Rowett Research Institute, Ørskov[40] and his group have demonstrated conclusively that, apart from the example cited, there is no real need to prepare feed grain by crushing, milling or in any other way. By using whole grains the problems are overcome of soft carcass fat and rumenitis with intensively-fed lambs.

Further, there is evidence, also from Ørskov,[41] which suggests that the depression in intake of roughage diets with cereals, due to an inhibition in the rate of cellulose digestion, is less with whole than with ground cereals. When wholegrains are fed, some of the grain will be seen in the faeces. There is a reduction of around 5 per cent in the digestibility of the grain but this loss will be more than amply compensated for by the saving in costs of processing.

Cereals are deficient in some nutrients and for pregnant and lactating ewes, early weaned and store lambs, dietary protein, or NPN, minerals and vitamins may be required.

When the allowance is controlled, concentrates are fed to sheep at around 0.5–1 kg (0.25–21 lb.) in the daily ration.

## Minerals

These are the inorganic elements remaining in the ash when vegetable or animal tissue is completely burnt. In animal tissue, over 90 per cent of the ash consists of the two elements calcium and phosphorus, which together form the hard constituents of bone. Almost 99 per cent of the calcium in the animal body is concentrated in bone, whereas only 80 per cent of the phosphorus is in bone, the remaining 20 per cent being in the soft tissues. Because these two elements are present in such large amounts and other elements in ash are present in much smaller quantities, in earlier times they received the greatest, sometimes the exclusive attention in the chemical analysis of food. Emphasis has changed somewhat within more recent years, and it is now well recognised that nutritional importance is not always on the side of the big battalions.

The literature of mineral metabolism in its application to the feeding of farm animals is enormous and cannot be reviewed comprehensively here.

Obviously, since mineral elements form an essential part of the animal body, unless they are present in adequate amounts, no animal can thrive for long. On most farms that are adequately cultivated and manured, however, there is no mineral deficiency. On most natural pastures the mineral content of herbage, though often low and always lower than that of cultivated pasture is, however, usually proportionate to that of other essential constituents of the diet and to the type of stock that normally graze them, the mineral requirements of sheep, particularly as regards phosphorus, being generally lower than those of cattle. There is therefore no justification for the indiscriminate use of mineral supplements or licks in sheep farming.

Some parts of the world have been shown to be deficient in certain definite mineral elements – phosphorus on the South African Veldt, iodine in certain districts of New Zealand and the United States, cobalt in the 'bush-sickness' areas of New Zealand and the 'pining' areas of Scotland, possibly copper in districts such as Derbyshire, where 'swayback' is unduly prevalent among lambs. In such cases supplementary feeding or manuring of pastures with the deficient element will both increase production and prevent disease.

## Cobalt

The case of cobalt is of especial importance in sheep husbandry. Nutritional pine in sheep is of sporadic distribution and has been long known to be preventable by the changing of the sheep periodically from 'unsound' to sound pastures. The condition arises because of low cobalt levels in soil and a deficiency in herbage. Cobalt levels below 0.1 mg/kg DM in herbage may indicate a possibility of deficiency. Cobalt is essential for the synthesis of vitamin $B_{12}$ by the rumen micro-organisms.

The clinical signs are loss of condition or general unthriftiness which are, of course, similar to those found with other diseases, for example parasitism.

A more precise method of diagnosis described by Dewey et al.[42] is to test samples taken from animals (for example, blood, urine, liver) using one of the several methods available. Cobalt deficiency in sheep can be prevented by various means including manurial application, drenching, the use of cobalt bullets, mineral licks or injection with vitamin $B_{12}$.

For prevention and more permanent treatment, the manuring of pasture with cobalt sulphate at the rate of 1–5 kg or even less per hectare will correct the cobalt deficiency of the soil for at least 3 years. The cobalt salt is mixed with superphosphate and applied as a top dressing. In New Zealand, aerial distribution of cobaltised superphosphate has been employed successfully in the treatment of cobalt-deficient hill pastures. There would seem also, both in Scotland and New Zealand, to be marginal areas of cobalt deficiency, where, although symptoms of deficiency are absent, sheep production can be substantially increased by supplying additional cobalt to the grazing sheep. However, on certain soils where the levels of manganese are above 1000 parts per million (ppm) this method may not be successful.

An alternative to manurial treatment is the direct administration of cobalt by what is called the 'cobalt bullet'. Obviously, repeated daily dosing cannot be carried out in farming practice. The ingenious device of the cobalt 'bullet' overcomes this difficulty. The method, as worked out by Marston and his colleagues in Australia, depends upon the known tendency of heavy objects swallowed by ruminants to become lodged in the reticulo–rumen or honeycomb and to remain there. Pellets called 'bullets' weighing 5 g and composed of 90 per cent cobaltic oxide and 10 per cent china clay follow this pattern of behaviour. Slow absorption of cobalt from the 'bullets' has been found to protect a sheep from all symptoms of cobalt deficiency for several years and probably for the lifetime of most sheep.

Whitelaw and Russel[43] have shown that 'bullets' can be administered successfully to 2–3-month-old suckling lambs, this treatment being superior to regularly monthly injection with vitamin $B_{12}$.

Although a few bullets may be regurgitated, and some may become encrusted with calcium phosphate, these workers recommend the use of bullets as the preferred form of prophylaxis although the direct application of cobalt salts to pasture and soil are also effective.

Because cobalt has become expensive, a recent approach is to spray pasture with a small amount of a cobalt salt. The application of 10 g cobalt sulphate in water, with a wetting agent, is adequate but has the disadvantage of having to be repeated at monthly intervals.

## Copper

Copper deficiency of soils is sometimes associated with cobalt deficiency, as in the so-called 'coast disease' of Southern Australia.

Uncomplicated copper deficiency of pastures occurs also in Southern Australia on pastures overlying certain types of sandy soil. In Merino sheep depastured there, the first and very remarkable evidence of copper deficiency is shown in the fleece. The wool loses the 'crimp' or closely-set transverse waves on the staple, so characteristic of the Merino fleece, and becomes quite straight instead. When the deficiency is corrected the crimp returns. Moreover, in the odd black Merino sheep the wool loses its pigmentation as well as its crimp. This loss of wool pigmentation in black-fleeced sheep of any breed forms a delicate and useful biological test for copper deficiency of pastures.

A more serious symptom of copper deficiency in sheep is a progressive anaemia. Copper is also concerned in the disease of lambs called 'swayback' in Britain and 'enzootic ataxia' in Australia, and which occurs locally in several other parts of the world. The disease, as the British name suggests, is a form of paralysis due to degenerative changes in the lambs' central nervous system produced before birth. It is a disease quite widely distributed in Britain, occurring not infrequently when hill land is improved with the application of lime to raise the spoil pH.

Injection of the ewe subcutaneously in late pregnancy, 10–16 weeks after mating, is a satisfactory means of preventing the disease. However, a deficiency in growing lambs has also been shown by Whitelaw et al.[44] to occur due to an increase in uptake by the plant of molybdenum and sulphur. This has the effect of rendering much of the copper in the ingested herbage unavailable to the animal and, in the circumstances quoted in the above work, causing pronounced hypocupraemia during suckling, some osteoporosis, making these lambs more susceptible to bone fractures and leading to a reduction in liveweight gain.

Routine injection with calcium copper edetate to maintain blood serum copper levels above 60 mg/100 ml proved to be a successful means of control. Very recent work has shown that a single dose of cupric oxide needles, as used successfully by Dewey,[45] will effectively protect lambs against induced copper deficiency as shown by Whitelaw et al.[46]

Breed differences have been found in blood copper levels where Scottish Blackface, Cheviot and Welsh Mountain ewes were grazed for an extended period on the same pasture. Weiner et al.[42] confirm that this may suggest that some breeds could be more prone to swayback. Copper toxicity is a hazard in housed sheep on intensive diets and care must be taken to ensure that copper levels of complete diet are less than 10 ppm. Hill[48] states that free access to copper-enriched minerals is not generally advocated for sheep due to the hazards of excessive copper intakes.

*Magnesium*

Deaths from 'staggers' or tetany (hypomagnesaemia) associated with low levels of magnesium in the blood, may cause losses especially on the best farms. Signs of the condition or its development are frequently not seen as death can be extremely rapid. Sheep seem to be most suceptible in the spring and the condition can be exacerbated by turning them on to lush pasture. The problem tends to be worse on intensively managed grassland, where potassium soil levels are high and when liberal applications of nitrogenous fertiliser are used.

Stress due to low temperatures, sudden disturbance, exercise, sudden change of diet or underfeeding may induce tetany.

The problem is a complex one and is dealt with more authoritatively in the veterinary section of this book. It has also been comprehensively reviewed by Kelly.[49] Several means of prevention are available. Adequately magnesium-enriched concentrates can be satisfactory, but self-help feed blocks containing magnesium may be consumed in inadequate amounts or, indeed, ignored entirely by sheep. Although less reliable, blocks may nonetheless offer the best practical solution for sheep on more extensive systems on upland and hill farms, when run on improved pastures.

Magnesium alloy bullets can give satisfactory protection but, as with those used with cobalt, suffer from the problem of regurgitation. The manurial treatment of pasture offers an alternative means of prevention, particularly on acid soils, but it is markedly less effective on alkaline soils. The application of calcined magnesite at around 650 kg/ha can be effective for up to 3 years and small amounts – 30 kg/ha – as a dust for foliar application, can afford protection for a few days after sheep are turned onto pasture.

## Selenium

Ill-thrift and infertility due to selenium deficiency have long been recognised in Australia and New Zealand. In the latter country, farms are recognised as being clearly deficient in selenium due to the occurrence of muscular dystrophy (white muscle disease) in new-born lambs. A drop in barrenness from 30 to 5 per cent from the supplementation of selenium to breeding ewes has been reported by Paynter[50], while growth responses in lambs have been obtained in Australia.

An extensive trial by Blaxter in Britain[51] indicated little evidence of a lamb growth response to selenium administration but this work suggested that a small percentage of farms could be mildly deficient. In recent years there has been a resurgence of interest in the possible occurrence of selenium deficiencies. Glutathione peroxidase, a selenium dependent enzyme, is more easily measured than the presence of the element in the blood and it is considered to be a sensitive predictor of selenium status. Indeed, selenium is now recognised as an essential micro-nutrient for health and for efficient animal production. A national survey in Britain has suggested that selenium deficiency may occur in some regions. Whilst responses to selenium and vitamin E administration have been reported, other studies have proved negative in British trials.

There is evidence of a functional interrelationship between vitamin E and selenium and the problem is not straightforward. Selenium can be extremely toxic at low concentrations and professional advice should always be sought before using selenium-enriched fertilisers or any form of animal treatment. Cobalt deficient lambs are, for example, more susceptible to selenium toxicity. Indeed, this counsel applies to all trace mineral problems since they tend to be complex in nature.

## Iodine

Iodine deficiency occurs locally in many parts of the world, one such notable locality being Montana in the United States. In sheep, iodine deficiency leads to the birth of lambs with enlarged thyroid glands and very little wool on their bodies. The enlarged thyroid is the origin of the popular name 'big-neck' for this condition. Iodised salt licks are an efficient preventive but since iodine deficiency in sheep is, so far as I am aware, unknown in Britain, there is no need for the provision of iodised salt licks for sheep in this country.

## Other Minerals

There are many other mineral elements of minor or localised importance in sheep nutrition. Thus, fluorine and selenium, in excess, are poisonous, although both in traces are essential for health. Sodium and chlorine, combined in common salt, should be accessible to sheep, and rock-salt blocks distributed over grazings, particularly in the spring of the year, are of definite value. For further and more detailed information on mineral metabolism, specialised manuals on animal nutrition should be consulted.

## Vitamins

In relation to sheep-feeding the subject of vitamins can be dealt with quite briefly.

While all vitamins have the common properties of being active in mere traces and of playing an essential part in the basic biochemical activities of the animal body, they have not necessarily any chemical relationship, one to another. It is most convenient, therefore, to treat them either as separate entities or as groups.

### Vitamin A

This vitamin is synthesised by sheep from the carotene pigment abundantly provided in all green food. The conversion of carotene to the vitamin is more efficient in sheep than it is in cattle. Consequently, the milk-fat of sheep is white, while that of cattle, especially Channel Island cattle, is yellow. The ability to store vitamin A in the liver is common to all animals, and in sheep the storage suffices to cover at least 6 months of dietary deprivation. It follows that provided sheep are fed green feed for any appreciable period of the year, there is no real possibility of the symptoms of vitamin A deficiency arising.

### Vitamin D

Under outdoor conditions this vitamin is formed by the ultra-violet rays of sunshine activating certain waxes or sterols in the sheep's own skin. Provided sheep are exposed to an adequate amount of sunlight, they therefore require no other source of vitamin D. Since sunlight is not very abundant in our climate, it has been suggested that vitamin D injections would benefit sheep production, especially under hill conditions. In general, experimental evidence has not confirmed this suggestion. It should be noted,

noted, however, that the liver storage of vitamin D is much less efficient than is the case with vitamin A, and that, in consequence, sheep confined to sheds during winter may well require a supplementary source. Fortunately, sun-cured hay, containing as it does a form of the vitamin ($D_2$) which can be utilised efficiently by sheep, will provide the supplement in sufficient amounts.

## Vitamin E

This vitamin, sometimes called the fertility vitamin, is of minor importance in sheep-feeding in Britain since it tends to be widely distributed in the common feeding-stuffs.

Deficiency causes a condition called 'white muscle disease', a form of muscular dystrophy in lambs, and 'ill-thrift', a type of unthriftiness in 6–12-month-old sheep and sometimes poor fecundity in ewes. Indeed, these disorders are normally associated with selenium deficiency and can be treated either with vitamin E, selenium, or both. The occurrence and effects of these deficiencies are extremely variable. Considerable economic loss has occurred in New Zealand from 'hogget ill thrift', where muscular dystrophy in new-born lambs is found and barrenness in ewes may also arise, due to an increase in embryo mortality.

Grain preserved by treatment with propionic acid will be low in vitamin E so a deficiency in sheep may occur, especially in intensive systems. A commercial trace element/vitamin product for oral dosing, containing selenium and vitamin E, other trace elements and vitamins, is available on the market.

## The B Group

This group of water-soluble vitamins includes a large number of unrelated chemical compounds with basic functions in cell metabolism. They are of considerable practical importance in the feeding of both pigs and poultry. This is not so, however, with ruminant animals, since rumen symbiotic bacteria are capable of synthesising all the B-group vitamins, making them available also for the metabolism of the ruminant host. The relationship of vitamin $B_{12}$ to cobalt deficiency in sheep has already been described.

Cerebrocortical necrosis (CCN) is associated with a deficiency of vitamin B – thiamine – due to some factor causing the vitamin to be unavailable to the animal. This condition, which results in characteristic nervous symptoms, can be cured by dosing with thiamine.

*Vitamin C*

In human nutrition this vitamin is of major importance, since a deficiency in human diet causes the well-known disease, scurvy. The case is, however, peculiar, since all mammals, with the curious exception of man, guinea pig, and monkey, are able to synthesise this vitamin in their own bodies. Symbiotic bacteria are not involved in this synthesis. It follows that this vitamin is of no importance in sheep-feeding.

This rather cursory summary of vitamins in relation to sheep-feeding, at first sight might appear to be inadequate in view of the vast literature and enormous amount of research that has been devoted over the last half-century to this branch of animal nutrition. It is, however, of minor importance in traditional sheep-feeding, although, undoubtedly, this new knowledge and its application is essential to the modern poultry industry in its fully intensified form. It is therefore certain that the more closely sheep production tends to intensification on the controlled environmental poultry pattern, the more necessary will it become to take the vitamins into careful consideration.

## Water

Drought, in arid countries such as Australia, has killed more sheep than has any disease, but since drought prevents plant growth, these periodical pastoral disasters are due to sheep starvation as well as to sheep thirst.

The sheep, of course, requires water to live, as does every other animal, but its requirements are definitely less than those of cattle, due partly, no doubt, to the fact that cattle are more dependent upon sweating in the disposal of surplus bodily heat. Dry sheep, that is to say sheep other than lactating ewes, may satisfy all their water requirements from the dew falling on pasture and from the moisture contained in pasture. There are green islands off the West Coast of Scotland and in the Shetland Islands where there is no source of fresh water and yet where sheep both live and thrive. Only where dews are absent and pasture withered do sheep require water to live, and under such conditions, fenced sheep have been seen, even in Scotland, dead of thirst alone.

Lactating ewes drink freely if able to do so, and there is no doubt that ewes and lambs thrive better if allowed access to clean and running water. The heavily beaten tracks to the drinking pools show how often they journey there in rainless weather. In a Scottish agricultural journal there was once a correspondence on the question as to whether or not sheep drink water. There was no need for any debate on the subject. Sheep of many breeds have been seen under all sorts of conditions drinking deeply and freely from streams, puddles, buckets and ponds. There is no dubiety on the matter. Nevertheless – and this in an arid climate is an important point in favour of sheep – they can exist far longer without water than cattle are able to do.

The water supply to hill sheep requires special consideration. Summer drought on hill grazing can be quite severe, and it is important to ensure that fencing is arranged so that access to streams is not inhibited. In winter the whole grazing may become covered with snow. Provided the snow is not completely frozen, sheep can satisfy their water requirements by eating it. Sometimes when eating hard and partially frozen snow the crackling sound is clearly audible, an unpublished observation made by Dr Joan Munro, a former colleague of Dr Fraser. In an unusually severe winter, however, such as that of 1947, snow may become too hard to be eaten, the streams are ice-bound and sheep may die of thirst. If, under such conditions, hay be fed to relieve their hunger they will die of thirst all the more quickly.

There is a certain amount of useful experimental evidence on the actual quantity of water needed by sheep under varying conditions of feed and climate. Bonsma[52] found that succulent fodders, such as saltbush or prickly pear, contain sufficient water to satisfy their requirements. Brauns[53] found that sheep on dry feed (maize endosperm and hay) drank between 1.15 and 1.5 litres daily under varying conditions of sunshine and temperature. Hindmarsh[54] gives the daily water requirements of sheep at 1 gallon (4.5 litres). At the Rowett Institute, Thomson and Fraser (unpublished results)[55] in experimental work on pregnant ewes, made the observation that in the later stages of pregnancy the water consumption of the ewes increased very considerably; indeed, in the last few weeks before lambing it was almost doubled.

## Feeding standards

The practice of sheep-feeding has not attained the exactitude possible, for example, in the feeding of dairy cows. On a scientifically-run dairy farm so much energy and protein is allowed for maintenance of the cow, and an addition is made – the production ration, according to yield or predicted yield of milk.

In the sheep-fold, conditions are quite different. The unit is the flock, or part of it – not the individual sheep. The production of the individual sheep is never quantitatively determined and thus rationing, as is possible in dairies, is out of the question.

Feeding standards have been formulated for sheep and, if intelligently applied, can contribute to more effective management and improved production. Thus, the recognition of the different nutrient requirements during the various physiological phases of the animal production cycle, and the adjustment of grazing management or regulations of the diet offered, can contribute to more economic use of feed resources and to high levels of production.

Requirements of the breeding ewe vary widely, between early and late lactation, post-weaning, pre- and post-mating, and mid and late pregnancy. Data are also available for the grazing lamb up to the stage of slaughter. More detailed information is available from the Agricultural Research Council.[55]

Under pastoral conditions the problem in the application of feeding standards is determining the amount and type of supplementary feed to offer. The exceptions are when pasture contributes but little to the diet, or when a sheep flock is housed. The alternatives then are whether to use traditional diets or to use 'complete diets' where roughage and concentrates are mixed.

Differential levels of feeding can be justified as, for example, lean as opposed to fat ewes, especially in mid to late pregnancy, and ewes suckling twins instead of a single lamb. By this means feed can be saved and more effective use made of it, especially in the more intensive systems, where a greater degree of control is possible.

### Practical feeding

When a flockmaster intends to feed sheep, his first consideration is to buy sheep that feed economically and well. He pays particular attention to the type of sheep he requires and to the district where they have been reared. In Scotland, feeding lambs are sold in successive sales throughout the autumn months. The breeder usually sells his lambs in successive drafts at these succeeding sales. He will, if his flock be big, usually have four drafts for sale: his tops, seconds, thirds, and shotts. These drafts are fitted to different purposes. Tops are for short keep and quick resale, seconds for longer keep. Thirds may be allowed to run on through winter and be finished as hoggs on grass the following summer. Shotts – a flock's refuse of undergrown lambs – are sometimes most profitable if fed indoors.

The point of this description of sale conventions is to emphasise the fact that the first important practical factor in sheep feeding is *time*. No matter how much or what food is fed, there is a minimum time required for every class of feeding sheep to attain finish. If that is not realised, it is easy to waste a great deal of food in attempting to push on feeding sheep more rapidly than is possible.

Once lambs are bought, it is a mistake to expect an immediate response to better feeding. Particularly if newly weaned, lambs do not behave like insensate mutton-making machines, putting on so much weight in proportion to so much energy ingested. They behave far more like home-sick children, restless for their homes. It is a waste of food to attempt to fatten lambs until they have settled down and have become accustomed to their new surroundings.

When lambs have settled, the serious business of feeding may commence. The essence of the business is gradualness. Sudden change of feed always upsets sheep, and unless they are introduced to new food gradually, combined in due proportion with the old, they may refuse new food altogether. This is particularly evident with sheep of older age-groups, such as cast hill ewes brought down to the lowlands. Some may take weeks before they will come to the troughs and eat grain.

Further, any sudden change of food upsets a sheep's digestion, and indigestion may result in death. If, for example, sheep that have been on pasture are to be folded on rape it may prove altogether disastrous to have them entirely on pasture one day, entirely on rape the next. It is far safer to put them first on rape for no more than a few hours gradually increasing the time on rape and decreasing that on pasture. Similarly with trough feed. It is a mistake, for example, suddenly to substitute maize for oats. It is better to start with a proportion of no more than one-quarter of maize to three-quarters of oats, gradually increasing the maize until that alone is being fed.

It is important to feed with an eye on the sheeps' dung. The proper consistency of sheeps' dung is formed pellets. The loose motions of the cow are unnatural to the sheep species. Sheep will not thrive to best advantage when scouring. This can be controlled. The cause is often worm infestation, or it may be more simply too laxative a diet. If, for example, sheep being fed roots scour badly, as they often do, the scour may be checked by substituting hay and grain for a certain proportion of roots.

It is also important to feed sheep at regular times. The hunger of an animal is a most punctual thing. Should a shepherd be delayed on his rounds, the fact is advertised by hungry bleating from the distant pastures.

Cleanliness is essential. Sheep are clean eaters and detest soiled or musty food.

Sheep should be called, never driven to their food, for a breathless and heated sheep newly chased by dogs is in no condition to make the best use of the food provided. Everything should be done quietly and in order. A fixed routine, regularly adhered to, ensures a contented flock.

Sheep are competitive eaters. Although all sheep are timid, some are more timid than others, and the more timid may be driven away from food by those that are bolder. Some weakly sheep may be so bullied that they hang on the outskirts of the flock, eating by stealth. These will never improve nor thrive. Ample space for feeding tends to prevent this happening, and having the feeding flocks carefully drawn as to age and size is also helpful.

All these practical points are of first importance in successful sheep-feeding. The fact that one man follows these simple rules with care while another ignores them explains why with the same rations and the same sheep one man will make a profit while another reaps a loss. They may all be included under the general title of Good Shepherding.

## References

(1)  Large, R. V. (1964) 'The development of the lamb with particular relevance to the alimentary tract,' *Anim. Prod.*, **6**, 169–78.
(2)  Ørskov, E. R. (1972) 'Reflex closure of the oesophageal groove and its potential application in animal nutrition,' *S. Afr. J. Anim. Sci.*, **2**, 169–76.
(3)  Blaxter, K. L. and Boyne, W. A. (1978) 'The estimation of the nutritive value of feeds as energy sources and the derivation of feeding systems,' *J. Agric. Sci.*, Camb., **90**, 47.
(4)  Armstrong, D. G. (1980) Some Aspects of Nitrogen Digestion in the Ruminant and their Practical Significance, 2nd T. Miller Memorial Lecture, North Scot. Coll. Agricul. Misc. Publ.
(5)  Ørskov, E. R. (197) 'Nitrogen utilisation by young and lactating ruminants,' *World Rev. Nut. and Diet*, **26**, 225–43.
(6)  Milne, J. A., Christie, A. and Russell, A. J. F. (1979) 'Effects of nitrogen and energy supplementation on the voluntary intake and digestion of heather by sheep,' *J. Agric. Sci.*, Camb., **92**, 635–43.
(7)  Agricultural Research Council (1980) *Nutrient Requirements of Farm Livestock: Ruminants.* HMSO.
(8)  Blaxter and Boyne, 'The estimation of the nutritive value of feeds.'
(9)  Agricultural Research Council. *Nutrient Requirements.*
(10)  Kammlade, W. G. (1947) *Sheep Science*, Lippincott, p.32
(11)  Nolan, T. and Black, W. J. M. (1970) 'The effect of stocking rate on tooth wear in ewes,' *Irish J. Agric. Res.*, **9**, 187–96.
(12)  Gunn, R. G. (1969) 'The effects of calcium and phosphorus supplementation on the performance of Scottish Blackface hill ewes, with particular reference to the premature loss of incisor teeth,' *J. Agric. Sci.*, Camb. **72**, 371–78.
(13)  Weiner, G. and Purser, A. F. (1957) 'The influence of four levels of feeding on the position and eruption of incisor teeth in sheep,' *J. Agric. Sci.*, Camb. **49**, 51–55.
(14)  Coop, I. E. and Abrahamson, M. (1973) 'Effect of tooth condition on intake of grazing sheep,' *N. Z. J. Exp. Agric.*, **1**, 58–64.
(15)  Sykes, A. R., Field, A. C. and Gunn, R. G. (1974) 'Effects of age and state of incisor dentition on body composition and lamb production of sheep grazing hill pastures,' *J. Agric. Sci.*, Camb., **83**, 135–43.

(16) Klinner, W. E. and Sheppersen. G. (1975) 'The state of haymaking technology – a review,' *J. Br. Grassld Soc.*, **30**, 259–66.

(17) Ferguson, J. A. and Black, W. J. M. (1974) 'Hay quality and the pregnant ewe,' Edinburgh School Agric. Ann. Rpt, 41–43.

(18) Wilkinson, J. M., Wilson, R. E and Barry, T. N. (1976) 'Factors affecting the nutritive value of silage,' *Outlook in Agric.*, **9**, 3–8.

(19) Demarquilly, C. (1973) 'Composition clinique, caracterisations fermentaines digestibilites et quantites lugerees des ensilage, modification par rapport ou förage vert initial,' *Ann de Zooteca*, **22**, 1–35.

(20) Wilkins, R. J., Hutchinson, K. J., Wilson, R. F. and Harris, C. E. (1971) 'The voluntary intake of silage by sheep, I. Inter-relationships between silage composition and intake,' *J. Agric. Sci.* Camb., **77**, 531–37.

(21) Dulphy J. P. and Michalet, B. (1975) 'Comparative effect of harvesting machine on intake of silage by heifers and sheep,' *Ann. Zootech.*, **24**, 757–63.

(22) Thomas, P. C., Kelly, M. C. and Wait, M. K. (1976) 'The effect of physical form of a silage on its voluntary consumption and digestibility by sheep,' *J. Br. Grassld Soc.*, **31**, 19–22.

(23) Thomson, D., Corner, H. H. and Cunningham, J. M. M. (1958) 'Silage for sheep,' *Scot. Agric.*, **33**, 155.

(24) Thomson *et al.*, 'Silage for sheep.'

(25) Tissier, M and Molenat, G. (1975) *Lènsilage de mais dans alimentation des brebis*, Centre de Recherches Zootech. et Vet. de Thiey: Bull. No. 20, 37–42.

(26) Nolan, T. (1974) 'Fattening of lambs on silage alone and with concentrates,' *Ir. J. Agric. Res.*, **13**, 137–46.

(27) Sheehan, W. and Fitzgerald, J. J. (1976) 'Lambs do better on hay than on silage,' *Farm and Food Res.*, **7**, 85–86.

(28) Reed, K. F. M. (1979) 'A note on the feeding value of grass and grass/cover silages for store lambs,' *Anim. Prod.*, **28**, 271–74.

(29) Wernli, C. G. and Wilkins, R. G. (1980) 'Nutritional studies with sheep fed conserved ryegrass, (2) Silage supplemented with dried grass or barley, *J. Agric. Sci.* Camb., **94**, 219–27.

(30) Marsh, R. and Murdoch J. C. (1978) 'Effect of some grain crop drying processes on the digestibility and voluntary intake of herbage by sheep,' *J. Br. Grassld Soc.*, **30**, 9–15.

(31) Osbourne, D. F. Beever, D. E. and Thomson, D. J. (1976) 'The influence of physical processing on the intake, digestion and utilisation of dried herbage,' *Proc. Nutr. Soc.*, **35**, 191–200.

(32) Greenhalgh, J. F. D., Ørskov, E. R. and Fraser C. (1976) 'Pelleted herbages for intensive lamb production,' *Anim. Prod.*, **22**, 148.

(33) Wernli, C. G. and Wilkins, R. J. (1980) 'Nutritional studies with sheep fed conserved ryegrass,' *J. Agric. Sci.* Camb., **94**, 209–218.

(34) Jackson, C. A., Orr, R. J. and Young, M. E. (1976) *A Comparison of Fodder Crops for Use by Grazing Lambs*, Grassld Res. Inst. Internal Rept. No. 348.

(35) Geisler, P. A., Newton, J. E., Sheldrack, R. D. and Mohan, A. E. (1979) 'A world of lamb production from an autumn catch crop,' *Agric. Systems*, **4**, 49–57.

(36) Keane, M. J. (1974) 'Use of forage crops in early fat lamb production,' *Irish. J. Agric. Res.*, **13**, 251–62.

(37) Jackson *et al.* 'A comparison of fodder crops.'

(38) Drew, K. R. (1968) 'Winter feeding of hoggets,' *Proc. N. Z. Soc. Anim. Prod.*, **28**, 94.

(39) Wedin, W. F., Hodgson, H. J. and Jacobson, N. L. (1975) 'Utilising plant and animal resources in producing human food,' *J. Anim. Sci.*, **41**, 667–86.

(40) Ørskov, E. R. (1976) 'The effect of processing on digestion and utilisation of cereals by ruminants,' *Proc. Nut. Soc.*, **35**, 45–52.

(41) Ørskov, E. R. (1979) 'Recent information of processing of grain for ruminants,' *Livest. Prod. Sci.*, **6**, 335–47.

(42) Dewey, W. D., Lee, H. D. and Marston, H. R. (1958) 'Provision of cobalt to ruminants by means of heavy pellets,' *Nature*, **181**, 1367–1371.

(43) Whitelaw, A. and Russel. A. J. F. (1979) 'Investigations into the prophylaxis of cobalt deficiency in sheep,' *Vet. Rec.*, **104**, 8–11.

(44) Whitelaw, A., Armstrong, R. H. and Evans, C. C. (1979) 'A study of the effects of copper deficiency in Scottish Blackface lambs on improved hill pasture,' *Vet. Rec.*, **104**, 455–60.

(45) Dewey, D. W. (1977) *Search*, **8**, 325.

(46) Whitelaw, A., Armstrong, R. H., Evans, C. C., Fawcett, A. R. and Russel. A. J. F. (1980) 'Effect of oral administration of copper oxide needles to hypocupraemic sheep,' *Vet. Rec.*, **107**, 87–88.

(47) Weiner, G., Field, A. C. and Wood, Jean (1969) 'The concentration of minerals in the blood of genetically diverse groups of sheep. 1. Copper concentrations at different seasons in Blackface, Cheviot, Welsh Mountain and crossbred sheep at pasture,' *J. Agric. Sci.*, Camb., **72**, 93–101.

(48) Hill, R. (1977) 'Copper toxicity,' *Brit. Vet. J.*, **133**, 365–73.

(49) Kelly, P. (1979) 'Hypomagnesaemia in sheep: a review,' *ADAS Quart. Rev.*, **34**, 151–66.

(50) Paynter, D. I. (1979) 'Glutathione peroxidase and selenium in sheep. II The relationship between glutathione peroxidase and selenium-responsive unthriftiness in Merino lambs.' *Aust. J. Agric. Res.*, **30**, 695–702.

(51) Blaxter, K. L. (1978) 'The effects of selenium administration on the growth and health of sheep on Scottish farms,' *Brit. J. Nutr.*, **17**, 105–115.

(52) Bonsma, H. C. et al. (1938) 'Drought rations for sheep,' *Fmg in S. Africa*, **13**, 53.

(53) Brauns, F. (1930) 16th Rept Div. Serv. and Anim. Indust. Onderstepoort, p.567.

(54) Hindmarsh, W. L. (1939) *Agric. Gaz.* NSW, **50**, 293.

(55) Agricultural Research Council (1980) *Nutrient requirements of farm livestock: ruminants.*

# 9 Sheep Pasture

Pasture is the wild sheep's natural food, and domesticated sheep are more dependent upon pasture – and therefore more independent of other foods – than are any other class of farm animal.

Indeed, in many systems sheep may derive from 85 to 90 per cent, or even all their energy intake from either fresh or conserved pasture herbage. Calculations by Greenhalgh[1] show that, in Britain, sheep may utilise around 2 kg of concentrates for each kilogram of carcass produced whereas cattle consume around 4 kg.

A vast body of knowledge has been accumulated concerning pasture, the species and strains of grasses and clovers most productive, persistent, digestible, and palatable to stock; methods of cultivation and manurial treatment. To discuss this subject would require a volume in itself, and there are indeed many such already available. Only those aspects relevant to sheep management need be mentioned here.

The efficient conversion of pasture into animal products depends upon its surplus leafage being eaten by cattle, sheep, or by both. Without ruminant animals capable of making that conversion, grass would be an entirely superfluous crop, lacking the slightest agricultural significance. Whether pasture is best exploited by sheep or by cattle is mainly a matter of economics and of the relative profitability of sheep and cattle products under given conditions. The two species can often be grazed together with advantage.

Under favourable conditions, the potential yield of nutrients from grassland in Britain is high, it having been calculated that a harvestable yield of 20 tonnes dry matter (DM) per hectare per annum is possible if water and nitrogen are supplied in adequate amounts. But weather, soil and other factors limit the best yields obtained to around 15 tonnes DM/ha in experimental conditions, and around 7 tonnes DM/ha is more usual under farming conditions, where little or no nitrogenous fertiliser is applied. Since yield of herbage increases progressively as additional nitrogen is applied up to 300–400 kg nitrogen per hectare, opportunities for greatly increased stocking rates are theoretically possible.

Herbage is only of value to the extent that the quality and amount ingested meets the needs of the class of grazing sheep at any particular time. The nutritive value of herbage is dependent primarily on its energy value and this is commonly expressed in terms of its digestibility or 'D' value (more precisely – the percentage of digestible organic matter (DOM) in the dry matter). The DOM of herbages contains an average 12.5–13.5 MJ metabolisable energy (ME) per kilogram but can vary widely, especially with indigenous herbage or rough grazing. Using this factor, it is possible to calculate the energy retention and animal production from a given weight of feed conserved, as illustrated by Greenhalgh.[2] Knowledge of the intake of herbage by grazing sheep and the factors affecting it is essential to the development of more precisely controlled and productive grazing systems.

The techniques used for estimating herbage intake include total collection of faecal output, the chromic oxide dilution technique, the use of faecal nitrogen/digestibility regressions or the use of *in vitro* D value of samples from sheep with oesophageal fistulae.

All of these are subject to large and sometimes unmeasurable errors and caution should be exercised in the comparison of absolute values between experiments.

Many factors influence herbage intake, some are intrinsic to the particular sward, while others are attributes of the sward and are amenable to management manipulation. Others, yet again, are of animal origin, such as size, breed, appetite drive due to physiological state, previous nutritional history and so on.

A grass sward is an extremely complex structure which is continually changing during the growing season as the proportions of leaf, stem, senescent material, and the ratio of grass to clover, are modified by age, grazing pressure or cutting. Although these sward characteristics, as well as others such as digestibility, height and density, species/strain composition, can affect intake, there are still many uncertainties as to the relative and absolute importance of any single, or combination of, factors, particularly with the diversity of grazing conditions under which sheep are kept throughout the world.

Thornton and Minson[3] state that the voluntary intake value of several pasture species appears to be closely related to the retention time in the reticulo–rumen. In general, improved intake tends to be associated with an increase in digestibility, but it is now recognised that quite different levels of intake of pasture can occur at similar digestibilities and also that, at comparable levels of intake, quite different responses in terms of animal performance may occur. Although the D value of leaf may be only marginally higher than that of the stem, voluntary intake of leaf can be 20 per cent higher for perennial ryegrass and, as Laredo and Minson[4] have shown for some of the tropical grasses, it may be as much as 36 per cent greater. The potential nutritive value of pasture species varies quite widely.

Amongst temperate species perennial ryegrass is superior to cocksfoot, and the legumes (white and red clovers) are more nutritious than grasses.

There are substantial differences between the main groups of indigenous hill plant species in Britain.

The broad-leaved hill grass, for example *Agrostis*, spp. tend to have a D value 5 units lower than perennial ryegrass throughout the season, while the fine-leaved grasses, for example *Festuca*, spp. are poorer again, by a similar margin. Hodgson and Grant[5] show that the sedges and rushes, which have a short growing season, decline in D value from around 60, in spring, to below 50 by autumn.

It has long been the view of experienced graziers that spring grass is infinitely more nutritious than that grown in autumn. This complex subject has been reviewed by Reed,[6] who indicates that, while there may be little change in D value throughout the main growing season, the lower net energy value and DM of autumn-grown herbage may be important factors in causing lower intake. This would appear to confirm the practical observations.

The presence of clover in swards has been shown to improve their nutritive value and consequently the liveweight gain of growing and finishing lambs[7] (Fig. 8.1). This occurs, state Rae *et al.*[8] whether the grass component is high-quality ryegrass or hill pasture, such as *Agrostis* or *Festuca* (see Armstrong and Eadie[9]). Greenhalgh[10] states that the superiority of clover compared with grass is considered to be due to its higher digestibility and greater voluntary intake (see also Fig. 9.1). This, suggests Ulyatt,[11] may arise in turn from differences in the structural carbohydrates of legumes, causing a more rapid breakdown in, and passage out, of the rumen, as well as higher rates of production of VFAs.

The optimum proportion of clover in swards of different types has yet to be established and the maintenance of vigorous growth in mixed swards presents a difficult management problem. Continuous stocking throughout the grazing season leads to a decline in clover content. High stocking rates can be equally damaging because of selective grazing. Indeed, it has been suggested by Marsh and Laidlaw[12] that clover will contribute more to a sward when rotationally grazed.

Unexpected complications from the improvement of pasture have occasionally arisen. Perhaps one of the most spectacular was the extraordinary effect of subterranean clovers on Western Australian sheep, as described by Bennets and his colleagues.[13] The story reads more like witchcraft than scientific research. Maiden unbred ewes showed copious

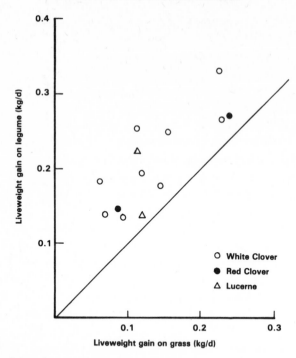

*Fig. 9.1* Liveweight gains of lambs grazed or fed on legumes or grasses (summary of results from Britain, New Zealand and Australia).

milk secretion; ewes that should have lambed were barren; wether sheep developed udders. These amazing happenings, due to an oestrogenic substance in the subterranean clover, were most dramatic in good clover years. On this and much other evidence that might be quoted it must be accepted that the final criterion of what is, or is not, good pasture cannot be properly assessed on its appearance, its botanical composition, nor on its chemical analysis. The productivity and health of the livestock grazing it is the only reliable proof of its actual value.

Nevertheless, under British conditions the proportion of clover in swards is likely to have fallen by the normal autumn mating period and even grazing of pure stands of red clover for up to 28 days pre-mating has not been found to be harmful in the West of Scotland.

## Utilisation

Seasonality of pasture is its most natural yet gravest fault. Were pasture available throughout the year, it would be the ideal food and the most economic basis for production in all flocks. All pastures, however, are apt to fail, either because of low temperatures in cold countries or low rainfall in hot countries. There are always periods of the year when grass is abundant and periods when it is scarce, so that if a regular stock is maintained on pasture throughout the year there must always be a time when the stock is underfed or a time when the pasture is undergrazed.

This problem has been met by the pastoralist in various ways. He may arrange his lambing to coincide with the probable date of grass being available. In this way the summer stocking of pasture is doubled by the sheeps' natural increase, and a better balance between stocking and pasture thereby achieved. When, because of adverse weather conditions, the growth of grass is delayed, grave difficulties may follow. The stock is increased at the very time when there is least pasture for them to eat.

The amount of herbage per unit area of pasture has been shown by many scientists to be an important factor affecting herbage intake (Fig. 9.2). In Australian investigations by Arnold and Dudzinski,[14] the weight of herbage below which intake by sheep has declined has ranged from 1100 to 4000 kg DM/ha but recent studies by Hodgson and Milne[15] with temperate pastures suggest that the critical level may be around 3500 kg DM/ha. Because of behavioural adjustment and other factors, such as the ratio of dead to green material, it is likely that satisfactory levels of performance can be obtained, especially under continuous grazing conditions, over a range of herbage weights. Nonetheless it is apparent that, depending on the circumstances, severe restriction on intake could occur once green material falls below 1300–1600 kg DM/ha.

Physical limitations to intake may also be due to factors such as height of the sward, including leaf length, available herbage per animal and sward density. With the animal, intake may be controlled by bite size, biting rate or daily grazing time. In Australian studies by Williams et al.[16] the intake of both suckling and reared lambs was reduced by 60–70 per cent when the sward was lower than 5 cm, but others have suggested 10–20 cm as plateau levels below which intake may decline.

Traditionally, sheep farmers favour high-density swards probably in the belief that a high amount of nutritious leaf is available. Yet Hodgson[17] has concluded that density does not appear to be an important factor and also that height and distribution of leaf and stem within the canopy are important. He has established that both bite size and biting rate are strongly influenced by grazing height, especially under strip-grazing conditions. Lambs also seem to be influenced by this factor under continuous stocking and therefore sward height is an important factor in determining intake.

Grazing time of adult sheep can vary from 3 to 13 hours daily. However,

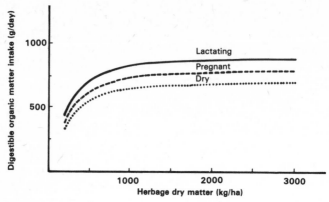

*Fig. 9.2* Relationship between the amount of herbage available (kg DM/ha) and the herbage intake of grazing ewes.

within a specific set of management conditions it is unlikely to vary greatly from day to day. The differences found are probably attributable to climatic factors and extremes in the fund of available herbage material.

In grazing management it is the adjustment of livestock numbers relative to size or weight of animal per unit area which can be manipulated by farmers. This in turn will have consequences on the amount of herbage (or herbage allowance) available per individual animal. This could be particularly crucial in late spring or during periods of slow growth.

Treacher and his co-workers at the Grassland Research Institute (GRI) have investigated the basic relation, under paddock grazing management, between herbage allowance and intake (see Penning and Gibb's findings in Fig. 9.3). Once lambs have reached a grazing age they will compensate for a reduced milk intake by increasing herbage intake but even the best-quality pasture does not replace milk, so causing a potential loss in liveweight gain. This explains why it is important to obtain high ewe milk yields in early fat lamb systems on grassland.

When the same group offered twin suckling ewes on perennial ryegrass allowances of herbage in the range 25–120 g DM/kg liveweight, lamb growth rate was not restricted until after the fourth week of lactation. Thereafter, allowance below 50 g considerably reduced lamb growth and, even on the highest allowance, ewes did not start to gain weight until after the third month. This is a clear illustration of the ewe's great ability, as discussed by Gibb and Treacher,[19] to continue milking while losing weight, given that she is in a body condition to do so.

In conditions of moderate herbage restriction and high milk yields, ewes can lose up to 20 per cent of their weight during lactation. Studies by Gibb and Treacher[20] comparing fat ewes (body condition score (CS) 3.2) and thin

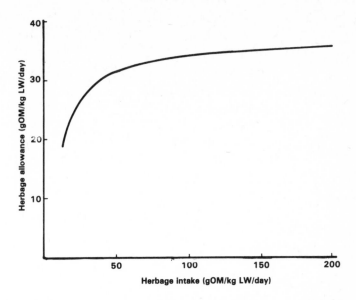

*Fig. 9.3* Relationship between the herbage allowance and herbage intake of grazing lambs (after Penning and Gibb[18]).

ewes (CS 2.4), and stocked to impose moderate to severe restrictions in intake, confirmed that fat ewes would tend to produce more milk, but mainly in the later stages of lactation.

Traditionally in Britain, lambs are rarely weaned under 16 weeks of age unless they are sold directly off the ewe for slaughter. In practice, this is no doubt aimed at deriving maximum benefit from milk even though yields have declined to very low levels.

When grass growth in the spring is late, or daily accumulation is slow, or when stocking rate is unavoidably high, herbage intake can be severely restricted. The farmer has the choice of accepting reduced lamb performance or supplementary feeding until adequate pasture is available. In the GRI experiments, milk yield over a 14-week lactation was reduced by 30 per cent at a herbage allowance of 30 g DM/kg liveweight (equivalent to 2 kg DM/68 kg ewe), compared with double that level. In the restricted group, weaning anytime from 8 weeks onwards was preferable to 14-week weaning, whereas weaning before that time in the more liberally treated group led to a reduction in lamb growth-rate.

However, when lambs in the '30 g' group were weaned at 17 weeks they subsequently consumed more pasture and reached similar weights at 25 weeks as their more liberally treated cousins.

The importance of pasture allowance for the individual ewe has also been suggested by other workers and it is clear that it is an important factor during the earlier phase of lactation, as shown in Table 9.1.

**Table 9.1**  Effects of different pasture allowances during lactation.

|  | Single suckling | | | Twin suckling | | |
|---|---|---|---|---|---|---|
| Pasture allowance (kg DM/ewe/day) | 2.0 | 4.4 | 7.4 | 2.1 | 5.0 | 6.5 |
| Pasture intake (kg DM/ewe/day) | 1.6 | 2.2 | 3.7 | 1.7 | 2.5 | 3.2 |
| Ewe milk production (1/6 week) | 78 | 85 | 80 | 86 | 107 | 124 |
| Lamb gain (kg/6 week) | 10.0 | 11.8 | 12.4 | 7.6 | 10.2 | 11.2 |

After Rattray and Jagusch.[21]

The particular values which may be appropriate under different grazing conditions will be influenced by factors such as pasture yield, proportion of dead material, sward structure and herbage digestibility. During lactation the demand for nutrients by the ewe is at its highest level and the intake of grazed pasture can be from 20 to 60 per cent greater than in the dry or pregnant ewe.

In many lowland and upland sheep-farming systems ewes in early lactation graze spring pastures before there is considered to be sufficient grass to provide a nutrient intake adequate for an acceptable level of animal performance. The customary practice is to offer ewes a cereal-based supplement, but the practical difficulty is to determine the appropriate amount.

The value of feeding supplements clearly needs to be carefully considered. Milne and his co-workers[21] at HFRO found supplements to have no effect on lamb liveweight gain over a 6-week period when the fund of herbage was 500–750 kg DM/ha. Where the threshold levels may be, and the relationship of fund of herbage, herbage allowance and type of concentrate, will need to be investigated before more precise and practical guidance can be offered.

Nonetheless, those sheep-farmers who prefer to stock lightly until sufficient pasture is available probably make the best choice.

Even though grazing management may not yet be ordered, in practice, on the basis of stocking rate adjustments related to available herbage per unit area, allowance per ewe, or both, more precise rules may ultimately emerge. Methods for the measurement of the fund of herbage using simple instrumentation are beginning to emerge and these could lead to more precise systems of management using feed budgeting techniques.

The seasonality of pasture is naturally much more extreme in some countries than in others and on some types of pasture than on others. It is less extreme in some pastoral countries of the southern hemisphere – New Zealand and Argentina, for example – one reason why the European sheep and cattle industries find difficulty in competing with the pastoral produce of these countries. It is very much more marked in northern countries – Scotland, for example – and in such northern countries is more extreme on natural than on cultivated pastures. At high altitudes as in the Scottish Highlands, seasonality of pasture is the real limiting factor in production. The gulf in productivity between the short season of rapid growth of hill pasture in summer and its long quiescence at other seasons of the year, particularly late winter and early spring, is very great. Consequently, if sheep are kept on such hill pastures all the year round, stocking is limited by what the pasture will carry at the leanest time of year. Even then there is consistent overgrazing in spring and undergrazing in summer, with inevitable progressive deterioration of the pastures concerned. On cultivated pasture the same problem arises, but in less acute form, and one of the most beneficial effects of modern grassland research has been the extension of the grazing season. This has been achieved by the use of earlier strains of pasture grasses and the use of fertilisers, notably nitrogen, to stimulate growth. The application of accumulated day temperatures in spring as a guide to the optimum date for nitrogen fertiliser application could lead to the more efficient use of fertiliser.

Stocking rate can be seasonally adjusted in association with conservation to meet flock requirements. Results from many grazing experiments have tended to show that at low grazing pressures continuous grazing – or set-stocking, as it is often called – gives higher animal performance than rotational grazing which, in turn, is usually superior at very high stocking rates.

Semi-arid conditions, where seasonal cycles of pasture growth are caused by lack of rainfall, have much in common with the Scottish hills in that they also experience cycles of variable nutritional provision from herbage.

Different breeds of sheep are of unequal value in converting various types of pasture into animal products. This aspect of pastures – owing to concentration on the botanical and chemical composition of the sward – is apt to be overlooked. Nevertheless, in sheep husbandry it is of considerable importance.

One of my earliest lessons in this connection was when I went round dairy farms renting winter grazing for ewe flocks. The usual bargain made was so much per day's grazing for a flock, the price agreed on to include the keep of the shepherd and his dogs. I discovered that all these dairy farms demanded a higher price if the flock was one of Blackface ewes, their reason being that from a somewhat bitter experience they had found that Blackface sheep grazed more closely than did other breeds.

The Scottish Blackface exploits a heather grazing better than most other breeds. The Romney makes particularly good use of pasture grown on marshy land. The Shetland, because of its wandering habits, is best suited to an island life. There are doubtless many other instances of variety of grazing habit and suitability to different types of pasture between one sheep breed and another. Further observation and study along these lines is most desirable, for in general it is more laborious and costly to alter a pasture than to change a breed.

All sheep are conservative. Sheep thrive and make better use of pastures to which they are accustomed. They thrive even better if born and bred on that pasture. Young sheep on extensive grazings learn where to go for food and water at different seasons of the year both by example and by experience. They also tend to stick to the stretch of ground they know. This habit, in Scotland, is called 'hefting' and at sheep stock valuations an added value is put upon sheep if they be properly hefted, that is to say, if the sheep, in groups which are essentially family groups, limit their foraging to a particular area of the hill farm.

Data obtained by Arnold[23] in Australian studies (Table 9.2) show that quite large differences in intake relative to liveweight can occur between breeds grazing the same pasture.

**Table 9.2** Comparison of herbage intake (g/kg liveweight) by ewes of different breeds on the same pasture.

| Breed | Intake | | | |
|---|---|---|---|---|
| | P | | L | |
| | (g/kg) | (%) | (g/kg) | (%) |
| Fine-woolled Merino | 17.0 | 100 | 20.2 | 100 |
| Dorset Horn | 16.6 | 110 | 22.2 | 110 |
| Corriedale | 13.2 | 78 | 16.8 | 83 |
| Border Leicester and Merino | 12.6 | 74 | 18.5 | 92 |

P = 14 pregnancy L = in lactation

After Arnold.[23]

This is an aspect which has been scarcely studied in Britain.

Differences in nutritional value among grasses and legumes appear to exist because of the variations in sites, rates and products of digestion.

Furthermore, the diet of the grazing sheep can differ appreciably in quality and quantity from that of sheep given cut herbage, which has been the main method of pasture evaluation. Techniques are being rapidly developed, as described by Corbett,[24] to allow determination of energy expenditure of sheep while grazing. This vast, complex and rapidly increasing body of information needs to be drawn together and computer modelling techniques, as described by Sibbald *et al.*,[25] may assist in reducing the complexities of science to a level comprehensible to the husbandman. In this way, it may become possible to apply feeding standards more realistically in the management of grazing sheep.

## References

(1) Greenhalgh, J. F. D. (1977) 'Present and future use of cereals as feeds for livestock,' *Proc. Nutr. Soc.*, **36**, 159–67.

(2) Greenhalgh, J. F. D. (1975) 'Factors limiting animal production from grazed pasture,' *J. Br. Grassld Soc.*, **30**, 153–60.

(3) Thornton, R. F. and Minson, D. J. (1973) 'The relationship between apparent retention time in the rumen, voluntary intake, and apparent digestibility of legume and grass diets in sheep,' *Aust. J. Agric. Res.*, **74**, 889–98.

(4) Laredo, M. A. and Minson, D. J. (1975) 'The voluntary intake and digestibility of sheep of leaf and stem fractions of *Lolium perenne*,' *J. Br. Grassld Soc.*, **30**, 73–77.

(5) Hodgson, J. and Grant, Sheila A. (1980) 'Grazing animals and forage resources in the hills and uplands,' *Br. Grassld Soc.* Occ. Symp. No. 12, 41–57.

(6) Reed, K. F. M. (1978) The effects of season of growth on the feeding value of pasture – a review, *J. Br. Grassld Soc.*, **33**, 227–34.

(7) Thomson, D. J. and Raymond, W. F. (1970) 'White clover in animal production, nutritional factors – a review,' *J. Br. Grassld Soc.* Occ. Symp. No. 6, 277–284.

(8) Rae, A. L., Brougham, R. W., Glenday, A. C. and Butler, G. W. (1963) 'Pasture type in relation to liveweight gain, carcass composition, iodine nutrition and some rumen characteristics of sheep: liveweight growth of the sheep,' *J. Agric. Sci.*, Camb., **61**, 187–90.

(9) Armstrong, R. H. and Eadie, J. (1973) 'Lamb growth on grass and clover diets,' *Proc. Br. Soc. Anim. Prod.*, **3**, 68.

(10) Greenhalgh, J. F. D. (1979) 'The utilisation of roughages,' *In:* The Management and Diseases of Sheep, British Council and CAB.

(11) Ulyatt, M. J. (1971) 'Studies on the causes of the difference in pasture quality between perennial ryegrass, short-rotation ryegrass and white clover,' *N.Z. J. Agric. Res.*, **14**, 352–56.

(12) Marsh, R. and Laidlaw (1978) 'Herbage growth, white clover content and lamb production on grazed ryegrass – white clover swards,' *J. Br. Grassld Soc.*, **33**, 83–92.

(13) Bennets, H. W., Underwood, E. J. and Simer, F. L. (1946) 'A specific breeding problem of sheep on subterranean clover pasture in Western Australia,' *Aust. Vet. J.*, **22**, 2.

(14) Arnold, G. W. and Dudzinski, M. L. (1967) 'Studies on the diet of the grazing animal. The effects of physiological status in ewes and pasture availability on herbage intake,' *Aust. J. Agric. Res.*, **18**, 349–59.

(15) Hodgson, J. and Milne, J. (1978) 'The influence of weight of herbage per unit area and per animal upon the grazing behaviour of sheep,' 7th Gen. Meeting Europ. Grassld Fed., Ghent, 431–37.

(16) Williams, C. M. J. Geytenbeek, P. E. and Allden W. J. E. (1976) 'Relationship between pasture availability, milk supply, lamb intake and growth,' *Proc. Aust. Soc. Anim. Prod.*, **11**, 333–36.

(17) Hodgson, J. (1981) 'Variations in the surface characteristics of the sward and the short-term rate of herbage intake by calves and lambs,' *Grass and Forage Sci.*, **36**, 49–57.

(18) Penning, P. D. and Gibb, M. J. (1979) 'The effect of milk intake on the intake of cut and grazed herbage by lambs,' *Anim. Prod.*, **29**, 53–67.

(19) Gibb, M. J. and Treacher, T. T. (1978) 'The effect of herbage allowance on herbage intake and performance of ewes and their twin lambs grazing perennial ryegrass,' *J. Agric. Sci.*, Camb., **90**, 139–47.

(20) Gibb, M. J. and Treacher, T. T. (1980) 'The effects of ewe body condition at lambing on the performance of ewes and lambs at pasture,' *J. Agric. Sci.*, Camb., **95**, 63–64.

(21) Rattray, P. V., and Jagusch, K. T. (1978) 'Pasture allowances for the breeding ewe,' *Proc. N.Z. Soc. Anim. Prod.*, **38**, 121–26.

(22) Milne, J. A., Maxwell, T. J. and Souter, W. (1981) 'Effect of supplementary feeding and herbage mass on the intake and performance of grazing ewes in early lactation,' *Anim. Prod.*, **32**, 185–95.

(23) Arnold, G. W. (1975) 'Herbage intake and grazing behaviour in ewes of four breeds at different physiological states,' *Aust. J. Agric. Res.*, **26**, 1017–24.

(24) Arnold, 'Herbage intake and grazing behaviour.'

(25) Corbett, J. L. (1980) 'Grazing ruminants – evaluation of feeds and needs,' *Proc. N.Z. Soc. Anim. Prod.* **40**, 136–62.

(26) Sibbald, A. R., Maxwell, T. J. and Eadie, J. (1979) 'A conceptional approach to the modelling of herbage intake by hill sheep,' *Agric. Systems*, **4**, 119–33.

# 10 Hill Sheep Husbandry

Much of the new knowledge which influences animal production is derived from analytical research, and traditionally it is the advisers and farmers who, drawing on this and their own experience, have combined these elements into a synthetic whole. The approach of the scientist to divide the whole into parts for the convenience of more intensive study is both justified and, indeed, necessary provided it is realised that ultimately it may be only a small contribution to a whole system. Indeed, the conduct of systems studies and analytical research in parallel has much to commend it since careful monitoring of systems creates an awareness of problems of relevance leading to meaningful analytical investigations.

In Scotland, hill sheep farming forms the first stage of the system of stratification on which the sheep industry of that country is based. The essential principle of the traditional system lies in the closest possible adaptation of a flock of sheep to an area of uncultivated hill pasture on which they subsist. The stocking is light because, apart altogether from the fact that the productive value of the pasture is itself low, much of the ground may be bare rock, bracken, water, or bog. Cunningham et al.[1] state that the stocking rate varies from 1 ewe to 0.5 ha to 5 ha or more on the poorest hills. Moreover, on any given hill farm, the stocking rate will vary from one part of the hill to another. The extreme variability of soil and vegetation, of altitude and micro-climate over the extent of the same hill farm is, most probably, the primary reason for the way Scottish hill-farming has developed. Breeding ewes are set-stocked, remaining on the hill land throughout the year and more or less totally dependent on pasture for their sustenance. Hunter and Milner[2] remind us that hill sheep are territorial in behaviour, establishing home ranges occupied by single animals or groups, these being developed on a daughter-dam or matrilinear relationship. These territories are maintained by a dominance interaction. Hunter[3] also points out that grazing intensity of hill pastures varies throughout the year due to seasonal availability of plant species, their palatability and the preference exhibited by sheep. The study by Eadie and Black[4] points out that, although most hill pasture species may have a short period when they have a relatively high nutritive value, they vary considerably in the extent to which this is maintained. Thus *Agrostis–Festuca* communities are usually heavily grazed and have a relatively high feeding value for a considerable period, while draw-moss (*Eriophorum vaginatum*), which is much valued for early spring grazing, is more or less worthless for the remainder of the year.

On the hills, around 80 per cent of pasture growth occurs over 4 months from mid May onwards, and by October or early November growth has ceased. Thus, there is undergrazing of pastures in the summer because of the low stocking rate, which is set at a level to allow preferential selection in winter. A study by Eadie[5] shows that utilisation on an annual basis of herbage grown is, at best, around 20–25 per cent. The consequence of this is the accumulation of a fund of dead material which at all times dilutes the diet.

Even on the best hill-grazings hill sheep perform at well below their potential, liveweight output being around 6–16 kg/ha. Weaning percentages vary between 50–120 per cent in unimproved systems and average flock weaning weights from 21–27 kg are usual. Individual ewe performance varies widely, but the extent to which this is attributable to genetic merit can be confounded by the disparity of home ranges occupied.

It is now quite clear from Russel's[16] study that it is primarily the level of nutrition obtained by the ewe throughout the year, not winter alone as is frequently believed, which limits ewe performance. There is a clearly defined cycle of ewe bodyweight which usually reaches its highest level by October, the start of the sheep year. By then the lambs are sold, the cast ewes disposed of, the rams purchased, the ewe hoggs off most farms away at their wintering. The ewes have about 2 months to recuperate from lactation before they are mated again towards the close of November. A mild and open autumn with some growth of herbage is of tremendous importance in building up the frames of hill ewes to withstand the hardship and nutritional strain of the succeeding winter and spring. Indeed, in adverse conditions, a ewe can lose up to 20 per cent of her body tissues, which may be acceptable if she goes into winter in good condition, but may be fatal if she is already lean in autumn.

Mating or tupping time is begun traditionally on the 28 November, although on better hills it may be somewhat earlier. Too-early mating could expose new-born lambs to severe weather at lambing, while a later date may give lambs insufficient time to grow before sale. Since the nutritional value of hill pasture has deteriorated long before the end of November, ewes will be slowly or even rapidly losing weight by that time. Doney and Gunn[7] have shown that reproductive performance is determined to a considerable extent by the body condition of the ewe, her weight and rate of change in weight at mating. This explains, in part, the poor lambing performance which is found on hill farms and losses at lambing time as described by Gunn and Robinson.[8] However, twins in any but small numbers, are unwanted on many hills grazings since there is inadequate feeding to allow the ewe to build up two full-sized lambs or to milk them adequately when born. In many cases, only sufficient to replace casualties are wanted but if special provision in the form of improved grazings is made, then twins can be handled satisfactorily. During the recovery period, post-weaning, and pre-mating phase ewes are particularly responsive to management and 'flushing' can be manipulated to an extent in improved systems to achieve particular lambing rates.

During December, January, February and March – the 4 worst months of the year – hill ewes are in lamb, and during March and early April are heavily in lamb. Their late pregnancy thus coincides with a low nutritional plane which sometimes comes perilously near starvation. All recent experimental work on sheep shows that full growth of the unborn lamb and of the ewe's udder depend on the ewe being upon a high nutritional plane in late pregnancy. Yet, in hill ewes, a loss of one-third in liveweight between tupping and lambing is not unusual. Can a system in which the nutrition of the ewe deteriorates as her pregnancy progresses be justified in practice?

The accepted principle in the management of hill ewes throughout the greater part of Scotland is that they should be given sufficient hay to keep them living through severe snowstorms and, apart from such a use of hay as iron ration to prevent starvation, the ewes should get no other food than what the plants growing on the hill pasture itself can provide. The arguments advanced to support this practice are partly economic, which is easy to understand. The most a hill ewe can be expected to produce in the course of a year is one lamb plus one fleece. Production being limited to this, or frequently even to much lower levels, so must expenditure be controlled very carefully if any profit is to be obtained. Some hill-sheep farmers also declare that supplementary feeding of hill ewes makes them lazy; that once fed they ask for and need more; that they find the primrose path to the hayrack or feed block easier than seeking out sparse grazing. On a good hill farm, lightly stocked, acceptable levels of production may be obtained without extra feed but the lambing percentages will vary from over 100 per cent in a good year to 75 per cent or so in poor years, without there being any variation from one year to another in breeding policy and management. On poor farms a lambing percentage of 85–55 would be more common. Lamb mortality can be a significant source of economic loss, state Gunn and Robinson[9] ranging from 9 to 17 per cent or more in hill flocks. It has been shown by Cunningham and Maxwell[10] that birthweight is an important factor, with losses beginning to rise steeply as weight at birth of typical Blackface lambs falls below 3 kg. Better nutrition in late pregnancy, that is the last 5–6 weeks of gestation, will ensure that the majority of birthweights will fall within a range likely to minimise losses. Furthermore, ewes in good condition will lamb down with a plentiful supply of milk and especially is this so in gimmers, which will also have a much stronger maternal instinct.

The benefits of supplementary feeding are nowadays more widely appreciated, and the advent of self-help blocks has contributed to this, by reducing labour demands. It probably also causes least disruption to grazing behaviour. The early experiments conducted at the Rowett Research Institute (RRI) by Orr and Fraser[11] some 40 years ago on the hand-feeding of hill ewes have been confirmed in subsequent experiments at the Hill Farming Research Organization (HFRO)[12] and in practice.

Lambing comes in late April and early May, sufficiently early to coincide with, and in late years, precede, the first new greenness of hill pastures. It is a period of hard work and acute anxiety, because there are so many risks – unseasonable weather, delayed growth of pasture, disease, vermin. Weather, being unpredictable and uncontrollable, leads to the widest variation in production. New-born lambs are particularly vulnerable to adverse weather – particularly wet, cold conditions. Survival appears to be greater in lambs with a hairy type of birth-coat and resistance to cold may well be an inherited trait. Lamb dysentery, louping-ill and other infective agents, are largely under veterinary control.

Losses from 'vermin are becoming more serious with the decline of game preservation and the extension of forests. Foxes, wild cats, and the hoodie, or grey crow, are all extending their range and becoming more common than they were earlier in the century, but the most deadly vermin of all is the killer dog, which, strangely enough, is so often the sheep dog.

The growth of hill lambs – as that of all lambs – is closely correlated with the ewe's milk yield. Compared with lowland lambs on productive pasture, a hill lamb is more dependent upon ewe's milk, and therefore the growth rate of a hill lamb is even more closely correlated with the milk yield of its dam.

The growth rate of a lamb on a good hill grazing will be around 230 g/day falling to 150 g/day by 8–9 weeks of age, yet growth rates of 325 g/day to 12 weeks have been recorded by Peart[13] for twin Blackface lambs. In increasing the milk yield of hill ewes it has been found that yield usually reaches a peak by the third or fourth week of lactation, with yields of up to 3 kg milk per day being recorded, where nutrition has not been a limiting factor. If, however, nutrition is inadequate over the first 3 weeks or so, peak yields are low and subsequent improvement in nutritional provision will do little to improve yield. Nevertheless, milk yield declines as lactation advances and good lamb growth can only be maintained if high-quality pasture is available to the lamb. Armstrong and Eadie[14] state that the cultivated grasses, and white clover in particular, are vastly superior in nutritional terms to even the best, bent-fescue, hill grasses. Under grazing conditions relatively small improvements in the quality of grazed pasture can produce remarkable improvements in performance. Eadie[15] discovered that an increase of 5 units of digestibility of grazed herbage led to hill ewes rearing single lambs producing 20 per cent more milk, with a similar increase in growth rate, while twin suckling ewes produced 70 per cent more milk and their lambs grew 40 per cent faster.

It will be apparent that there is a considerable amount of scientific knowledge on hill sheep (see Lucas[16] for an account). A synthesis of this has led to a proposal by Eadie[17] for a radical approach to the management of hill sheep. This recognises the fundamental importance of the nutrition of the ewe, not only during the winter period of starvation, but more importantly in summer and late autumn. It is the latter periods which determine the weight and quality of weaned lambs and the size of the future lamb crop. The use of improved pasture in summer and autumn is the key to better nutrition. This pasture may be provided by fencing off an area of *Agrostis–Festuca*, 1 ha for 12–15 ewes, or where this is not available, by reseeding, employing the appropriate technique from the many available. The 'two-pasture' system, as it is now widely named, is based on the use of the enclosed paddocks after lambing, through to weaning, when the ewes return to the open hill to join the dry ewes and hoggs. Depending on seasonal growth, ewes will return to the paddocks sometime before mating and will remain there as long as feed is available, before returning to the hill. Many hill farmers are now lambing in large hill parks, and in a late spring, when there may be insufficient growth in the improved areas, some of the ewes with singles may graze the hill until marking, when grass growth is usually adequate for them to return to the improved areas. Because of better utilisation of the enclosures and greater production from reseeds, an increase in stocking rate is possible. Experience has shown that impressive increases in production are possible. At the Redesdale Experimental Husbandry Farm (EHP), Walters[18] describes how in one hirsel, production has increased by 251 per cent and ewe numbers have risen from 450 to 1001, under the care of one shepherd with some assistance at peak periods. At Sourhope HFRO production has risen by 123 per cent, thus yielding an additional 34 kg lamb per hectare. Most hill sheep farming systems depend on the maintenance of a regular stock of breeding ewes on the farm. The term 'regular' means that the different ages of ewe up to casting age are roughly equal in number. The success of the enterprise depends, essentially, upon the sheep, both in type and in numbers, being suited to the land. Hill sheep farms vary widely in soil, vegetation and climate and the choice of breed is determined by its ability to thrive in the particular environment. Of the three hill-breeds in Scotland, there can be little doubt that the Blackface, Cheviot and North Country Cheviot have the necessary qualities as do the Welsh Mountain in Wales and Swaledale in the North of England.

Having selected the most suitable breed for a particular hill sheep farm, the next step is to get the ewe stock hefted and acclimatised to the ground; this is often difficult and expensive particularly in areas where tick-borne diseases are prevalent. Therefore, once a ewe stock is acclimatised and hefted to a hill, the basic stock is often maintained for a century or more. Sudden changes in stocking are hardly ever made, the risk of probable loss being too great. Suppose, for example, a breeder of Blackface sheep desires to change over from the Lanark to the Newton-Stewart type or vice versa, he does not sell off his old ewe stock and replace it with the new type of sheep. He changes his stock over from one type to another by the safer though more gradual method of successive crosses with the new strain of ram. The change is made in some five generations without any sudden disturbance of the hefting and acclimatisation of the ewe stock. When the change-over from Blackface to Cheviot stocks or vice versa was made in the past, successive top-crossing with rams of the opposite breed was the method most frequently employed.

Hefting and acclimatisation – which includes acquired resistance to prevalent diseases – are so important that in most hill districts of Scotland these factors are taken into account when a hill sheep-stock changes hands, their value amounting to a considerable sum of money. The conditions by which such valuations are made are now governed by the Hill Farming Act of 1946. Because of the proved value of having the ewe stock hefted and acclimatised to the hill to which their ancestors were first introduced perhaps 150 years or more ago, such ewe stocks are customarily taken over by valuation on the ground when one tenant or owner succeeds another.

The hill-sheep farmer has four main products to sell. These are:

(1) Wether lambs – a few fat, but mainly store.
(2) Ewe lambs – surplus to stock replacement requirements and therefore few in number and of secondary class.
(3) Cast ewes – sold, usually at 5–6 years of age as being too old for the hill but still capable of rearing crossbred lambs on marginal or low ground.
(4) Wool.

(1) Wether lambs. Modern market trends, with an ever-growing preference for lightweight carcasses and leaner meat, favour the hill wether lamb. It should be possible to dispose of a greater proportion of them for immediate

slaughter than was practicable in the past. Nevertheless, the majority of wether hill lambs will probably continue to be sold as stores. In this connection it must be remembered that the feeder who buys such lambs is extremely concerned with their quick response to better feeding, and not at all with the ability of the breed to which they belong to live and thrive under hill conditions. That is why the fleshing ability of hill sheep breeds can never be ignored nor be entirely lost sight of in the search for hardiness and milk. The balance, always a delicate one, cannot be allowed to tip too far either way.

Hill wether lambs are bought largely on 'character'. Character is something based on the reputation gained by lambs off a particular grazing for the way in which they thrive, grow, and fatten when transferred to lowland keep. The lambs off some grazings do much better than others and, although the fact is common knowledge among sheepmen it is not always at all easy to suggest a satisfactory scientific explanation to fit the facts. No doubt factors of breeding, parasitism, previous nutrition, are all bound up in it.

The Blackface breed and similar types are preferred for finishing before the end of the year, whereas the Cheviot types are frequently purchased for carrying through for slaughter in late winter, early spring. On many hill farms, although not on all, there may be a small area of cultivated land or area suitable for reclamation, on which forage crops can be grown. Crops which are most favoured on upland and hill farms are the catch crops such as rape, Dutch (stubble) turnips, fodder radish and radicole (Raphanobrassica). Rape is the most popular crop, but yield is unpredictable. Under poor soil conditions, Dutch turnips and the turnip hybrids have proved to be more adaptable and consistent yielders, but wastage can be high, as also for rape, where up to 35 per cent of the crop may be spoilt. The digestible organic matter (DOM) content (7–9 per cent) of the bulbs of Dutch turnips is low and supplementary feed of these up to 0.25 kg/day is necessary. Liveweight gains of lambs on forage crops can be very variable, from 20 to 150 g per day having been recorded. In the first instance, this may be due to the need for lambs to adapt to a new diet, and to assist this, they are frequently run on to pasture. Lambs are sometimes folded in 'breaks' and when moved to fresh fodder, intake has been shown, for example by Drew,[19] to increase dramatically. The differences in the amounts of forage available and the morphology of the crop may be an explanation.

Good crops of rape can finish at least 50 lambs per hectare. Fodder radish is prone to run to seed head, and must be grazed over a short period.

The technique developed for the intensive finishing of early weaned lambs, using cereal-based diets, has been used successfully to finish hill lambs as the study by Sheehan and Lawler shows.[20] Great care is required during the introductory phase, since catastrophic losses can occur if inadequate roughage is not included in the feed. Diets based on maize are superior to those with barley, which can be included whole, but conversion ratios are invariably poor, ranging from 5.5 to 9 kg feed per kilogram liveweight gain. Cost of feed ingredients is important, but more critical is that slaughter lamb price should rise or be high during the finishing period. It is usually tail-end lambs that have not been fattened off greencrop by December, which are housed for finishing on this system.

They tend to finish at a time when market prices are rising.

(2) Ewe lambs. While all the wether side of a hill sheep stock is destined for sale, usually as lambs and nowadays only exceptionally as mature wethers, the majority – and these the best – of the ewe lambs are retained as stock. Usually about one-quarter of a hill ewe stock requires to be recruited each year to replace casualties and those ewes that have reached casting age. Since in a hill sheep stock a lambing percentage of 80 is considered satisfactory, and since only half the lambs born are female, there may be only 40 ewe lambs, or fewer, from which to select the 25 or so needed to replenish the numbers for every 100 ewes kept. That means that only on good farms in good seasons is there much scope for selection. On poor farms in had seasons there is none.

Hill sheep farmers have been criticised for the method, or rather lack of method, by which such selection is made; in most cases it is less a matter of selection than of culling. A certain number of ewe lambs will be undersized because of being poorly suckled or late-born. Some will be too far off the desired breed type for retention for stock if any reasonable uniformity in sale produce is to be retained. By the time these rejections are made there is generally an insufficient number of lambs left for any more subtle selection. A study of selection of ewe lambs for hill flocks has been done by Dunlop.[21] Where numbers are adequate, selection is frequently based on the family history of the lamb, as well as upon its own appearance. A good hill shepherd will know the parentage of most of the lambs and in some cases by memory alone without written record, tracing their ancestry back over several generations. I have, myself, heard a shepherd plead for the retention in the flock of a rather unimpressive-looking ewe lamb because, 'Her grannie was one of the best-doing ewes we ever had on the place.'

Because it is only the ewe lambs surplus to stock replacement needs that are sold, the number of hill ewe lambs on the market is always much less than that of wethers. That is one important reason for their greater sale value as compared with wethers.

Carrying hill ewe lambs through their first winter is always difficult and often expensive and methods of doing so vary widely with district. In some parts of the Scottish Border country the ewe lambs retained for stock are never weaned, they stay beside their dams on the hill all winter. Where practicable, this method has several substantial advantages. There is no check due to weaning; there is a strengthening of the family group on which the system of hefting is based; there is the advantage of the ewe's example in teaching the lamb to make the best use of a hill grazing at all seasons of the year.

Under any system by which ewe hoggs are at liberty on the hill during tupping time there is a risk of the rams serving them. Unless they are particularly forward and well-grown the risk is slight since they are unlikely to come into season. A precaution adopted on Border farms where ewe hoggs were home-wintered was to 'break' the hoggs by tying a piece of sacking over their rumps, thus affording a mechanical obstacle to service. A cynical student once condemned this practice asserting that he had seen a Blackface ram returning from the hill with a piece of sacking on either horn! Another Border method, which became obsolete, was to retain one part of the farm for the wintering of all the ewe lambs. This area was called the 'hogg' hill, the word 'hogg' being applied from about the month of October onwards to sheep born in the spring of the same year. Recently the practice has been revived on part of Sourhope HFRO where on a 203-ha unit, 18 ha have been reserved for wintering 160 hoggs. They are fed, on average, 13 kg each of hay and sugar-beet pulp, plus 13 kg grass nuts.

The most common system of ewe hogg-wintering, taking Scotland as a whole, is 'away-wintering' where the hoggs are sent to lowland rented grazing from the beginning of October until the end of March. The price asked has now reached a fairly high figure – some £6–£9 per head – and hill flockmasters often grudge this cost, calling it a second rent. In fact, the cost is less uneconomic than it may seem. Sending the ewe hoggs away to wintering lightens the stocking of the hill in its barest season, thereby permitting a larger ewe stock to be kept. Therefore, provided the profit per ewe covers or exceeds the wintering cost of a ewe hogg, there is no actual loss involved. In addition to the cost of away-wintering, there is, however, the further difficulty of obtaining suitable lowland grazing. In former times dairy farmers, once the cows were off grass, were willing enough to let their winter grazing to hill flockmasters, but they are much less eager to do so today. The high cost of concentrates and the greater profitability of milk production, provided home-grown feeding stuffs are mainly used, have encouraged dairy farmers to save their winter grass with the object of getting their cows earlier on to pasture in spring. These difficulties have, in many instances, compelled flockmasters to seek alternative methods of wintering their hoggs. It was never in any case a method entirely free from objection since the hoggs always returned to the hill too soon, in most cases at least a month too early. There is little growth of hill grasses until late April, so that the hoggs suffer a check, often accompanied by casualties. If they could be kept away until 1 May, results would be much better.

Storm risk is an important element in any commonsense approach to the hogg wintering problem. Hill ewe hoggs are farming capital, not annual income, so that if the ewe hoggs are spoilt or lost, the effects are evident in the annual balance sheet for at least 4 years. In many a severe winter the hill sheep farmer who has sent his ewe hoggs away, battling to save his ewe flock, must often feel thankful that his flock recruits, at least, are reasonably secure. A variant which was once practised, namely to send ewe lambs off to foggage at weaning in September, and to return home to the hill in early January, has been shown to be as successful as any other.

A further method of dealing with the hogg wintering problem is the 'hogg-house' or in-wintering method once widely used in the North of England described by Wilson.[22] Hoggs may be confined for a short period and taught to eat hay and concentrates, thereafter being allowed out to a field or a hill paddock for the remainder of the winter with either free access or being brought in at night. In other cases, once housed they are kept inside until returned to the hill in March or April, the date being determined by the prevailing weather conditions and the amount of growth on the hill.

Considerable argument has centred on the relative merits of home, away and in-wintering systems and experiments have frequently given different conclusions. However, many of these have failed to observe that it is the specific levels of nutrition from weaning, throughout the winter and succeeding summer which influences growth and development, so causing variation in subsequent breeding performance, but Gunn[23] has provided some material.

The size of the lamb at weaning, relative to mature weight, the quality and amount of grazing available on an away-wintering or 'tack', or on the hill farm, and the amount of food offered to housed hoggs, all lead to quite large differences in the growth and development of hoggs during winter. Hoggs, which are in good condition and weight when they return to the hill, will make good breeding stock. On the other hand, hoggs which have made only moderate winter growth, say 2 kg, will normally show considerable compensatory growth during summer and will end up little lighter in weight than those which started heavier in the spring. In their first breeding year the latter will tend to have a higher percentage of twins, which is understandable, but differences will be minimal in succeeding years. Philips et al.[24] have shown that severe dietary restriction over the winter will lead to poor breeding performance of gimmers, due to inadequate development of the genital organs as well as to a reduction in reproductive potential throughout their breeding life.

In-wintering permits easy manipulation of nutrition. During the decline in the growth phase in hill sheep from November to January, small hoggs will not respond to liberal feed, while this, later in winter, will cause them to fatten rather than to grow. The aim should be the management of wintering systems to sustain growth from weaning to November and again in late winter and early spring, so as to avoid excessive body condition when the hoggs return to the hill; they should reach about 80 per cent of adult weight by 16 months.

(3) Cast ewes. Among mankind, the high hills in winter and rough weather are best left to the young. So also is it with sheep. There comes a time when a hillbred ewe will live with greater certainly and produce to greater advantage under lowland conditions. This fact is acknowledged in the cast ewe trade, an important aspect of commercial sheep farming. In Scotland the common practice is to cast the ewes at a certain fixed age, usually at 4, 5 or, exceptionally nowadays, 6 years of age. Often the younger the casting age, the higher the sale price. The ewes in the regular annual cast are warranted sound above and below, which means that they have their full complement of 8 incisor teeth and both halves of the udder are functioning. Ewes defective in either respect are sold separately. Cast ewes, as in the case of lambs, are sold on character just as much as upon size and appearance. Undoubtedly, there is a big difference in the value of the cast ewes off various grazings and when a buyer has had good or ill fortune in his purchase in one year, his experience is likely to be reflected in his eagerness or otherwise to purchase the same cast in subsequent years. Thus, it comes about that for particular casts there is the keenest competition because of the 'character' casts off the same hill stock have gained in former years. The underlying biological reasons why some casts have such a much better character than others is very complex. Disease, without a doubt, is one factor. Cast ewes infested on their own hill grazings with fluke or lungworm may – far from throwing off their infection under better nutritional conditions as might be expected – find the best lowland pasture an ovine cemetery. The death rate among cast ewes may be heavy but at least one flock of cast ewes known to Allan Fraser – they were North Country Cheviots off the then famous stock of Lynegar in Caithness – refused to die; like old soldiers, they only faded away! Breeding is clearly an important factor and character may be made or marred on the same grazing by a change of breeding policy and indeed the good character of cast ewes in any hill sheep stock is one of the best possible

advertisements for the soundness of the breeding methods employed and, in particular, the choice of rams.

In addition to character, uniformity of type and a good size of frame for the breed are important selling points in the cast ewe trade. Uniformity of type in a flock can be achieved only by a consistent policy in the purchase of bought-in rams. Size of frame in the ewe depends very largely on good wintering of the ewe hoggs, a further reason why hill flockmasters pay such careful attention to their method of hogg wintering.

At most of the annual cast ewe sales, ewes are sold either as 'uncrossed' – when they have always been mated with rams of the same breed – or as 'crossed' – when they have been mated at least once with a ram of some other breed. On average, a higher price is paid for uncrossed than for crossed ewes, even when they are of the same age and type. It is of some interest to speculate why this should be so. Clearly, it must arise in the first place because the majority of sheep farmers believe that crossing lowers the value of a cast ewe, but why should it be so? On reason advanced by a surprisingly large number of flockmasters is based on the scientifically discarded hypothesis of 'telegony', which means that the influence of the sire upon the offspring is carried over in some obscure manner into the offspring of a different sire, used subsequently upon the same dam. To take a crude hypothetical example, it implies that a white woman first married to a Negro and having had children by him might, on a second marriage to a white man, give birth to an occasional faintly coffee-coloured baby. There is no scientific evidence for this belief, nevertheless it dies very hard. Yet, even supposing telegony were true, the belief would, in the case of cast ewes, seem irrelevant since the majority are crossed in their first mating after purchase. In most cases they are not bought for the purpose of pure breeding, so why should they be presumed to lose value?

A second reason frequently advanced is that a cast ewe that has been crossed has had 'more pulled out of her' than she should have had if the lamb had been purebred. Cast ewes of the two Scottish hill breeds, Blackface and Cheviot, will have been crossed with a Border Leicester ram if they have been crossed at all and the Border Leicester-cross lamb is certainly, nutritional conditions of the ewe being equal, slightly heavier at birth. The difference is, however, a small one and without doubt the extra strain on the ewe's resources in producing it cannot be of the same order as that of a ewe rearing twins of its own breed. A further interesting suggestion often made is that a cross-lamb from a ewe crossed for the first time has greater vitality and vigour than lambs born of the same ewe by later crosses with the same breed of ram or even the same ram. There is no scientific evidence to support this idea but it is, in a way, an extension of the hypothesis of hybrid vigour, known to be important, but also difficult to interpret on theoretical grounds. It is a speculation, no more, and yet one which I am unable to dismiss as altogether untrue because I have been inclined to speculate along similar lines. There this interesting but obscure problem must rest meantime. On the one hand we have undoubted preference of sheep farmers for the uncrossed ewe, a preference for which they are prepared to pay money. On the other hand there is no satisfactory theoretical explanation, but it would be wrong because of this to assume too dogmatically that there is nothing to explain.

Should a cast ewe have any of her incisor teeth missing – 'broken-mouthed' is a trade expression – she will sell at a price considerably lower than that of a 'whole-mouthed' ewe of the same age and quality. The reason – a fairly obvious one – is that broken-mouthed ewes can be handicapped in their grazing.

(4) Wool. The economic importance of wool production from a hill flock is well recognised, for it can be as high as 20 per cent of the total flock revenue. Walters[25] confirms that the recent increased value of lambs, when combined combined with a substantial improvement in the lamb crop can, however, reduce wool sales to around 11 per cent of flock income. The comparative poor quality and low fleece weights – 1.5–3 kg being the usual range – are often a matter for expressed surprise by visitors from wool-producing countries, who consider wool a more easily-won product of poor hill land than well-grown lamb.

For a short period in British sheep history – roughly during the time of the Napoleonic Wars – there was a diversity of view as to whether the industry should concentrate on mutton or on wool. The victory went to mutton, but not without a struggle. There is no doubt that the rapid expansion of hill sheep farming in Scotland at that time was due, in the main, to the relatively high price of wool and that the incentive provided by the English woollen industry was the economic force behind that expansion.

The conclusion of the Napoleonic Wars, the resumption of fine wool imports from Europe, the subsequent development of the Merino sheep industry in Australia – all combined to direct sheep production in Britain, including the Scottish Highlands, towards mutton production which, in the absence of refrigeration, remained a naturally protected home market for the home producer during the greater part of the nineteenth century. The Merino breed which had been tried without success in Scotland, although well established elsewhere, eventually became extinct in Britain, although it has been reintroduced recently for experimental purposes. The position was reached in which the main business of hill sheep farming became the supply of store stock for lowland mutton and lamb production, a position which remains unchanged today.

Doney[26] states that wool growth in hill breeds is seasonal to the extent that 80 per cent of the fleeces are produced in the 6 months from June to November. In contrast, three-quarters Merino one-quarter Cheviot sheep do not exhibit this seasonality of wool growth, suggesting that this is a function of the genetic background of hill breeds (see Doney[27]). Improved nutrition during lactation will increase wool growth, but has a negligible effect during pregnancy.

'Hardiness' in hill sheep is frequently considered by farmers to be of great importance, but it is a concept which has no precise scientific definition. Undoubtedly the fleece is a weather protection coat and its properties of thermal insulation in different weather conditions can be considered a component of 'hardiness'. Although the extent to which fleece types differ in efficiency as a protective coat is not wholly understood, it is clear that it is the depth of wool, irrespective of its length, which is the important factor in insulation, as also may be volume of wool per unit area of skin. It is wind which can substantially increase the loss of heat by breaking the structure of the fleece. Even moderate wind speeds of 12–20 km/hour will cause an increase of heat loss of 3–5 times that of still conditions. Heat loss is of more importance in the new-born lamb, and the tolerance of cold is considerably greater in those born with a predominantly hairy birth-coat. The fleece of the adult sheep enables it to tolerate an exceedingly wide range of thermal environments.

### References

(1) Cunningham, J. M. M., Smith, A. D. M. and Doney, J. M. (1971) *Trends in Livestock Populations in Hill Areas in Scotland*, Hill Farming Res. Organ., 5th Report, p. 88.
(2) Hunter, R. F. and Milner, C. (1963) 'The behaviour of individual, related and groups of South Country Cheviot Hill Sheep, *Anim. Behav.*, **11**, 507
(3) Hunter, R. F. (1954) 'The Grazing of Hill Pasture Sward Types,' *J. Brit. Grassld Soc.*, **9**, 195.
(4) Eadie, J. and Black, J. S. (1968) 'Herbage utilisation on hill pastures, Proc. Occ. Symp. Br. *Grassld Soc.*, **4**, 191.
(5) Eadie, J. (1967) *The Utilisation of Grazing Hill Sheep: Utilisation of Hill Pastures*, Hill Farming Res. Organ. 4th Rep. p. 38.
(6) Russel, A. J. F. (1971) 'Relationships between energy intake and productivity in hill sheep,' *Proc. Nutr. Soc.*, **30**, 197.
(7) Doney, J. M. and Gunn, R. G. (1974) *Progress in Studies in the Reproductive Performance of Hill Sheep*, Hill Farming Res. Organ., 6th Rep., p. 69.
(8) Gunn, R. G. and Robinson, J. F. (1963) 'Lamb mortality in Scottish hill flocks,' *Anim. Prod.*, **5**, 67.

(9)   Gunn and Robinson, 'Lamb mortality.'

(10)  Cunningham, J. M. M. and Maxwell, T. J. (1979) 'Improved sheep production on hill farms,' *Vet Ann.*, **19**, 69.

(11)  Orr, J. N. and Fraser, A. (1932) 'Restoring the fertility of Scottish sheep grazings,' *Trans. High Soc. Scot.*, 5th Ser., V, **434**, 64.

(12)  Hill Farming Research Organization, 1st Rep., p. 88.

(13)  Peart, J. N. (1967) *Lactation and Lamb Growth*, Hill Farming Res. Organ. 4th Rep., p. 69.

(14)  Armstrong, R. H. and Eadie, J. (1977) 'The growth of hill lambs on herbage diets,' *J. Agric. Sci.*, Camb., **88**, 683.

(15)  Eadie, J. (1967) *The Nutrition of Grazing Hill Sheep*, Hill Farming Res. Organ., 4th Rep., p. 38.

(16)  Lucas, I. A. M. (1974) 'The contribution of science to the improvement of hill sheep production,' *Proc. Brit. Soc. Anim. Prod.*, **4**, 45.

(17)  Eadie, J. (1970) *Hill Sheep Production Systems Development*, Hill Farming Res. Organ., 5th Rep., p. 70.

(18)  Walters, B. R. (1978) 'Farm Unit Reports: 2 Dargues Hope,' *Redesdale Exp. Husb. Farm Annual Rev.*, Min. of Agric. and Fish.

(19)  Drew, K. R. (1968) 'Winter feeding hoggets,' *Proc. N.Z. Soc. Anim. Prod.*, **28**, 94.

(20)  Sheehan, W. and Lawlor, M. J. (1974) 'Studies on the indoor finishing of Blackface lambs,' *Ir. J. Agric Res.*, **13**, 33–38.

(21)  Dunlop, G. (1947) Selection of ewe lambs for hill flocks, *J. Min. Agric.*, **54**, 222–27.

(22)  Wilson, W. (1946) 'Wintering hill hoggs,' *J. Min. Agric.*, **52**, 350.

(23)  Gunn, R. G. (1964) 'Levels of first winter feeding in relation to performance of Cheviot hill ewes,' *J. Agric. Sci.*, Camb., **62**, 92 and 123.

(24)  Philips, R. W. McKenzie, F. F., Christensen, J. V., Richards, G. S. and Petterson, W. K. (1945) 'Sexual development of range ewe lambs as affected by winter feeding,' *J. Anim. Sci.*, **4**, 342–46.

(25)  Walters, B. R. (1979) 'Components of Sheep Income,' Redesdale Exp. Husb. Farm, Ann. Rev., p. 14. Min. of Agric. and Fish.

(26)  Doney, J. M. (1961) 'The fleece of the Scottish Blackface breed 1. seasonal changes in wool production and fleece structure,' *J. Agric. Sci.*, Camb., **56**, 365.

(27)  Doney, J. M. (1966) 'Breed differences in response of wool growth to annual nutritional and climatic cycles,' *J. Agric. Sci.*, Camb., **67**, 25.

# 11  Lowland and Upland Systems

Going back into history, into medieval times, it is clear that methods of keeping sheep were very different from those currently practised.

In those days, the shepherd drove the village flock to their common by day, folding it within movable hurdles on the village arable ground at night. The enclosure of the commons made this system impossible, so that the sheep, perforce, had to remain folded on the arable land both by night and by day, just as the pig-sty, from being a night shelter only, became the prison cell of the English cottar's pig. It was in this way that the classical system of folding sheep in England began.

Many years ago, Mr J. F. H. Thomas wrote an excellent little book called *Sheep Folding Practice.* In it he defined the folding system as 'the daily confinement of the flock on an area of food crop defined by the use of a suitable form of portable fencing'.

In its most intensive form, 'the closed form' of folding, the flock is always moving on to clean fodder and clean ground, with most beneficial results on worm control. Sometimes in the 'run-back system' the flock, although given access to fresh fodder, is not fenced off from that already eaten when, owing to the tremendous concentration of sheep on a small area, worms may be a serious problem.

In the folding system's most intensive form, lamb-creeps are provided so that lambs may creep through the hurdles ahead of the ewes, with excellent effects on the nutrition and health of the lambs.

The folding system reached its greatest development and prosperity in the later eighteenth and throughout most of the nineteenth century. It was combined with corn-growing, often barley-growing, on light land. The folded flock was of great benefit in consolidating the land by treading and in the even manuring of such light land, with good results on the white crop succeeding. In fact, on much of the lighter arable land of England the folded flock was the pivot of farm management.

By far the fullest description of this or other methods of the past is to be found in the closely documented work by Trow-Smith, [1] *A History of British Livestock Husbandry to 1700.*

In this century, sheep husbandry methods have been too wasteful of land

and labour. There is now a realisation that sheep can be managed intensively, but incorporating a greater proportion of grass in the system. One ewe to 2–3 ha on the hill, 6–7 ewes per hectare on lowland, hardly makes sense, while one full-time shepherd to care for 400–600 ewes, at modern wage rates does not make sense either.

It is, then, only right, sensible, and proper that new methods of sheep husbandry aiming at higher productivity per ewe per hectare should be explored energetically, and at the time of writing many such explorations are in progress. Some may suceed, some will fail. All aim at what has come to be called 'intensification', and some are much more revolutionary than others.

Because of the adaptability of sheep, the wide variety of breeds and crossbreeds available and the diversity of resources used, the systems of sheep production are so numerous that detailed discussion of them all is not possible.

Food resources include grassland, either short-term leys or permanent pasture; forage crops; arable by-products, for example sugar-beet tops; stubbles, etc. and concentrates. These, along with elevation and climatic conditions, the locality and its market outlets, for example fat, store or breeding stock, and the availability and the price of breeding stock, combined with the interest of the farmer, are amongst the main factors which dictate the choice and type of system. Also, time of lambing is an important variable, since it has a direct influence on pasture utilization and, hence, grassland management. It is also of importance in relation to management at lambing and the risk of neonatal lamb mortality.

The majority of breeding ewe flocks are based on crossbred ewes, for example Mule, Scotch or Welsh Halfbred, Greyface, etc. or breeds such as the Clun or similar types, usually crossed with a meat-type sire for fat or store lamb production. Flocks of this type are frequently dependent on grassland to a considerable extent and especially in upland areas. In fact, 85–90 per cent, or thereby, of the annual food energy requirements of the flock are derived from this source, either grazed or with part conserved as hay or silage for winter keep.

Since profit is understandably the main aim of a production enterprise, it is essential to appreciate the relative importance of the factors which are likely to influence this objective. The size of the lamb crop, that is the lambing percentage, and the value of the lambs sold account for around 80 per cent of the gross margin per ewe, while stocking rate, combined with these are all important when gross margin per hectare is considered. This is not to say that feed costs, flock replacement costs, veterinary costs or labour can be ignored but profitability in Britain could be greatly improved by producing larger and better lamb crops from a reduced area.

One of the most important management decisions to be made is deciding the time of lambing for the flock. In a mild climate, with early grass growth and an ample supply of conserved foods, early lambing in December or January for the production of spring fat lamb may be considered. Although a system of this type will inevitably involve higher feed costs, these may be offset by the greater market value of the fat lamb. With early lambing, the appetite of the ewe is at a peak when she is dependent on conserved feed. To achieve high returns a large proportion of lambs have to be slaughtered at around 12–16 weeks of age, and before prices fall. This requires high liveweight gains of up to 300 g per day, which demands high-quality forage and liberal concentrate use to stimulate lactation, or alternative early weaning and the use of concentrate diets for the lambs.

The favoured choice of many flockmasters is to lamb the flock at a time to coincide with the start of the spring flush of grass. Understandably, there is a trend for later lambing with increasing elevation, from lowland to upland. Not only is the commencement of grass growth later, but adverse weather at lambing can have a profound effect on lamb losses, unless ewes are housed at lambing.

Mating time will therefore vary from August to October. On many upland farms the shorter growing season can mean that grass growth is minimal by the time the rams are joined with the ewe flock. Whatever the conditions, grazing management should be organized to ensure that a fund of 2000–2500 kg dry matter (DM) per hectare or more of pasture is available for several weeks before and during mating. The aim will be to improve body condition, or alternatively to hold it at the desired level (condition score (CS) 3–3.5) if that has already been reached. Where pasture is scarce supplementary feeding may well be justified.

In Britain the average lambing percentage of crossbred ewe flocks, as recorded by the Meat and Livestock Commission (MLC)[2] is around 130–140 per cent, yet the best flocks achieve around 180–190, with some over 200 per cent lambs born with 170–180 or so, per cent lambs reared.

The technical knowledge is undoubtedly available to allow a considerable improvement in lamb output, if it were applied. The MLC in its excellent booklet *Feeding the Ewe* shows quite clearly that when high lambing percentages are sought, ewes should be at a CS of 3.5 at mating, and maintained at this level for a month after service. This is a difficult target to reach. Much will depend on the initial condition of the ewes and the quality and amount of pasture available and on the period of time required to reach it. As the season advances, pasture quality declines as does the intake of the ewes, probably because of the effect of shortening daylength, so allowance must also be made for these seasonal factors.

A practice which I have found flockmasters to use fairly regularly is to deliberately reduce ewe body conditions after mating, and to 'flush' before, and over, the mating period. Admittedly, this could well have an effect when ewes are below the target condition score (CS) of 3.5 but there seems to be little purpose in this practice, if shortage of pasture before and at mating time does not allow improvement to this level and for its maintenance throughout the mating season. However, there is some evidence to suggest that crossbred ewes which improve in condition from around CS 2.5–2.75 to over CS 3 and then maintain weight could produce more lambs.

Within any flock, body condition can vary considerably and leaner ewes should be drawn out for preferential treatment, some 6–8 weeks before mating. It is highly probable that these may be the most productive ewes in the flock and they certainly merit special attention.

In most grassland systems, particularly those based on spring lambing, ewes can well fend for themselves on pasture, after the end of mating and until the end of the year or thereby, depending on weather conditions. In this mid pregnancy period some weight loss can be allowed, but certainly no greater than CS 0.5–1.

The traditional method of wintering the halfbred ewe flock in the Scottish Border country was based on winter pasture, roots, concentrates and hay. To that list there may now be added silage, straw, liquid and block urea-based supplements.

No one with practical experience of sheep can doubt the value of hay in keeping ewes in healthy conditions. Whatever may be said against hay on the grounds of nutritive losses in winning and so on, there can be little question that good hay, well made, suits sheep very well.

Either allowing free access or providing an allowance of 0.5–1 kg hay per ewe are customary practices.

Although it is many years since the early pioneers obtained successful results in using silage it is only in recent years that is has increased in popularity. Its greater flexibility in grassland management, allowing conservation in early summer or in autumn and its ease in mechanisation, are decided advantages. It should have a digestibility (D) value of 65–70, a (DM) content of 25 per cent or so and be well preserved with a sweet smell. Intake of precision-chopped silage can be 20–33 per cent higher than long material. The benefits of chopped silage may be less apparent during pregnancy, since concentrate allowances can be adjusted but, as Apolant and Chesnutt[3] point out, advantages post-lambing can occur in enhanced milk yield and consequently in lamb growth. An allowance of 4–5 kg silage per ewe daily is adequate, but sheep, being fastidious eaters, may not this amount of inferior silage, thus needing larger amounts of concentrates.

Roots, particularly turnips before lambing, and swedes thereafter, are one of the traditional bulk feeds. Folding the flock for 2–3 hours each day, with a run back on to grass can, by eliminating harvesting and storage, markedly reduce costs. However, some stored roots may be needed for feeding in the lambing field, and for ewes with lambs-at-foot, until adequate pasture is available.

Barley straw, clean and free from moulds and supplemented with 300 g or so suitable compound, containing at least 15 per cent crude protein, has been shown by Broadbent and Jacklin[4] to be satisfactory, with additional concentrates being provided in late pregnancy.

The use of concentrates in steaming-up ewes before lambing was common practice on the Scottish Borders long before the scientific evidence discussed in an earlier chapter of this book had provided theoretical chapter and verse to support the practical argument. About a month or so before lambing, the troughs were put out and, starting with 200 g or so per head, the ration was increased to 650–800 g, or even 1 kg by the time the lambs were born. At 8–10 weeks before lambing, lean ewes should be drawn out and given 100–200 g of a high-energy concentrate, later joining the main flock.

When rams have been raddled, or fitted with harness, ewes can be later marked – 'smotted' – so that previous to lambing they can be segregated into groups to allow more precise nutritional management with a saving in feed costs.

Bruised oats is the traditional cereal used in concentrate/meal mixes, but barley fed either whole or rolled, dry or wet stored, is nowadays more popular and entirely satisfactory. Dried sugar-beet pulp, loose or as nuts, can also be used to contribute to part of the meal mixture but, because of its fibre content, it should be restricted to 10–15 per cent of a mix or to around 400–500 g per ewe, if given separately. Other energy feeds are also used, including wheat, maize, flaked maize and others, as supply and cost allow.

In recent years, feed-blocks and liquid feeds dispensed by ball lickers, or in troughs, have become more common. Most contain some non-protein nitrogen (NPN), usually urea. They are suitable supplements for moderate-to low-quality roughages from mid pregnancy until 3–4 weeks before lambing. Most blocks contain salt or some constituent to limit intake, but they vary widely in composition, and not all ewes will eat them or consume a similar amount. When upland ewes are run on to rough grazing or when the roughage, that is hay or straw are of low digestibility, then supplements of this type can justify their relatively high costs.

By 4–5 weeks before lambing, concentrates should include sufficient protein and particularly adequate undegradable protein which can be provided by including, say 5–10 per cent fishmeal or 5 per cent linseed cake.

Allan Fraser wrote, 'Fishmeal, if price permits is a commendable addition to a concentrate mix, as has also been the view of many shepherds in the Borders.' Science has now demonstrated the reasons and confirmed the accuracy of these observant men.

Examples of suitable concentrate mixtures are given by the Meat and Livestock Commission (MLC).[5] Apart from the draft hill ewe, which will require a total of about 16–20 kg per head of concentrate, most ewes will consume around 50–55 kg in grassland systems, while the level may rise to 70–80 kg in the earlier lambing flocks. In areas with early growth of grass, or when high-quality roughages are used, the requirement for concentrates may be considerably lower.

In the winter management of his flock in Britain, the farmer is faced with the choice of running the flock on pasture – possibly intensively stocked – on a sacrifice area, or mob-stocking at 100 or so per hectare, with frequent moves, or housing when winter feeding commences.

Stocking rates at up to 30 ewes per hectare may not necessarily detrimentally affect the botanical composition of the pasture but herbage yields, up to mid June, will be reduced, at high stocking rates, says Black.[6] On many soils, severe poaching may occur even at moderate stocking rates of around 10–14 per hectare.

Where early spring grass is valued and high summer stocking levels are sought then winter housing has to be considered.

In his book, referred to earlier, Trow-Smith[7] wrote:

> The modern flockmaster is only now re-discovering the advantages of sheltering his animals from the winter weather and is building once again houses for his flocks which cannot be dissimilar to those which went out of use when the great wool boom died down – although they survived on the Continent. On the big Wiltshire farms there were bercaries of stone or timber with thatched roofs and equipped with hay racks and mangers; furthermore, the different classes of stock – ewes, wethers, and shearlings were housed separately.'

Some of the modern attempts to intensify sheep husbandry are not, in fact, so novel as their protagonists suppose.

Nevertheless, respect for tradition should never degenerate into conservative complacency. The old ways of sheep husbandry have served wonderfully well in their time, but times change and methods will have to change with them.

Once again there is, with returning profitability, a resurgence of interest in sheep housing, which is an essential and integral part of systems in Iceland

and Scandinavia. A reduced labour requirement, easier supervision, pleasanter working conditions, more precise control over nutrition, a reduction in lamb mortality, avoidance of poaching on heavy land, and earlier grass in spring are among the benefits of housing a breeding flock. Against these must be set the high capital cost, if new buildings are erected, higher feeding costs and the greater risk of disease.

It is possible to modify existing buildings cheaply and a sheep house may be used for other purposes such as calf- or turkey-rearing, grain or forage storage, to spread costs. An alternative mentioned by Speight and Cunningham[8] is to use unroofed yards.

Essentially, housing should be seen as a tool in intensification, and the Edinburgh College of Agriculture has developed a simple computer program to allow the economic evaluation of introducing sheep housing on individual farms.

Experience has shown that housing will be successful, given that ewes have adequate trough and floor space, that there is good ventilation without draughts at floor level, and these are combined with sound nutritional and health management.

I have found that housed ewes may appear to be thriving; their wool dry and their faces clean, yet when handled they can be much leaner than appearances would suggest.

Because the flock is entirely dependent on hand-feeding, rationing must be carefully regulated so that a weight loss of not more than CS 1 is allowed to occur in mid pregnancy. A sound programme of preventative medicine is needed and veterinary consultation is desirable. Foot rot can wreak havoc, particularly if bedding is damp. Several years ago in collaboration with my then colleague, the well-known Dr J. A. A. Watt, we eliminated foot rot from a flock of Greyface ewes. But failing this, it is essential to segregate all lame sheep before housing and cure them before they join the flock.

Shearing of March/April lambing ewes soon after housing is now being practised. The evidence suggests that heat stress, if it occurs, is reduced, food intake is greater and heavier lambs at birth are generally obtained. From experience other practical advantages include easier inspection, both of body condition and udders, and there is a reduced space requirement.

Perhaps the main disadvantage is the risk of cold stress when ewes are turned out in spring, hence the relevance of clipping date. Later, the number of cast of 'couped' ewes will be greatly reduced.

It is, I believe, important to assess carefully all the economic and management consequences before venturing into a housing system.

The specialised skills of the shepherd and the flockmaster come to the fore at lambing time and have been well documented.

Regrettably, neonatal lamb mortality represents a major loss to the sheep industries in many countries. As the value of the individual lamb increases, so also can the justification be made, welfare aspects apart, for increased supervision over the lambing period. The reasons for these losses are likely to vary depending on climatic conditions, management practices, incidence of disease and other factors. Surveys in lowground flocks in Scotland show that of total losses, 30–40 per cent are due to abortion and stillbirths, 20–30 per cent to starvation and exposure, 15–20 per cent to infectious disease, 5–10 per cent to congenital defects and 5–10 per cent to predators and misadventure. Clearly, better nutrition and management of the in-lamb ewe has much to contribute.

If new-born lambs are weak and slow to suckle or to stand or when milk let-down by the ewe is delayed, they can be effectively fed using a stomach tube. A first feed of 50–100 ml colostrum, preferably from a ewe, is best. Cow colostrum can be used although it is less satisfactory. If further feeding is required, 100–200 ml per lamb per feed, at least three times daily depending on its size, is needed. This technique is potentially one of the important advances in practical sheep husbandry, but it is vital that the technique is understood, and is learned by demonstration from a skilled operator.

Losses due to exposure and the chilling or hypothermia which will occur in wet, sleety and cold weather can also be reduced by another relatively new technique. I refer to the 'Moredun' method developed by Andy Eales and colleagues[9] and developed from basic research in Australia and the Moredun Institute.

Hypothermia can arise because of excessive heat loss or depressed heat production. The first will occur in adverse weather conditions and the second in small lambs with inadequate reserves of fat at birth. These small lambs may be unable to stand or they may have dams which are short of milk. Lambs so affected develop hypoglycaemia – a deficiency of glucose in the bloodstream. Identification of the hypothermic lamb is dependent on establishing its rectal temperature, which, if below 37°C, is indicative of the condition. This is considerably more reliable than depending on observation of the behaviour and appearance of the lamb. Even skilled shepherds with an acute perception of lambs so afflicted find this method of considerable benefit. Simple and easy-to-operate electronic thermometers are available. Young lambs under 5 hours old should be dried, warmed in air at 35–40°C and given a feed. Similar treatment for lambs over this age is suggested but the addition of 25 ml for small lambs, to 50 ml for large lambs, of a 20 per cent solution of glucose. This is given directly into the abdomen, that is interperitoneally. Because of the complexities of this technique described by

Eales *et al.*[9] it is best to seek veterinary guidance before adopting it.

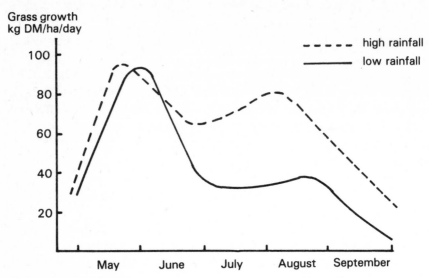

*Fig. 11.1* Grass growth rate variation over the season (from Ministry of Agriculture and Fisheries[11]).

## Grazing management

The challenge to the flockmaster, if he wishes his sheep enterprise to compete in economic terms with others, is to produce high yields of grass and to utilise it efficiently. This implies high stocking rates and matching flock requirements to grass growth – combined with obtaining lamb growth rates consistent with marketing objectives.

Grass growth is highly seasonal and varies considerably due to rainfall, soil and elevation (Fig. 11.1). It can be manipulated by the strategic use of nitrogen fertiliser, and the carry-over effect of winter grazing can also affect growth in the spring. Additional flexibility can be introduced by the integration of grazing and conservation, while the use of earlier-maturing ram breeds can allow earlier fat-lamb sales.

The intake of herbage has been shown by Hodgson[10] to be dependent upon the weight of herbage per unit area, that is herbage mass. On a mixed grass–clover sward the intake of herbage will be maximal at around 1200–1800 kg DM/ha. In practical terms this will be a sward height of 2.5–6 cm.

Because of the spring/early summer flush of growth it is possible to achieve satisfactory levels of intake at the lower end of the desired herbage mass, 1200–1400 kg DM/ha. This allows high stocking rates over this period, with 25 ewes and lambs per hectare, or even more being possible. Other than for early lambing flocks an allocation of grassland for conservation, preferably silage, can be incorporated. Indeed, as an insurance against unforeseen conditions, part of this should be regarded as a buffer to be grazed if need arises. The use of lightweight electric fencing within a grazing field is a convenient means for isolating surplus to grazing requirements.

As lambs grow the overall flock requirement increases and this may continue beyond the point when the mid season decline in grass growth occurs.

It is at this time that a silage aftermath is invaluable and (4–6 cm in height) during mid to late summer herbage mass of 1600–1800 kg DM will ensure high lamb performance.

Traditionally, sheep farmers have been reluctant to use nitrogen fertiliser on grass for sheep. In experiments involving up to 100 kg nitrogen per hectare in a fat lamb system using Suffolk-cross twin lambs, I obtained lamb growth rates of 365–380 g daily over the period April to the end of June, with a high proportion of lambs being sold fat by that time.

There is clear evidence that levels of nitrogen fertiliser application, grass yield and stocking rate potential are related. Thus, at low levels of nitrogen input, 0–75 kg nitrogen per hectare grazing stocking rates of around 10–14 ewes per hectare, depending on weight, or 650–750 kg ewe weight per hectare may be possible. At inputs of 150–225 kg nitrogen per hectare, stocking capacity can be significantly increased to 20–25 ewes, that is 1200–1500 kg/ha or more, depending on conditions.

It is now possible using a computer program and incorporating data which takes into account site, climate, monthly rainfall, level of nitrogen, winter forage requirements, breed of ewe, lamb crop and growth rate to estimate grazing and overall stocking rate as a guide to planning.

An early application of nitrogen of around 70–80 kg/ha will be beneficial in stimulating early growth. No matter how adequate and nutritious hand-fed diets may be, young grass will stimulate lactation, a fact well known to stockmen, who nevertheless are frequently reluctant to use nitrogen to aid this objective.

When to cease offering supplementary concentrates in spring is a difficult decision. It would appear that at least 750 kg DM/ha, that is around 1.5 cm of herbage, should be allowed to accumulate before reducing or eliminating concentrates. Flockmasters frequently stock highly at this time, and decision-making is as much an art as a science, as yet, and depends on acute observation and awareness of all the important factors.

Individual animal performance tends to fall as the season advances and also as stocking rates increase. At high stocking rates where sheep are the only enterprise only a limited area of conservation aftermath may be available, but it is essential in an intensive system as a buffer. The use of nitrogen to stimulate mid-season growth can also be an effective way of providing more mid-season forage. If around 50 per cent of the nitrogen to be applied is used in the June–July period, then the yield profile can be altered (Fig. 11.2).

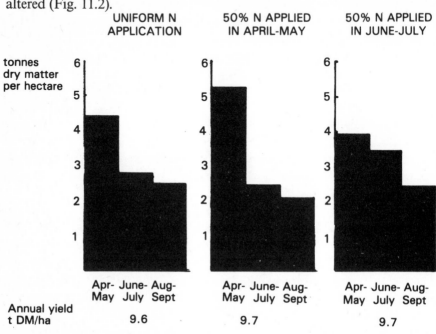

*Fig. 11.2* The effect of the pattern of fertilizer N application on the pattern of grass growth and annual yield at 300 kg N/ha[12].

In this way, nitrogen application throughout the season will be proportionally: 20–30 spring, 50 midsummer, and 20–30 per cent autumn. The late dressing is applied with the aim of providing adequate herbage for mating.

Flexibility can be introduced by early weaning lambs at 10–12 weeks of age, by using aftermath of conservation areas where silage will be used in other livestock enterprises and by growing forage crops. The so-called follow-nitrogen system, which is based on a sequential application of nitrogen at 7–10-day intervals, in 3 or 4 blocks across a field, allows some flexibility in amount and timing of fertiliser use to adjust growth to flock requirements. Given that stocking rates are effectively adjusted, nitrogen use, whatever the overall level from 0–300 kg per hectare, should be around 10–12 kg per ewe.

It is well known that clover is more nutritious than grass, but in practice it suffers from a late start to growth in the spring, and a highly seasonal pattern of growth. The amount of clover eaten appears to be directly related to the proportion of clover present in the sward horizon which the sheep grazes. In mixed swards a balance has to be struck, in which there is adequate grass for early growth, yet sufficient clover to have a nutritional impact in mid season. Curll[13] has shown that the effects of grazing pressures, intensity of defoliation, preferential selection – if it occurs – and intermittent grazing, all need to be investigated before practical guidelines will become available to ensure stable proportion of grass clover in swards.

The general maxim that lenient and infrequent grazing favours grass whilst hard grazing, and some rotational systems, encouraging clover is widely appreciated but is a somewhat unsatisfactory guideline. In the absence of better information I would suggest 30–35 per cent of the overall DM should come from clover in mid season.

The control of parasitism can be tackled either by using a scheme of routine dosing appropriate to circumstances, or by the use of clean grass and grazing systems coupled with dosing.

## Grazing systems

The farmer has a bewildering choice of grazing systems. Confusion can arise because these so-called systems are not all analogous, since some have implications within the overall farm plan, while others are tactical, including manipulation during the grazing season. Amongst the systems which have to be considered are clean grazing; set-stocking, or continuous, grazing; split paddock; rotational with or without creeps, with its several variants such as forward lateral or sideways creeps.

Some of the various systems are illustrated in Fig. 11.3, reproduced from Spedding's book *Sheep Production and Grazing Management.*

One of the major constraints to increasing stocking rate has been the hazard of a greater severity of gastro-intestinal parasitism – 'worms'. The value of 'swedes' or first year ley in this respect, is well recognised, (see, for example, Rutter[15]) but the innovation of the clean grazing system allows the principle to be extended to larger leys or permanent pasture.

At its simplest, grazing areas allocated to cattle and sheep are alternated annually, but in practice the system is more likely to include conservation, taken from within each grazing area, or as part of the overall rotation, as illustrated in Fig. 11.4. An important principle is not to allow lambs to run on grass to be grazed by ewes and lambs the following year. It is also essential to dose ewes before they go on to clean grass and perhaps before lambing as well. For a variety of reasons, some of which are not yet well understood but could include the effects of climate on overwintering of larvae and infection from peri-parturient ewes, a build-up in parasitism in lambs may occur late in the season. This is not a criticism of the system, but rather the more we understand how to use it to control worms effectively, the more valuable it becomes.

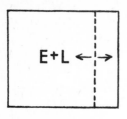
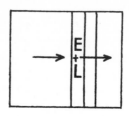
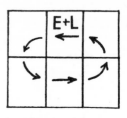

1. Set-stocking with a movable fence to adjust the stocking rate of ewes (E) & lambs (L).

2. Continuous folding of ewes (E) and lambs (L).

3. Rotational grazing of ewes (E) and lambs (L).

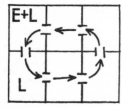
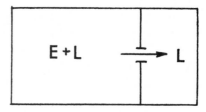

4. Forwards creep-grazing of lambs (L) ahead of their ewes (E).

5. Set-stocking with a creep.

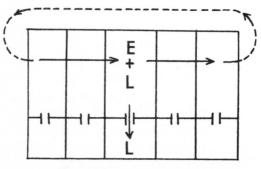

6. Sideways creep grazing.

Fig. 11.3 Systems of grazing (after Spedding[14]).

YEAR 1                      YEAR 2                      YEAR 3

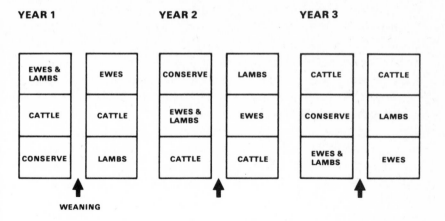

*Fig. 11.4* A system of clean grazing for permanent pasture with conservation (source: Rutter, 1983).

Comparison of systems is extremely difficult and scientifically complex, and perhaps the definitive experiments have still to be undertaken. As biological knowledge increases (see, for example, Maxwell[16]) it should become possible to determine more precisely the operational guidelines on which to base meaningful comparisons.

The practical choice of the majority of farmers is still to use a mixed grazing system.

Unless the clean grazing principle is used, then mixed grazing will invariably improve individual sheep performance and overall animal output per hectare.

If it is true that intermittent grazing, at any stocking rate, does not increase cut herbage production, then any benefits ascribed to rotational grazing must arise for other reasons. When rotational grazing and continuous stocking have been compared, as described by Nolan,[17] at equivalent stocking rates any differences found have been minimal, other than at very high stocking rates. The level of adoption of rotational grazing is limited. The main advantage to be gained from its use is the greater flexibility for adjustments between the grazing and conservation areas and for making short-term decisions during periods of rapid growth. As indicated earlier, continuous grazing does not imply a fixed stocking rate for the whole of the grazing season. The regulation of grass supply, by one of more of the many means available, is an integral part of a rotational system. The use of an elaborate layout of paddocks will not appeal where the enterprise is small or on arable farms with short-term leys where benefits may be marginal.

The choice of system is much dependent on the farmer and his objectives, since there is no overwhelming evidence to suggest that there is a 'best' method of grazing sheep.

The achievements of the most successful farmers are always a guide and stimulus. In the arable areas in particular sheep have come to be seen as the poor relation. However, high output can be achieved with a sheep enterprise and the key to this is a high overall stocking rate of 15 ewes per hectare being an essential target with lambing percentage of 175 or higher being desirable.

That this can be achieved is without doubt and Henry Fell[18] illustrates how sheep can compete with other enterprises. I believe he has clearly established the important principles for a profitable intensive system. Apart from a high stocking rate, he indicates the need, especially on heavy land, to remove the flock from the grassland area during winter. This may involve housing the flock or using some other method. He also proposes an alternative or addition to silage aftermaths for overcoming the 'summer gap' which is the use of suitable root crops. Nonetheless, it is productive grassland, intensively stocked at 25–30 ewes plus lambs per hectare at the height of the grazing season which contributes significantly to the competitiveness of the sheep enterprise.

In many of the arable areas with lighter soils, such as on the Chalk Downs, the inclusion of shorter leys in the cropping system is still considered to be worth while.

A sheep flock offers a considerable degree of flexibility since it can be rotated round the farm and, in economic terms, it is likely to be more profitable than suckler cows and equal to, or better than, finishing beef cattle.

Where grassland is intensively managed and arable crops and by-products are consumed, excellent results can be obtained, as Harry Ridley has shown at Chimbolton Down in Hampshire. The farm carries a large ewe flock which is lambed in the open, with straw bale 'houses' for shelter, and accordingly lambing occurs in late March–April when the weather is kinder.

Grass is sown down in 60 ha blocks which are subdivided to approximate 20 ha units, so that each flock is around 250–300 ewes, stocked at 15 ewes (mules) and lambs per hectare. Starting in February with 67 kg/ha nitrogen dressing, a total of 280 kg/ha is applied in 5 dressings in each month of April to July and again in September.

Hay is the basis of conservation and around 20 per cent of the ley area is shut up, but this would be greater were it not possible to utilize arable products in winter. Since each flock has its own conservation area, the heavier stocking (18–19 ewes per hectare) in early summer can be reduced when the flocks go on to the aftermaths in July and August.

Stubble turnips are sown after the winter barley is harvested, to be used along with grass during tupping in November, and this is adequate to bring ewes into good condition.

The new leys, undersown spring barley, also become available in autumn

Table 11.1  Sheep results for Chimbolton Down Farms 1977–80.

|  | 1977 | 1978 | 1979 | 1980 |
|---|---|---|---|---|
| Ewes | 2172 | 2569 | 3000 | 3100 |
| Lambs | 3795 | 4567 | 5240 | 5424 |
| Lambing (%) (lambs sold/ewe to tup) | 174 | 175 | 173 | 174 |
| Grass acres (leys) | 442 | 500 | 524 | 540 |
| Output/acre (£) (lambs and wool) | 209 | 240 | 275 | 300 |
| Stocking rate (ewes/acre) | 4.9 | 5.2 | 5.8 | 5.74 |
| Average price/lamb (£) | 22.86 | 25.00 | 23.63 | 27.65 |
| Price of shearlings (£) (ewe replacements) | 50 | 65 | 55 | — |

as harvest in early August. These young seeds, along with others available on adjoining farms, are grazed, first by the store lambs, rarely more than a third of the crop, weaned in early September. Later, ewes follow the lambs in October for flushing.

Stubble turnips usually last through until February along with grass from the last year of the ley, which is ploughed-out so that any damage occurring to the pasture is of no consequence. An allocation of 100 kg hay per ewe is sufficient to provide supplementary fodder although grazing of winter-proud cereals can add a few days' additional keep.

The production achieved on this farm has been consistently maintained at a high level (Table 11.1). Especially interesting are the lambing percentages and stocking rates achieved with modest labour inputs – one full-time shepherd and part-time assistants.

On upland farms a much shorter growing season is experienced and lambing is invariably later than in the lowlands. Under poorer conditions the proportion of lambs sold as stores will increase, these being sold during August and September. In recent years there has been an upsurge in the in-wintering of ewes to avoid poaching on heavy land and to reduced labour inputs. The potential for grass production in upland areas can be very high because of the better distribution of rainfall than occurs at lower elevation.

The potential of difficult conditions is well illustrated by the achievement of Edward Moir at Courance in Aberdeenshire. On a farm 170–300 m above sea-level, with limited shelter and steep fields, a flock of 445 Greyface ewes is lambed in April. The overall annual stocking rate averages 12.8 ewes per hectare, with a lamb weight of 947 kg/ha being produced from 1.48 lambs reared per ewe and all lambs being sold fat. Ewes are wintered on silage and are housed at lambing for ease of management.

Over a wide range of conditions the sheep flock can be integrated with other enterprises or run as the dominant enterprise, and to perform at high levels, so that production per unit area is extremely high and probably even better than the extremely intensive but labour-demanding systems of a bygone age.

## References

(1)  Trow-Smith, R. (1957) *A History of British Livestock Husbandry to 1700*. Routledge & Kegan Paul, p. 113.
(2)  Meat and Livestock Commission (1981), *Feeding the Ewe*, Sheep Improvement Services.
(3)  Apolant, S. M. and Chesnutt, D. M. B. (1982) An evaluation of silage for pregnant and lactating ewes, 55 Ann. Rep. Agric. Res. Inst. N. Ire., p. 33.
(4)  Broadbent, J. S. and Jacklin, D. (1983) 'An evaluation of straw as the sole source of roughage in the diet of pregnant housed ewes,' winter meeting: paper 150, *Brit. Soc. Anim. Prod.*
(5)  Meat and Livestock Commission, *Feeding the Ewe.*
(6)  Black W. J. M. (1978) 'Winter grazing of pasture by sheep,' *Irish J. Agric. Res.,* **17**, 131–40.
(7)  Trow-Smith, *A History of British Livestock Husbandry.*
(8)  Speight, B. R. and Cunningham, J. M. M. (1965) 'Methods of wintering ewes,' *Brit. Soc. Anim. Prod.,* **7**, 283 (abst.)
(9)  Eales, F. A., Small, J. and Gilmour, I. S. (1982) 'Resuscitation of hypothermic lambs,' *Vet. Rec.,* **110**, 121–23.
(10) Hodgson, J. (1977) 'Factors limiting herbage intake by the grazing animal,' Proc. Int. Meeting Anim. Prod. from Temperate Grassland (ed: B. Cilsenan) A. N. Foras Taluntais.
(11) Ministry of Agriculture and Fisheries (1978). *Focus on Grazing Systems for Beef and Sheep.*
(12) Doyle, C. J. and Wilkins, R. J. (1983) 'Grassland production: realising the potential,' Br. Grassl. Soc. Occasional symposium No. 15 (Ed.: J. Corrall).
(13) Curll, M. L. (1982) 'The grass and clover content of pastures grazed by sheep,' *Herb. Abstr. and Rev.* **52**, 403–411.
(14) Spedding, C. W. R. (1965) *Sheep Production and Grazing Management*, Baillière, Tindall.
(15) Rutter, W. (1983) 'Grassland management for the lowland ewe flock,' *In:* (ed.): W. Haresign. *Sheep Production.* Butterworth, pp. 207–218.
(16) Maxwell, T. J. (1978) 'Management decisions in grazing systems,' *In: Sheep on Lowland Grass,* Brit. Soc. Anim. Prod. Summer Meeting, pp. 21–27.
(17) Nolan, T. (1980) Research on mixed grazing by cattle and sheep in Ireland, *In:* (ed.): T. Nolan and J. Connolly Proc. Workshop on Mixed Grazing, Agric. Inst. Eire.
(18) Fell, Henry (1979) *Intensive Sheep Management*, Farming Press.

# 12  Intensive Systems

Because of economic factors and the application of technology the trend in advanced agricultural systems has been towards increasing intensification. This has led to greater control being exercised over animal reproduction, use of food and of environmental conditions, and has been most successfully applied in pig and poultry systems.

The efficiency with which animals convert feedstuffs into food for man has been assessed in numerous studies. These can be confusing because of the different criteria used and their application over different time-scales.

In assessing feed efficiency for a single animal from birth, beginning of lay or lactation, the individual lamb slaughtered at a young age is relatively efficient, producing 3.3 g protein/MJ of metabolisable energy (ME). In contrast, Holmes[1] has shown 3.2, 2.6 and 2.8 g protein/MJ ME being obtained in egg, bacon and milk production (1650 kg concentrates per cow), respectively.

However, breeding populations have also to be maintained, and the ewe following a 5-month gestation period may lactate from nil to a few days, up to 14–16 weeks. Thus, with an annual lamb crop, the ewe can be unproductive for at least 3 months or more.

When the same measure of feed efficiency (g protein/MJ ME) is applied to include breeding populations, the figures for sheep are 0.5 g for 1.4 lamb per ewe per year and 0.9 g for double that number. For egg, bacon and milk production coupled with 18 months beef, they are 2.9, 2 and 2.1 respectively.

The more efficient systems are dependent on high-quality forages or large inputs of concentrates, but sheep utilize an extremely wide range of food resources. Indeed, as illustrated in previous chapters, the utilization by sheep of vast areas of natural grasslands is one of the great merits of the species.

One of the limitations of the sheep is its modest reproductive performance. An increase in litter size, in the frequency of lambing, or both, can have a profound impact in production efficiency as illustrated in the Table 12.1.

If the efficiency factor shown in Table 12.1 is divided into 100, the amount of digestible organic matter (DOM) in food required to produce 1 kg of carcass will be obtained. On this basis, 1 lamb per year from Scottish

**Table 12.1** Effect of litter size on efficiency of production of lean and on ratio of obligatory products to meat.

| | Efficiency† | | Calculated on 1 crop/annum | | |
|---|---|---|---|---|---|
| Litter size | 1 crop/ year | 3 crops in 2 years | % total lean which is mutton meat* | Kg wool per kg lean from lamb* | Kg cereals required to produce 1 kg lean meat† |
| 1 | 3.9 | 5.3 | 36 | 0.86 | 59 |
| 2 | 6.4 | 8.2 | 19 | 0.35 | 38 |
| 3 | 8.2 | 10.0 | 12 | 0.22 | 30 |
| 4 | 9.5 | 11.3 | 9 | 0.16 | 27 |

† From Large[2] $E = \dfrac{\text{Weight carcass}}{\text{Weight DOM consumed}} \times 100$

\* From Blaxter[3]

Halfbred ewes by a Suffolk requires 25.6 kg DOM/kg carcass while it is about half, 12.2 kg DOM for triplets.

The data in the table are derived from calculations which are theoretical, in that they assumed similar lamb growth rates, irrespective of different litter sizes, and no increase in lamb mortality associated with higher ewe fecundity. Nonetheless, the principles are clearly established.

## New systems

The synthesis of frequent or accelerated lambing systems, as they are commonly described, has reached the stage of practical application in several countries, notably the United States, France, Eire and in Britain.

This has been made possible because of the tremendous advances in the technology of sheep production and, particularly, knowledge of reproduction, nutrition and management. The fundamental science on which frequent lambing systems are based has been comprehensively reviewed by the Scottish Colleges[4] and more recently by Dr John Robinson[5] and his colleagues. In their scientific work they have made an exceptional contribution to many of the relevant components of intensive systems.

### Breed of ewe

Frequent lambing systems are more likely to prove successful if breed types are used which have an extended breeding season. This will exclude, therefore, the majority of British breeds. The Dorset (horned or polled), the

Finnish Landrace and their crosses have been found to be suitable, and in France, the Romanov and Pre-Alpe-du-Sud have been successfully used, while in the United States, Outhouse[6] says the Rambouillet appears to adapt successfully.

Based on his practical experience of running a commercial system at the Harper Adams Agricultural College, Dr Tempest[7] has found the Cambridge and the Friesland × Dorset Horn to lamb successfully at continuous 8-month intervals. He suggests the Suffolk and its crosses should be considered as Suffolk × Mule and Suffolk × Welsh Halfbred ewes have been lambed down in September.

The suitability of breeds and crosses for intensive systems obviously needs to be investigated. It should be apparent, however, that a frequent lambing system ought not to be embarked upon unless the breed type chosen has shown a proven potential to adapt to out-of-season breeding.

Apart from choosing a breed of ram which will produce the type of carcass and growth pattern required, breeds known to have adequate libido over spring and early summer should be used when natural mating is practised. The Suffolk and other Down breeds have been used successfully but there are some doubts about the suitability of the Texel.

*Breeding intervals*

Theoretically, with a 5-month gestation period 2 lamb crops per year should be possible and some individual ewes can successfully achieve a lambing interval of 6 months. Because uterine involution may take from 28 to 45 days to complete following parturition, and at least 2 oestrus cycles are required to ensure adequate levels of conception, the minimum lambing interval that can be reasonably expected on a flock basis will be 7 months. This has been successfully achieved at the Rowett Research Institute (RRI),[8] while an 8-month cycle is used at Harper Adams Agricultural College. So 3 crops in 2 years seems to be a realistic practical target.

An important objective in frequent lambing systems is to avoid extended lambing periods. In Europe the customary practice is to synchronise oestrus using impregnated polyurethane vaginal pessaries. More recently, subcutaneous silastic progesterone implants have found favour in the United States and it is likely that basic research will lead to improved and more reliable methods.

Conception rates tend to vary from year to year and to be lower during the April–July period of natural cyclic anoestrus. It is desirable to identify those ewes which are not in lamb, and pregnancy diagnosis, at 18 days using blood plasma progesterone concentrations or ultrasonic methods, is an integral part of the system.

By running more than one flock, a considerable improvement in overall efficiency can be obtained. A simple system involving 2 flocks, 'out of phase' as illustrated in the diagram below allows those ewes which do not conceive to be moved, or 'slipped', to the next flock giving them an opportunity to mate earlier than the next scheduled mating.

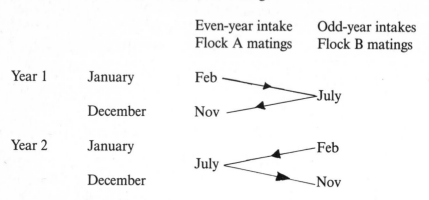

|  |  | Even-year intake<br>Flock A matings | Odd-year intakes<br>Flock B matings |
|---|---|---|---|
| Year 1 | January | Feb | |
|  | | | July |
|  | December | Nov | |
| Year 2 | January | | Feb |
|  | | July | |
|  | December | | Nov |

show the transfer of ewes which have not conceived. The transfer of ewes would depend on success at each mating period. (Scottish Agricultural Colleges[9].)

In contrast to the mating times above, Tempest[10] mates his flocks in August, December and April. A more sophisticated system has been practised at Nouzilly in France involving 7 flocks, which are therefore separated in physiological state by about 7 weeks.

In his early studies under relatively ideal and carefully controlled conditions, Robinson[11] used an artificial daylength period to induce oestrus out of season. Subsequent studies have shown that satisfactory induction of oestrus, conception rates and litter sizes can be achieved by a combination of oestrus synchronisation with pessaries, followed by an injection of 400–500 international units (IUs) of pregnant mares serum gonadotrophin (PMSG), when pessaries are withdrawn.

*Mating and lambing procedures*

It is essential to bring ewes to mating in good body condition, at least condition score (CS) 3. Following sponge removal, mating or artificial insemination (AI) should be delayed for 48 hours. When individual hand-mating is practised teaser rams can be used to identify ewes in heat, but even so, 1 fertile ram to 5–12 ewes will be needed. Alternatively, multi-ram groups of up to a maximum of 30 ewes have been successful. The difficulty of providing sufficient rams suggests that AI, using fresh semen, has an important role in intensive systems and it has been used successfully by Robinson.

Even when oestrus is synchronised, lambing in a flock may extend over a period of 6 days or so. Induction of parturition is possible using an injection of 16 mg betamethasone. Tempest[12] has obtained a 24-hour lambing period by this means. Clearly this brings a new concept to the organisation of a lambing which, of course, need not be at a weekend!

The reproductive cycle can therefore be very closely controlled as shown in Table 12.2.

**Table 12.2** Control of the reproductive cycle.

| Day | Operation | Weekday |
|---|---|---|
| − 14 | Sponges implanted | Thursday |
| − 2 | Sponges withdrawn }<br>  PMSG injected } | Tuesday |
| 0 | Rams joined | Thursday |
| +140 | Ewes penned | Thursday |
| +142 | Ewes induced | Saturday 6 p.m. |
| +144 | Ewes lambed | Monday 6 a.m. to<br>Tuesday 6 a.m. |

(After Tempest[13])

*Nutrition of the ewe*

It was originally thought that maintaining the weight of ewes lambing frequently would pose a serious problem. If properly applied, the knowledge gained from research will ensure that liveweight will be maintained from one cycle to another, and indeed increased in time, as has been shown by Robinson et al.[14]

The pattern of food intake should be adjusted to obtain the desired level of body condition at appropriate phases. Maximum ovulation requires a high level of body condition (CS3) at mating and this should be maintained month of pregnancy. Maintenance of a high level of body condition throughout pregnancy, especially with ewes carrying a large number of foetuses, can lead to inappetance and hypoglycaemia in late pregnancy. This can best be prevented by restricting food intake in mid pregnancy in order to gradually reduce body condition to about CS 2.5 or slightly lower. In practice, this would involve an increase in stocking rate or the use of poor-quality pasture during the grazing season. With housed ewes, it is relatively easy to adjust feed allowances. To obtain high birthweights, an increase in weight in the last 2 months of pregnancy of around 20 per cent is a reasonable target. Once means of identifying foetal number accurately becomes commercially realistic, even greater precision in nutritional provision will become possible.

The level of energy intakes (Table 12.3) which have produced good results when used by Robinson,[15] can be used as a general guide. When housed, ewes can be grouped according to body condition. Routine handling to assess changes in condition, particularly in late pregnancy and when ewes are lactating, is an important aspect of management.

**Table 12.3** Mean daily metabolisable energy intakes (MJ) for highly productive ewes maintained on a seven month lambing interval.

| Weaning to end of first month of gestation | 30–90 days of gestation | No. of lambs | Mean from 90 days gestation to parturition | At 90 days | At parturition | During lactation |
|---|---|---|---|---|---|---|
|  |  | 1 | 12.4 | 9.4 | 15.4 | 18.8 |
| 15.0 | 9.4 | 2 | 14.5 | 9.4 | 19.6 | 24.5 |
|  |  | 3 | 15.8 | 9.4 | 22.2 | 26.2 |
|  |  | 4 | 16.6 | 9.4 | 23.8 |  |

Depending on the length of the breeding cycles, lactation may vary from 4 to 10 weeks. When maximum lamb growth rate is sought, a high level of milk production is needed, and energy should be supplied at around three times the maintenance requirement. Concentrates will need to be fed to ewes and should contain 170 g CP/kg DM with an inclusion of a protein source of low degradability. The incorporation of 20 per cent fishmeal would meet these requirements. Since the removal of a high-quality protein will lead to a consequent decrease in milk yield and a decline in fat mobilisation, it can be excluded from diets after 3 or 4 weeks. This has the benefit of encouraging an increased intake of food by the lambs, allows the restoration of live body condition, and begins the drying-up process, which is necessary when lambs are abruptly weaned.

The food resources used are basically the same as for many other systems – grazed pasture, conserved forages, arable by-products and concentrates. The management procedure adopted at Harper Adams in the use of farm resources is illustrated in Table 12.4. A specification for the farm integration of a 2 flock, 3 crops-in-2-years system is presented by the Scottish Agricultural Colleges.[16]

Robinson has calculated the apportionment of food resources on the basis of energy requirements and as affected by frequency of lambing and litter

**Table 12.4** Integration of an intensive ewe flock lambing at eight month intervals with farm resource at Harper Adams Agricultural College.

| Stage of Cycle | December mating | | August mating | | April mating | |
|---|---|---|---|---|---|---|
| | Month | Location | Month | Location | Month | Location |
| Flush | Nov. | Cow leys | July | Sheep ley | Mar. | Sheep ley |
| Tup | Dec. | Cow leys | Aug. | Sheep ley | Apr. | Sheep ley |
| Preg. No. 1 | Jan. | Cow leys | Sep. | Sheep ley | May | Sheep ley |
| Preg. No. 2 | Feb. | Housed | Oct. | Cow ley | June | Sheep ley |
| Preg. No. 3 | Mar. | Housed | Nov. | Cow ley | July | Sheep ley |
| Preg. No. 4 | Apr. | Housed | Dec. | Housed | Aug. | Sheep ley |
| Lamb | May | Housed | Jan. | Housed | Sep. | Housed |
| Lactate | June | Sheep ley | Feb. | Housed | Oct. | U/S ley* |

\* U/S ley = Undersown ley
After Tempest[17]

size. The effect of these factors on the biological efficiency of meat production is very clearly shown in Fig. 12.1.

### Ewe replacement policy

Experience indicates that the longevity of ewes in intensive systems is as great as in others. Production has been successfully maintained at the RRI for five pregnancies, with many ewes successfully achieving 8 pregnancies which is in accord with American results. Tempest has retained ewes in his flock for as long as they are highly productive, and has a replacement rate varying between 20–25 per cent per annum.

Replacements can be either ewe lambs or 2-year-old ewes. The optimum age to mate ewe lambs would appear to be around 8 months at about 60 per cent of mature weight. Ewe lambs should be well-grown, since ewes lambing frequently, unlike those lambing annually, have less opportunity to increase body size.

### Physical performance

Using mainly the well-proven breed type, that is the Finn × Dorset, the average conception rates and the ranges of these, in parentheses, have been 88 (81–98) 81 (73–88) and 89 (76–98) per cent at Aberdeen by Robinson *et al.*,[18] Edinburgh by Speedy and Fitzsimons,[19] and Harper Adams by Tempest,[20] respectively. There is a distinct seasonal effect in the results, with the lower levels of performance occurring with matings in the February–May period, and the highest performance in autumn. Using Ram-

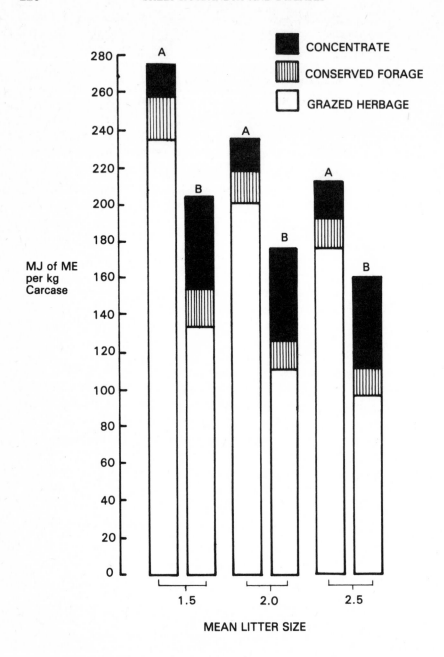

*Fig. 12.1* The effect of mean litter size and breeding frequency on the total food energy (MJ of ME) and of the individual components of the diet required to produce 1 kg of carcass[9].
(A = breeding once a year, B = breeding three times in two years.)

bouillet ewes, a similar result was obtained at Purdue, Indiana. The average conception rate was 86.3 per cent and it varied from 83 per cent in winter to 92 per cent in autumn in Outhouse's[21] study.

There is no doubt that satisfactory litter sizes can be obtained, these having averaged 2.1, 2.33 and 2.21 at Aberdeen, Edinburgh and Harper Adams respectively.

Even though an increase in litter size frequently leads to an increase in perinatal mortality, this need not occur under the controlled management conditions of intensive systems. At Harper Adams the overall mortality from birth to sale has been 16 per cent, which by any standard is an excellent achievement. Losses at Edinburgh were much higher, at 27 per cent, but new techniques which have recently been developed can contribute to reducing losses. For example, Robinson routinely administers approximately 60 ml of either ewe or cow colostrum, by catheter, to all lambs from larger litters and his lamb mortality is now well below 10 per cent.

As a result annual lamb production per ewe can be relatively high – 3.5 RRI, 2.7 Harper Adams and 2.13 Edinburgh – leading to a significant improvement in carcass production per ewe, when combined with more frequent lambing (Table 12.5 and Fig. 12.1).

**Table 12.5**  Relationship between breeding frequency and carcass production.

| | Frequency of breeding | | | | | |
| | Once/year | | | 3 times in 2 years | | |
| --- | --- | --- | --- | --- | --- | --- |
| Litter size | 1.5 | 2.0 | 2.5 | 1.5 | 2.0 | 2.5 |
| Carcass production (kg/ewe/year) | 24.7 | 31.7 | 36.8 | 34.7 | 44.6 | 52.2 |

After Robinson[22]

## Other systems

Many of the principles and practices appropriate to frequent lambing can be applied to once-a-year lambing flocks, particularly so when exploiting highly prolific breeds. Lambing can be organised to occur outwith the normal season and with lambs being produced for specialised markets when prices are high. Lambs may be weaned at, or soon after, birth and reared on milk substitutes for specialised markets demanding very light carcasses. Alternatively, lambs may be weaned at 4–5 weeks on to concentrates, or suckled until 10 weeks or later. On more conventional systems the aim would be to obtain a maximum number of suckled twins by cross-fostering and only rearing surplus lambs artificially.

*Lamb finishing*

A component of frequent lambing systems is the early weaning of lambs at a stage when they are capable of thriving satisfactorily on solid food. Even though lambs may show little interest and consume very little solid food until 2–3 weeks of age, creep feed should be made available from that stage or earlier. Given early access to creep feed, lambs can be weaned at 4–5 weeks with a minimum check to growth. In general, the earlier the weaning, the greater the check, which is usually minimal when lambs are 6–7 weeks old. Age and level of solid feed intakes are better guides to weaning than weight, which is not a particularly good measure of a lamb's ability to survive on solid food. With a high intake of milk, solid food consumption may be poor, rumen development can be delayed, yet the lamb can be well-grown and thrifty and still suffer a check at weaning.

Because lambs abruptly weaned at around 4–5 weeks have rumens which have not attained mature proportions; the diets used have to be highly digestible and are generally based on cereals.

Research by Orskov[23,24] has shown that lambs can digest whole cereals, apart from oats, which are less suitable and which ought not to be included in diets until lambs are at least 30 kg liveweight.

The protein concentration in cereals is inadequate for early-weaned lambs and supplementation is required. Theoretically, the amount of supplement included could be reduced as the lambs mature. However, slaughter weights are usually reached at a relatively young age, 10–16 weeks, so that changes in dietary composition are scarcely worth while. As changes in diet composition sometimes cause a decline in intake with sheep, the risk is not justifiable. With individual gains of 350–500 g per day or more, growth rates are rapid so that lambs are finished at a young age. The protein supplement should contain a source of low degradable protein, such as fish- or meatmeal. An example of a supplement is: 80 per cent fishmeal, 15 per cent ground limestone, 4 per cent molasses, 1 per cent trace minerals pelleted to about 4 mm diameter to avoid segregation when mixed with whole grain. It should be used at a 10 per cent inclusion rate.

When roughage is used it should be of high quality, preferably with a high legume content, and the amount offered should be restricted. Roughage can also be incorporated into pellets (6–12 mm diameter). Indeed, Tempest reports obtaining better lamb performance with complete diets incorporating a proportion of roughage than with cereal-based diets. This is in accord with my own experience with conventionally-reared lambs weaned at 12–14 weeks and finished on complete diets.

When processed grain is the sole energy feed, the proportion of propionic acid formed in the rumen is increased, and this leads to the deposition of soft fat, which is rejected by the meat trade. This problem is largely overcome either by feeding unprocessed grain or by the inclusion of roughage in the diet.


## Economics of intensive systems

As very few commercial flocks produce 3 crops in 2 years there are very few data available to determine the circumstances in which these systems could be profitable.

A study by a Working Group[35] at the University of Reading, which had to base their calculations on several assumptions (whereas more concrete data is now available) concluded that the system did not merit a development study. A later study by the Scottish Colleges using a computerised model with data then available, indicated no clear economic advantage to frequent lambing compared with an annual lamb crop. Frequent lambing systems were shown to be sensitive to lamb mortality, to changes in purchased feed costs – which are approximately doubled when lambs are intensively reared to slaughter on concentrate diets – and also to lamb prices, which vary seasonally. This model could be used to assess the economic viability of the system as economic conditions change.

The real test is to assess new systems in a commercial context. A very acceptable economic performance has been obtained at Harper Adams, the gross margin per ewe of around £50 being almost three times the average figure of £17.50 recorded by the Meat and Livestock Commission (MLC) in lowland flocks. It is noteworthy, however, that Tempest has skilfully integrated his intensive flock with other farm enterprises to reduce finishing costs, in particular. In addition to the use of pasture for ewes, lambs born in May are weaned about 10 weeks old and stored at grass, to be finished on beet tops at 8–10 months of age. September–born lambs are reared indoors and kept on moderate-quality diets to finish at 6 months old for the Easter market, while December-born lambs are intensively finished on concentrate-type diets to finish at 4 months old for the same market.

Although the undoubted improvement in biological efficiency is attractive, the crux of the matter is that proper planning in the availability and use of cheap food resources, is essential. Unless a flockmaster already possesses a high level of technical knowledge and management skills, coupled with experience in sheep husbandry, he would be ill-advised to embark on running a complex system. If skilled labour able to contend with three lambings each year is not available, or suitable housing has to be erected, the risks and economic constraints may be unacceptably high.

Changing tradition systems of sheep production, towards the pattern more common to pigs and poultry, is now technically feasible. The final question which might be posed is whether intensive sheep production, even though possible, is desirable. It may be preferable to use grain, surplus to human requirements, for naturally granivorous species of domestic animals such as pigs or poultry. Also, the extent to which producers may confine or impose restriction of movement could give rise to welfare issues, largely of an ethical nature, but none the less ones about which people feel deeply. Criticisms on welfare grounds rarely arise in conventional systems.

As has been shown, pasture and its utilization has, and probably will continue to have, an important role in sheep systems of all kinds. The sheep in the future will have an important role in converting a diversity of forages and other materials into human food and apparel, in a wide range of systems.

## References

(1)  Holmes, W. (1977) 'Choosing between animals,' *Phil. Trans R. Soc. Lond.* B. **281**, 121–37.
(2)  Large, R. V. (1970) 'The biological efficiency of meat production in sheep,' *Anim. Prod.*, **12**, 393–401.
(3)  Blaxter, K. L. (1969) Proc. 2nd World Conf. Anim. Prod., Maryland, USA pp. 31–40.
(4)  Scottish Agricultural Colleges (1977) *A Study of High Lamb Output Production Systems*, Technical Note No. 16.
(5)  Robinson, J. J., Fraser, C. and McHattie, I. (1977) 'Development of systems for lambing sheep more frequently than once per year,' *In: Sheep Nutrition and Management*, US Feed Grains Council, London, 5–33.
(6)  Outhouse, J. B. (1974) 'Ewe productivity on accelerated lambing pregnancies,' Stat. Bull. No. 49. Agric. Exp. Stat. Purdiu, Indiana.
(7)  Tempest, W. M. (1982) 'Management of the frequent lambing flocks,' *In:* (Ed.) W. Haresign, Proc. 35th Easter School in Agric, Sci., Butterworth.
(8)  Robinson, Fraser and McHattie.
(9)  Scottish Agricultural Colleges, *A Study of High Lamb Output Production Systems*.
(10) Tempest, 'Management of the frequent lambing flocks.'
(11) Robinson, Fraser and McHattie, 'Development of systems for lambing sheep.'
(12) Tempest, 'Management of the frequent lambing flocks.'
(13) Tempest, 'Management of the frequent lambing flocks.'

(14) Robinson, Fraser and McHattie, 'Development of systems for lambing sheep.'
(15) Robinson, Fraser and McHattie, 'Development of systems for lambing sheep.'
(16) Scottish Agricultural Colleges, *A Study of High Lamb Output Production Systems.*
(17) Tempest, 'Management of the frequent lambing flocks.'
(18) Robinson, Fraser and McHattie, 'Development of systems for lambing sheep.'
(19) Speedy, A. W. and Fitzsimons, J. (1977) 'The reproductive performance of Finnish Landrace × Dorset Horn and Border Leicester × Scottish Blackface ewes mated 3 times in 2 years,' *Anim. Prod.*, **24**, 189–96.
(20) Tempest, 'Management of the frequent lambing flocks.'
(21) Outhouse, 'Ewe productivity.'
(22) Robinson, Fraser and McHattie, 'Development of systems for lambing sheep.'
(23) Robinson, J. J. and Orskov, E. R. (1975) 'An integrated approach to improving the biological efficiency of sheep meat production,' *Wld Rev. Anim. Prod.*, **11**, 63–76.
(24) Orskov, E. R. (1977) *Nutrition of Lambs from Birth to Slaughter: Sheep Nutrition and Management.* US Feed Gains Council, London, pp. 35–46.
(25) Working Group – Reading University and Grassland Research Institute (1973) An Assessment of Continuous Lamb Production in the United Kingdom, (ed.) J. G. W. Jones.

# 13   Introduction to the Diseases of Sheep

Disease may be defined as any deviation from good health, good health being that state in which the animal is in complete adaption to its surroundings so that it can thrive and give of its best. Unfortunately, in animal husbandry the understanding and appreciation of what is meant by good health can vary very considerably from area to area and from farmer to farmer depending on the experience of the individual concerned. The major portion of this book is about the breeding of sheep, the proper environment in which they should be kept, and the management methods which should be employed to achieve the optimum production of mutton and wool, therefore the question of what constitutes good health has already been dealt with adequately. The present section concerns itself with those factors which interfere with good health and which are commonly referred to as the diseases of sheep. It is felt before proceeding further that the reader should be warned of two pitfalls in reading about disease. One is that so much emphasis is placed on the diseases caused by the living parasites, commonly known as germs, that it is apt to be forgotten that much ill health is due to errors in breeding and husbandry, and secondly, one tends to become disease-minded and forgets that the majority of animals do in fact enjoy good health.

Disease consists in departures from the normal in the structure and functions of the different organs of the body due to one or several of the numerous causes which are to be discussed in this chapter. In other words, the body gets out of tune and the upset may be so severe as to cause the death of the animal.

The causes of diseases may be classified as follows:

(1)  Living agents
(2)  Nutritional deficiencies
(3)  Genetic abnormalities
(4)  Upsets in metabolism
(5)  Poisonous substances

# Living agents

Under this heading are included the bacteria, viruses and fungi (commonly known as microbes or germs) and the parasitic animals such as worms and flukes. All the infectious diseases of sheep fall within this group.

## Classification of living agents

(1) Bacteria. The science of bacteriology began over 100 years ago from the observations of Pasteur on the fermentation of wine and it was these observations which gave the first real glimpse of the teeming world of unseen living organisms that abound in nature in numbers beyond all comprehension. They are present in the water we drink, the food we eat, the air we breathe, and on almost everything we touch. Many of these very small living organisms are essential for our existence and it is only the comparative few that can be incriminated as the cause of disease. Among these few may be recognised the most potent killers of life.

Bacteria are very small single-celled organisms varying considerably in shape and size but all can be seen quite readily with an ordinary microscope, while the majority of disease-producing ones, as well as many others, can be grown in the test tube in the laboratory. It is in this way that the disease-producing bacteria are recognised one from the other as well as differentiated from those which are harmless.

(2) Viruses. So far as we understand them, viruses also appear to be living agents but they are even smaller than bacteria, many of them being below the limit of vision of the ordinary laboratory microscope so that a special electron microscope is used. Unlike bacteria, viruses will not grow on anything but living tissue, and this, combined with their smallness, make them relatively difficult to handle except in specially equipped laboratories. It is now well known that both plants and animals are capable of developing diseases due to virus infection. There are not only viral diseases of man and animals such as smallpox, foot-and-mouth disease, louping-ill (and a host of others) and viral diseases of plants, such as potato viruses, the mosaics, and the virus yellows, but bacteria themselves are frequently destroyed by virus-like agents.

At one time, all known viruses were the cause of one or other disease. With the development of better culture techniques it has been possible to grow a number of viruses in the laboratory which have not as yet been related to any sickness, while some disease-producing viruses can readily be found in healthy animals. These facts complicate specific diagnosis considerably. Recently, viruses have been discovered which are the cause of some forms of cancer both in man and animals.

Until the development of the electron microscope most viruses had never been seen, so that it was impossible to determine their appearance, but we now know that, despite their smallness, they vary considerably in shape: some are elongated particles, while others are short rods, spherical, or many-sided, and a large number of the viruses which destroy bacteria are tadpole-shaped with a 'head' and a 'tail'. All contain nucleic acid which becomes incorporated in the host cells and initiates replication of the virus.

(3) Parasitic animals. A parasite is not a particular species but is any animal which has adopted the parasitic way of life. An animal which forms an association with another animal in order to obtain from it an adequate food supply without giving any benefit in return is a parasite. Many different animals have adopted the parasitic method of living during the course of evolution, and while some may do little harm to their hosts, others bring much suffering to domestic stock and great economic loss to man. The sheep in Britain or in any other country, is no exception to this suffering and the parasitic animals which the sheep may carry are numerous. The liver-fluke, the tapeworm, the lungworm, the stomachworms, the tick, the blowfly, and the mange mite are but a few.

## The communicability of microbes and parasites

So far, in discussing the diseases produced by living agents, most emphasis has been placed on the appearance of the microbes but their shapes and sizes are only a small factor in their production of disease. When an animal – as for instance the sheep – becomes infected by a disease-producing living agent one naturally wonders where the agent came from and how it managed to be where it is. In other words how do these agents move? Paradoxically, one can start by saying that in a number of microbial diseases the germs do not move at all since they may be present in healthy sheep and yet do no harm until environmental change in the sheep upsets this equilibrium and allows the microbes to multiply and produce disease. This is what happens in pulpy kidney and braxy. On the other hand, the living agent may be brought to the sheep from some reservoir of infection. This may be sheep or other animals suffering from the disease or harmlessly carrying the agent, or it may be soil or pasture. In both cases it is necessary for the living agent to be taken from the source to the new host. Bacteria and viruses are transferred by contact of healthy animals with infected ones either directly or by the sexual act, by the ingestion of contaminated food or by unhygienic handling, biting insects, such as ticks. With most parasitic animals, such as worms and flukes, the infecting organism has to undergo what is termed a life-cycle, often involving an intermediary host before it can transfer from sheep to sheep.

Fortunately for ourselves and our animals, the mere transfer of disease-producing organisms to a new host does not necessarily lead to illness, for certain other factors determine whether this is so or not. In the first place, the number of organisms transferred is usually of importance and except for very infectious conditions, such as foot-and-mouth disease, the weight of infection needs to be fairly considerable. Secondly, most animals have good mechanisms for repelling a potential invader and these have to be broken down before the organism produces disease.

*Properties of bacteria, viruses, and parasitic animals which enable them to cause disease*

By this day and age most people know that bacteria, viruses, and parasites can cause illness but it is perhaps not so well known how such small, living organisms can bring about debility and death of both man and his stock. This is perhaps not so surprising, for scientists themselves are not very clear on some aspects of this subject but nevertheless many harmful effects of these organisms which cause disease have been traced with considerable precision.

In the course of their multiplication in the host many pathogenic bacteria produce an array of substances that poison the tissues of the host and among them are the most formidable poisons known. Surprisingly enough, many of the bacteria which produce the most potent of toxins have very little invasive capacity for the host tissues and remain localised, the powerful toxin diffusing from a circumscribed site of infection to be carried throughout the body; this is the case in pulpy kidney, lamb dysentery, tetanus, etc.

At the other extreme there are organisms, such as the causal agent of anthrax, which are highly invasive and do not cause death until there has been an enormous multiplication throughout the body.

In between these two extremes is a third category in which the organism is both invasive and also capable of producing poisonous substances, but possessing neither characteristic to great degree. These latter organisms do not kill so rapidly, taking days, weeks, or months to do so, or they may not kill at all, only causing illness until destroyed by the infected host.

In such bacterial infections, the invaded tissues are usually damaged and the host reacts in defence by what is termed 'inflammation'. Depending on which organs are invaded various names are given to this, for example pneumonia (lung), hepatitis (liver), encephalitis (brain), metritis (womb), etc. The invading parasite proliferating in the host and reacting with it may interfere with the functions of vital organs and so bring about serious harmful effects, as, for instance, invasion of the bowel wall along with its reaction of inflammation (enteritis) which may seriously interfere with the nutrition of the host. Invasion and damage of the lung giving rise to pneumonia will have serious effects on the exchange of oxygen and carbon dioxide which upsets the whole metabolism of the animal. Disturbances of this sort play an important part in many diseases of sheep.

Viruses have much in common with bacteria as regards producing illness but nevertheless differ considerably. A virus, for instance, can only bring about damage to those tissues of the body which it can actually invade, so that, unlike bacteria it cannot cause remote effects by diffusable toxins. As with bacteria, however, considerable numbers of virus particles may be present in the bloodstream and spread to sites far distant from the original point of invasion. The actual disease effects are caused by the incorporation of the virus into the cells of one or more tissues of the body, which interferes with the normal function of such cells (in many cases to the extent of killing them). Generally, considerable inflammation follows and this in itself, although often ultimately beneficial, can upset the vitality of the infected animal for a time.

The effects of parasitic animals on their hosts are not essentially different from those of bacteria but differ somewhat in detail. The reaction of the host's tissues to the parasite is again that of inflammation, stimulated either by direct injury by the parasite when the latter feeds on the tissues or moves through them or by poisonous substances produced by the parasitic animals. These act either locally around the invader or throughout the body, having passed into the host's blood or lymph stream. Parasitic animals may, however, injure the host in other ways than those common to themselves, bacteria and viruses. For instance, they may exert mechanical effects on various parts of the host, such as by pressure on organs or by blocking natural channels such as blood vessels, bile ducts, and air passages. Some during their feeding processes remove essential substances from the host as, for example, parasitic animals living in the host's gut may utilise its food supply or feed upon its blood. Other parasitic animals which may be relatively harmless in themselves introuce dangerous microbes into the host, and so are often the indirect cause of serious trouble. The sheep tick is known to transmit the causal agents of louping-ill, tick-borne fever, tick pyaemia, and redwater, while the liver-fluke carries into the sheep the microbe causing black disease. Lastly, but by no means of least importance, is the effect which parasites have in lowering the host's resistance to bacterial or viral infection, but this relationship may act in reverse, in that some chronic microbial diseases lower the host's resistance to the invasion of parasites. This complex relationship is commonly seen in sheep, as, for instance, the frequent association of stomachworms with Johne's disease and lungworms with bacterial pneumonia.

## Resistance of the sheep to living agents

So far most emphasis has been placed on the parasite and its harmful effects on the host, but the question of resistance of the sheep to infection must be considered. It is obvious that the host's resistance to infection is of great importance when discussing sheep husbandry and clearly merits description. A proper understanding goes far towards keeping livestock healthy.

Immunity may be either innate or acquired. The former is the resistance which is determined solely by the inherited qualities of the animal and is an expression of its genetic constitution. The most obvious example is when a species has a complete natural insusceptibility to an infection – sheep are never affected with swine-fever or with myxomatosis. Innate resistance may not be so complete and lesser differences could well be of great practical importance. This relationship between infection and genetic status of the host is of course a familiar idea among flockmasters and shepherds. It has been discussed widely at agricultural meetings and there can be no doubt that differences in innate resistance to various infections do occur between breeds of sheep and even between individual animals in a breed or flock. Little is known of the extent of these differences or of the laws that govern their inheritance, for experimental techniques are difficult and expensive, but evidence has been obtained that susceptibility to one strain of scrapie in one breed of sheep depends upon a single dominant gene.

It has long been accepted that animals which have suffered from certain infectious conditions and recovered do not usually take the same disease again and this is due to their development of acquired resistance. It is also the case that this immune state may be acquired when animals are in contact with disease-producing agents and yet withstand them without obvious signs of illness. The degree and speed of development of acquired immunity is under the genetic control of the host but information on this particular aspect is scanty in all species and it is not known whether sheep vary to any practical extent in their capacity to react.

Acquired immunity which the animal develops after being ill with the infectious disease, being latently infected, or after vaccination with any one of the many vaccines, now in use, depends on the production of what are termed 'antibodies' (or immune bodies) which are highly specific for the particular organism in question. These are not present in animals which have never been in contact with the disease, have not been vaccinated, or have not sucked immune colostrum. Immune bodies work in a number of different ways, some neutralise the poisons or toxins which various bacteria produce, some actually kill the invading bacteria, while others prevent the organisms invading the tissue of the animal. Similar immune bodies are also known to protect animals against viral diseases, although their action tends to be less complete, since the virus spending most of its time within the cells of the body is partly protected from the defences of the host, only being susceptible to attack when it is living free in the body fluids such as in the blood or lymph.

Whether a sheep is susceptible or resistant to a given infection depends in the main on such specific immunity but other less clearly defined factors also operate and are undoubtedly important in both individual and flock immunity. On occasion these factors make the understanding of disease more difficult than one imagines. There is, of course, little difficulty in such virulent infections as anthrax and foot-and-mouth disease where it would seem that the mere presence of the organism among non-immune stock is sufficient but with other diseases consideration must be given to predisposing factors which cause alterations in the susceptibility of the host. For instance, most sheep carry throughout their lives a variety of parasites potentially dangerous for them but under good husbandry conditions they remain harmless and only manifest their presence when environment or breeding becomes unsatisfactory. Despite the probable importance of these non-specific factors, it must be realised that both the scientist and stockman are all but ignorant of their exact nature so that it is much easier to talk about them than to define them. In general we do know that sensible breeding for a given environment, adequate and correct nutrition, a good physical environment and a high standard of hygiene are very important for keeping livestock healthy but it is nevertheless easy to think of a number of instances where disease arises in sheep kept under what appear to be ideal conditions. There is convincing evidence that strong, well-grown animals are more susceptible to blackquarter, braxy, black disease, and louping-ill, while an outbreak of pulpy kidney disease in sheep often follows the transfer of a flock from poor to good feed.

Resistance of the host to parasitic animals such as stomach and intestinal worms seems to be similar in many ways to that already discussed in connection with bacteria and viruses. It is generally agreed that sheep can develop an acquired resistance to worms which is largely dependent upon antibodies. The precise mechanisms that operate have not as yet been defined but the immune response is in most cases incapable of removing all the worms present so that in ordinary, practical husbandry terms, a certain level of worm infestation appears inevitable. Although a great deal still remains to be discovered, an effective vaccine against lungworm has been developed which stimulates active acquired immune responses and is widely used in cattle but not, so far, in sheep. Other factors which are thought to operate in the epidemiology of parasitic diseases since they are known to alter susceptibility are genetic make-up, age, and standard of nutrition of the host. Dormancy of larvae during their development in the host is common in worm infestations of sheep and probably accounts for some of the carry-over of worms during winter months. Carry-over in other species and over-wintering in soil is also important.

In the following chapters it will be shown time and again that immunity can be conferred by inoculation of a vaccine. The vaccine used differs very considerably from disease to disease but in every case it will contain the specific organism, either living or dead, or some product obtained from its growth. The discovery and manufacture of many of the vaccines in common use in the prevention of sheep diseases have often been difficult and unfortunately there still remain a number of conditions of sheep known to be caused by a bacterium or virus against which no useful vaccine has yet been produced. Further research is necessary to this end.

A certain amount of confusion often arises in practical disease prevention as to the difference between vaccine and immune sera, both of which are used. The two products are quite different, for immune serum is obtained by bleeding horses which have been previously vaccinated and hyperimmu-nised against a particular disease-producing microbe and/or its toxins. When injected into sheep, such serum contains the immune bodies of the horse and these are capable of protecting the sheep against that disease. Unfortunately (for such sera are expensive to produce), the immunity lasts for only a few weeks after which the sheep are susceptible once more. On the other hand the immune bodies that sheep produce following inoculation with a vaccine remain for a considerable time so that active immunity can persist for months or even years and can be further stimulated when the sheep face natural infection. The main advantage of giving serum is that the immune bodies already present protect immediately, whereas it often takes several weeks for an animal to make adequate quantities of immune body after vaccination. In other words, with immune serum, protection is immediate; with vaccination, protection occurs only after several weeks when active immunity has developed. Immune serum is occasionally used for cure as well as prevention since the immune bodies which it contains have the capacity to destroy or neutralise the infecting organism.

Newly-born animals sucking the first milk of their mothers take in many of the immune bodies of the parent and during the first hours of life these penetrate the gut and enter the bloodstream so making the newly-born animal immune to the same diseases as its parent. Such transferred immune bodies, like those transferred artificially by the inoculation of serum, only persist for a matter of weeks after which time the young animal once more becomes susceptible. Practical use is made of this biological phenomenon in protecting lambs against lamb dysentery, pulpy kidney disease, tetanus, etc. when vaccination of the mother protects the newly-born against the disease.

## Chemotherapy of living agents

For many years it has been one of the aims of science to find substances that are more poisonous to microbes and parasitic animals than to the host animal so that when given to the host animal invading organisms are destroyed. Since ancient times it has been customary to treat wounds and illnesses with various herbal mixtures but it was not until Lister introduced his ideas of antiseptic surgery that any logical use was made of antiseptic chemicals. Following Lister's discovery, such things as carbolic acid, chloride of lime, iodine, and a host of other substances have been found to be highly effective in killing bacteria, viruses, and parasitic animals but unfortunately these substances are also very efficient killers of tissue so that their value is strictly limited. The damage of tissue by the chemical may be often more serious than the original disease and shepherds should be very wary of using what they term 'strong' medicines, lotions, or ointments since they invariably do more harm than good.

In 1906 the German biologist Ehrlich did, however, produce a chemical, a synthetic arsenic compound, which on intravenous injection into the infected host was capable of destroying the causal organism of syphilis and yet was harmless to the host. In the early 1930s a further chemical was found which, although prepared in the manufacture of dyes for wool, was seen to be very active in the prevention and treatment of infections of man and animals caused by the bacteria known as streptococci. So started the great era of sulphonamide drugs of which there are now many varieties widely used. The discovery in 1940 of penicillin, a product of the yeast *Penicillium notatum*, opened an even greater field of therapy than did the sulphonamides and the search for new antibiotics has produced such drugs as streptomycin, chloromycetin, aureomycin, and many others. In a similar way considerable research is being done in the sphere of parasitology where the search for chemicals active against such parasites as flukes, worms, flies, and ticks has led to the discovery of a number of very efficient drugs and dips, the latter replacing those previously based on DDT, gammexane, and dieldrin which may no longer be used because of their potential danger to human health and wildlife. The new drugs have indeed caused a revolution in human and veterinary medicine but it must be realised that each drug is not a panacea for all ills and they must be used intelligently against a background of accurate diagnosis and good animal management.


## Nutritional deficiencies

Errors in diet can cause specific debilitating diseases of animals no less spectacular or economically important than those due to germs and other organisms. Deficiency diseases may be classified as follows:

(1)  Deficiency of food commonly called starvation or malnutrition.
(2)  Vitamin and allied deficiencies.
(3)  Deficiency of major elements.
(4)  Deficiency of what are termed 'trace elements'.

Whenever a vital function or series of vital functions in an animal is interrupted, owing to continued failure of the diet to supply the above substances in sufficient quantity, disease develops. The symptons depend on the particular functions of the body that are affected.

It is important to realise that virtually any dietary restriction leads to some change in one or more tissues but in diagnosis it is important to know whether the changes seen are caused by a deficiency of one or more dietary essentials or whether the changes are merely due to malnutrition. In the former case there are frequently special tissue alterations which are dependent on the absence of one or more essential nutrients from the diet. Some of these lesions may only be produced by a lack of a single nutrient, others are common to deficiencies of more than one material. In malnutrition there is usually a marked decrease in fatty tissue not only under the skin but also about the intestines, kidneys, and uterus, as well as a number of non-specific microscopic changes.

Although the absence or deficiency of a dietary factor from the food is the obvious reason for the development of disease, a number of other contributory factors often play an important part. There may be interference with absorption, even though adequate amounts of a nutrient are ingested as, for instance, in Johne's disease, or there may be an increased excretion from the body. In other circumstances, although the intake of essential nutrients is adequate for the normal body, the needs may be increased as for example during pregnancy, and unless the special requirements are met, a deficient state develops. Finally, certain materials may block the action of other substances, as sometimes occurs with copper in the causation of swayback.

Since the nutrition of the sheep has already been discussed in previous chapters, reference should be made to them for further details as to normal requirements.

## Genetic abnormalities

Several abnormalities of sheep are due to inherited anomalies and these occur when there is some defect in the genetic make-up of an animal derived from either or both of the parents. The essential defect is present in the fundamental units (genes) of the egg or sperm. Since these units control the development and function of the growing and adult animal any alterations in them may result in inherited changes which cause death or abnormality of the animal carrying such defective genes. It is known that most species of animal carry abnormal genes but it is only when in-breeding or line-breeding is practised intensely that such abnormalities really become important. The advent of the atomic era and the possible ill effects of radiation upon genetic development may indeed enhance the importance of this group of diseases.

## Upsets in mineral metabolism

There are several important diseases of sheep, for example, lambing sickness, hypomagnesaemia, swayback, etc. which are known to be associated with characteristic deficiencies in the mineral content of the blood and tissues, but it is generally agreed that these disorders are not caused entirely by deficiencies in the diet, the cause of the deficiency in the blood often being unknown. The symptoms are nevertheless caused by the abnormally low content of the specific mineral in the blood and/or tissue fluids. Despite our ignorance as to why these deficiencies arise, it is possible in many instances to prevent or cure the disease by giving the deficient mineral either in the food or by injection.

## Poisonous substances

Animals may be poisoned by plants, contaminated grazings or by injudicious or wrong dosing. In all cases the poisonous substance is taken into the body and there causes its ill effects by damaging or destroying different vital tissues in the body. Many of these poisons cause damage to the gut when they are swallowed but others are absorbed and damage the internal organs.

## Diagnosis of a disease

Diagnosis may be defined as the art of recognising a disease and of distinguishing it from other diseases and it is the first and most essential step in dealing with any outbreak of illness. It is only by accurate diagnosis that the appropriate and proper treatments and preventive measures can be applied. It is obviously useless, for example, to inject any one of the many specific vaccines or sera into sheep until it is known definitely that the particular condition is present or threatens to be present in a flock.

Similarly a change of management methods may be successful only when applied against certain diseases. There is no wonder-drug available which will cure and prevent all or many diseases, there is no short cut to bypass accurate diagnosis which alone decides the procedure to be adopted in any outbreak.

Diagnosis in the sheep is often difficult and although clinical examination of a sick animal may reveal the cause of the illness, other methods are frequently employed. For example, in worm infestations of sheep an examination of the faeces to estimate the number of worm eggs present is often necessary for proper differential diagnosis, while blood samples from one or more sheep may have to be examined in different ways before certain diseases may be distinguished. Probably more important is diagnosis by post mortem findings followed by microbiological and chemical tests. It is for this reason that it is often advisable to sacrifice an ailing animal before beginning expensive or time-consuming methods of prevention for the rest of the flock.

# 14 Reproduction and Genital Diseases

A major improvement objective in sheep production is to increase the number of lambs weaned per ewe which of course is dependent on the number of lambs born per ewe lambing. This in turn depends on the number of ova and foetuses which die during pregnancy and during the birth process. This chapter is concerned with the mortality aspects since other sections of the book deal with the management, nutrition and genetics of reproduction.

Total losses are estimated to vary from about 10–40 per cent of fertilised ova and most of these losses occur during the first month of pregnancy. If two or more ova are fertilised in a ewe and only one embryo dies lambing will occur normally, but if only one egg is fertilised and dies the ewe will return to service or will be barren at lambing time. Since the dead embryo is absorbed nothing unusual is seen.

The causes of embryonic death early in pregnancy are not well understood but, in general, ewes in good body condition at mating have lower rates of embryo mortality than ewes in poor condition. There is some evidence that high levels of feeding during the month after mating leads to greater losses as also does severe starvation at this time even if the lack of food is only for a few days. In tropical countries a very high atmospheric temperature during early pregnancy is known to increase embryo mortality. It must be admitted, however, that the specific cause is often unknown but it is possible that on occasions the fertilised ova are defective rather than are the management and nutrition of the ewe, and in this case the ram may be involved as well as the ewe.

### Abortion and foetal mortality

Abortion has long been recognised as a cause of major loss when sheep are bred. Until comparatively recently such things as foot rot, cold winds, feeding of unripe, or too many, turnips or swedes and rough handling were blamed for much of the trouble, and even today the causes of outbreaks of abortion are often not diagnosed. However, it is now appreciated that a wide range of infectious agents such as viruses, bacteria, parasites, etc. can cause abortion and foetal mortality. If more than 1–2 per cent of a lambing flock abort, it is probable that a specific infection is the cause. In Britain, about a million lambs are lost from infectious abortion.

The clinical signs can vary considerably, not only may infection result in abortion late in pregnancy but the foetus may die in the uterus early in pregnancy and be absorbed so that the ewe appears barren at lambing time. Infection with the same organisms may also result in the birth of stillborn lambs at full term or in the birth of weakly lambs which fail to survive. Clearly, a specific diagnosis based on the clinical signs is not usually possible so that veterinary and laboratory help should be sought.

The three most important diseases of this type in Britain are: vibriosis, enzootic abortion of ewes and toxoplasmosis.

*Vibriosis (campylobacteriosis)*

This disease has been reported from many parts of the world and it is a different disease from vibriosis of cattle. The infection seems to appear from nowhere since it often occurs in closed flocks where no sheep have been purchased. Unlike in cattle, the infection is not spread venerially from sheep to sheep by the ram but by the eating of food or pasture contaminated by discharges from the womb or droppings of carrier and infected sheep. When infection arrives in a closed flock for the first time it is thought that the disease is brought on to the farm by various wild birds such as crows and sparrows, etc. Vibriosis is a very contagious disease and infection spreads quickly and widely in a flock even in the last weeks of pregnancy so that, management-wise, little can be done to prevent an outbreak occurring.

Nevertheless aborting sheep should immediately be removed from the rest of the flock. Heavy losses only happen in the lambing season when the infection first becomes established and rarely, if ever, occurs 2 years running mainly because all the sheep become immune. For this reason such sheep should be kept for future breeding.

Abortion usually occurs in the last 6 weeks of pregnancy and this is followed by the birth of stillborn or weakly lambs born at normal time and these do not survive. The loss may be as high as 70 per cent of the lamb crop but usually it is about 15–20 per cent. Ewes of all ages are affected which is not generally the case with other infections. Diagnosis can only be carried out in the laboratory to which aborted or dead lambs along with their placentas should be sent.

Treatment and prevention of vibrionic abortion is not very satisfactory. The feeding of antiobiotics in the face of an outbreak has been tried but it is very expensive and not totally effective, while the sporadic occurrence of the disease makes widespread vaccination wasteful although it does give reasonable protection. Vaccination of a flock even after an outbreak has commenced can be of value in reducing losses but unfortunately the vaccine has to protect against several different strains of the organism. The commercial production of such vaccines is difficult and expensive and at least in Britain this and its limited use has not justified their manufacture up to the present time.

## Enzootic abortion of ewes (EAE)

This is often called 'kebbing' and is caused by a microbe of the chlamydial group, others of which cause psittacosis, encephalitis, joint-ill and pneumonia. Abortion of this type occurs mainly in low ground flocks especially those where new breeding sheep are purchased each year. It is rarely seen in true hill flocks kept on the higher ground. At one time 'kebbing' was considered to be a disease of sheep of the 'Borders of England and Scotland' but it is now known in most parts of the world.

*Plate 14.1* The placenta from a ewe infected with the virus of sheep abortion.

Although venereal transmission from ram to ewe may occur, the commonest method of spread is from aborting sheep at lambing time. Any age of sheep can be infected including new-born lambs and the most likely route is by the mouth. The use of lambing pens, sheds and in bye fields for lambing enhances the rate of spread. The infection remains latent in the newly infected lamb or ewe until they themselves are well advanced in their first pregnancy after infection when the organism begins to multiply in the foetal membranes to bring about abortion. The organism is discharged from the vagina in the foetal membranes and then in the vaginal discharges for about 3 weeks after abortion.

Most of the abortions occur in the last 2 weeks of gestation when fully formed foetuses are aborted, but is is very characteristic of this type of abortion that ewes may carry their lambs to full term but they are stillborn or born very weakly. When twins or triplets are born 1 or 2 of the lambs may be stillborn or weakly while the other appears healthy. The placenta or 'cleansings' have a characteristic appearance being thickened and brown and often covered in places with a pink cream like 'pus' which contains masses of the causal organism. Shepherds experienced in the disease readily recognise the characteristic appearance of these placentas but it is better to send one or more away to the laboratory for a specific diagnosis. It is essential to send the placenta to the laboratory and not the foetus or aborted lamb. A test on the blood of ewes can also be carried out if the disease has not been diagnosed from aborted material.

In newly infected flocks the abortion rate may be very high, as many as 30 per cent of all ages of pregnant ewes aborting. In following years the rate generally drops to about 5–10 per cent and it is the younger animals which abort, since the older ones are immune because of previous infection. Management practices designed to keep young sheep away from the lambing area and from aborted ewes helps to limit spread of infection but care in the purchase of new breeding lambs and sheep can keep a flock clean. Fortunately, the development of an effective vaccine some years ago has played an important part in reducing the incidence of this one-time scourge. The vaccine must, however, be used correctly and since the correct use varies from farm to farm depending on the sheep management practised it is always better to seek veterinary advice. A newly introduced vaccine contains two or more different strains of the abortion organism and this makes vaccination much more efficient. If vaccination has not been practised or has not been effective for one reason or another treatment of an affected flock with the drug tetracycline can be of considerable help in reducing the number of abortions. Again veterinary advice is essential.

It has been shown that the sheep abortion chlamydia is a risk to pregnant women helping with difficult lambings in infected flocks and great care should be taken, protective clothing and gloves must always be worn. Infection causes serious illness and abortion. Ideally pregnant women should not work with lambing sheep.

### Toxoplasmosis

*Toxoplasma gondii* is a parasite similar to the coccidia which cause coccidiosis of poultry although the latter has no connection with sheep abortion. Sheep abortion of this type occurs in all parts of the world and in some years it is the commonest form of abortion in Britain. Similar to enzootic abortion, toxoplasmosis occurs mainly in low ground flocks.

It is known that man, dogs, foxes and cats are commonly infected with *T. gondii* and it is now evident that the cat is the major source of infection in the environment. Spread of infection can be by birds, rodents, flies and earthworms but cats are probably the commonest source of infection of sheep since they contaminate food, pasture, hay and straw, etc. Once infection has gained entry to a flock further spread occurs at lambing time, ewes and lambs eating food contaminated by the discharge of aborting ewes and by newborn lambs licking infected vaginal discharge. Vertical transmission of infection from ewe to lamb is not as important as environmental contamination.

The effect of toxoplasmosis on the pregnant ewe is related to the stage of gestation at which infection takes place so that foetal reabsorption or birth of stillborn or weak lambs at full term are all symptoms to be expected. Normal lambs are, however, often born to ewes when they are infected late in pregnancy. After abortion, the ewe is immune and can be used for future breeding. The appearance of an aborted placenta can be characteristic of the disease in that the cotyledons are covered in white gritty flecks while the membrane between the cotyledons is normal. However, the aborted placenta may appear normal, hence the need for laboratory diagnosis. Both aborted placentas and foetuses should be sent.

Although no vaccine is available a practical method of immunising a flock is to mix lambs and their ewes in amongst the aborting sheep along with non-pregnant and newly purchased sheep so that they acquire an infection and develop immunity. Unlike enzootic abortion, *T. gondii* will not remain latent to cause abortion at the next pregnancy. However, before practising this method of immunisation the flockmaster must be very sure that no other form of abortion infection is present in the flock and, particularly, that enzootic abortion is not present and unrecognised – which is not uncommon.

Until recently, drug treatment of toxoplasmosis was of little value but current work has shown that the oral feeding of Monensis to pregnant sheep halves the losses from abortion. Further, the lambs which survive are heavier at birth.

## Salmonella

Several sero types of salmonella have been found as causes of abortion in sheep. Until recently the commonest was *Salmonella abortus ovis*, especially in the South-West of England, but *S. dublin* and *S. typhimurium* are now more common while *S. montevideo* is also a frequent cause of abortion in many flocks, particularly in Scotland.

## Listeria monocytogenesis

This also causes abortion in pregnant ewes but will be considered in more detail later, since it causes other forms of disease in sheep.

There is good evidence now that two or more infections of specific abortion-producing organisms occur in flocks; the enzootic abortion of ewes/toxoplasma combinations is the most common but vibrio and listeria may also be involved.

In addition to the specific abortion-producing organisms, a number of diseases of the mother will cause death of the lamb in the uterus. In Britain tick-borne fever and louping ill are particularly important in this context when pregnant, non-acclimatised sheep become infested with infected ticks. In other countries this is also true of foot-and-mouth disease, Rift Valley fever and blue tongue. Border disease, Q fever and brucellosis can also occasionally cause abortion.

**Non infectious neonatal mortality**

This accounts for about half the deaths of lambs before they reach the age of a few days and it is estimated that about a million lambs die in this way in Britain each year. Foetal stillbirth, parturient stillbirth and early death after lambing are the clinical signs. The factors which predispose to this loss have only recently been defined by researchers at the Moredun Institute, Edinburgh and as a result methods of diagnosis, prevention and cure are now well established. Hypothermia or fall in body temperature, is the main clinical symptom and this causes death of the lamb unless the fall can be reversed by suitable treatment by the shepherd. There are two features which causes hypothermia, one is low production of heat within the lamb and the other is high heat loss of the lamb to the environment. The latter is probably the easier to understand, small lambs lose heat faster than large lambs, some birth coats have low insulation values compared with others and wet also decreases the insulation values of the coat. Low air temperature and high wind speed increases heat loss. Hence, the value of shelter and the drying of the birth coat. Variations in the capacity of a lamb to produce heat begins during pregnancy, the most important factors being the size of the placenta and the nourishment of the ewe. Good nutrition in early pregnancy helps the placenta to develop adequately and allows good growth of the foetus or foetuses. Good nutrition in late pregnancy is essential, particularly when twins and triplets are involved so that decent-sized lambs will be born while milk production with adequate colostrum is also encouraged.

Prevention of hypothermia is mainly a question of common sense and good shepherding and the understanding of the causes. Diagnosis is by using a thermometer to measure body temperature and the Moredun lamb thermometer is designed for shepherds for this purpose (the normal temperature of a lamb is 39°C or more while a temperature of less than 37°C indicates severe hypothermia).

Treatment of mild hypothermia (37°C or 39°C) is straightforward. The affected lamb should be sheltered, dried, fed colostrum by stomach tube and, if possible, left with the mother. Severe hypothermia is very dangerous and glucose injections are indicated in the first instance followed by active warming preferably in a 'Moredun lamb warming box', and again colostrum should be fed by stomach tube. Shelter should be provided for the recovering lamb preferably in company with the mother.

# 15 Diseases of the New-born Lamb

There are several specific diseases which affect new-born lambs and, because the central nervous system is affected, cause interference with movement. Since the clinical symptoms may be similar, considerable confusion may arise over differential diagnosis so that it is advisable to seek professional and laboratory help before commencing preventive methods.

## Swayback

Swayback, which affects new-born lambs, occurs in various parts of Britain. Its incidence varies considerably, large numbers of lambs being affected on some farms while on others only an occasional lamb shows symptoms. The symptoms of swayback are due to a progressive destruction of the white matter of the brain while the lamb is in the womb and is associated with low amounts of copper in the ewe. These low levels of copper in the ewe are not always associated with copper deficiency of the pasture and the deficiency is then termed a 'conditioned' one; it is recognised that an excess of molybdenum and sulphate in the pasture can interfere with the utilisation of copper. Recently, it has been shown that the utilisation of copper is under genetic, as well as environmental, control, and it is now known that some breeds of sheep are more susceptible to swayback than others. In Britain, swayback occurs most commonly in areas where the soil type is of peat, limestone or clay. It is also thought that heavy liming of pasture brings about the disease. The incidence of the disease in Britain is variable from year to year and is higher in those years when the winter weather is mild. (Plate 15.1.)

The symptoms are those of incoordination of gait, stumbling and blundering movements but often the lamb is quite unable to walk.

The name 'swayback' has been given to this disease because of the characteristic swaying movement sometimes seen when the lamb walks. Severe cases never recover, although they can survive for some time especially if the lambs are fed and sheltered. Milder cases may live and grow into adult sheep. On occasion, symptoms may not develop until the lamb is a few weeks old and then only if the animal is hurried or chased; these are called 'delayed' cases of the disease.

Diagnosis usually depends upon microscopic examination of the brain. In some cases, obvious cavities and spaces filled with gelatinous material are seen when the brain is examined and this makes diagnosis easy. Copper assay of the blood of the lamb and its mother, or copper assay of the liver of the lamb if it is dead, can also be helpful.

*Plate 15.1* A characteristic gait of a swayback lamb.

Treatment is generally of no value and severely affected lambs should be destroyed; less affected lambs can be fattened. Prevention of swayback is very satisfactory and is commonly carried out on many 'swayback' farms. The pregnant ewe should receive a copper supplement during the last 3 months of pregnancy but it is very important to remember that comparatively small amounts of copper can kill sheep by copper poisoning. For this reason the copper supplement must be controlled and a safe way is to give the extra copper by injection of a proprietary form between the tenth and sixteenth week of pregnancy. Recently good results have been obtained from oral dosing with a bolus of cupric oxide needles which lodge in the abomasum or with a glass bolus containing copper. The latter method is claimed to be free of the danger of causing copper poisoning. Mineral licks, top dressing of pasture and drenching are not nearly so satisfactory.

## Border disease

Border disease was first described as a result of many outbreaks seen in flocks on the central Welsh–English border area; it almost certainly occurred early in the century. It is now recognised in many parts of the world where sheep are kept.

The main clinical feature is the birth of weakly lambs of poor conformation, lambs born with a fuzzy halo of long hair in breeds with pure wool, lambs affected by a tremor or with marked contractions of the legs, lambs with kempy coats coloured brown or black. Growth rate is poor, scouring occurs and death is frequent. The whole flock appears to be poorly doing and early abortion is common so that many ewes appear barren. A proportion of hairy shaker lambs survive but thrive badly.

Diagnosis on clinical grounds is comparatively simple when hairy shaker lambs are numerous. Confirmation of diagnosis is based upon laboratory examination of the brain and by the isolation of the causal virus or by the demonstration of the viral antigens in the lamb tissues. It is important to distinguish Border disease from the common abortion infections and from swayback.

The cause of Border disease is infection of the ewe in early pregnancy with a placenta crossing pestivirus which is related to the virus causing mucosal disease of cattle and the European swine fever virus although these infections do not appear to spread to sheep and cause Border disease. The introduction of Border disease infection to a clean flock is by the purchase of sheep from infected farms since the causal virus remains persistent in infected sheep which spread virus over long periods of time and which infect susceptible pregnant ewes so carrying on the disease. Non-pregnant ewes when infected become carriers and immune to the same type of virus. Abortion material is rich in virus. Affected lambs can also transmit the disease by lateral contact (contagion).

The control of Border disease is not easy since vaccination is as yet not a practical procedure. If the disease has been recently introduced and the level of infection in the flock is low, segregation of affected lambs with their mothers is indicated, the lambs being slaughtered as soon as economically possible. In more heavily infected flocks all lambs should be slaughtered before the next breeding season. Susceptible ewes being retained for breeding can be mixed with affected lambs during the summer period in the hope that they will become immune before the next breeding season at least to the same strain of virus.

## Daft lamb disease

This is a relatively uncommon condition of lambs and the clinical signs are usually seen at birth. It is mainly confined to the Border Leicester breed and its crosses and is thought to be due to an inherited recessive factor causing defects in the developing brain. The lambs show severe incoordination of movement but the most characteristic feature is that the head is carried high, with the mouth pointing backwards over the neck or to one side, so-called 'star gazing'. Some lambs are incapable of walking or may stagger in circles as though blind or senseless. In less severe cases the animal may be able to suck and survives to become a normal adult sheep except that symptoms recur if the animal is excited for any reason.

Accurate diagnosis is possible only by laboratory examination of the brains of affected lambs.

Prevention of the condition depends upon eliminating the genetic defect from the flock by severe culling of parents and lambs. New rams should not be purchased from affected flocks. There is no doubt that these methods have reduced the incidence of daft lamb disease very considerably over recent years.

## Watery mouth

Watery mouth is a shepherd's name for a condition which is ill-defined but on occasion causes heavy losses in young lambs in the immediate postnatal period 12–48 hours after birth.

Affected lambs do not suck and quickly collapse into a coma. The mouth is wet with saliva and stomach contents while the abdomen is swollen. Scour is not seen, indeed the lamb is constipated while the guts gurgle with wind. Death follows within a few hours. Post mortem shows the presence of uterine secretions (meconium) and mucin in the abomasum. The bacterium *Escherichia coli* can be found in the intestine, blood, spleen, liver, etc. Predisposing causes include inadequate colostral intake, hypothermia, prenatal infections while the penning of lambs with their mothers encourages postnatal infection. Early castration and docking with rubber rings appears to encourage the onset of watery mouth.

Treatment to overcome the constipation with enemas and laxatives is useful and the giving of antibiotics often reduces losses. *Escherichia coli* antiserum has been used as have various vaccines, but veterinary advice is necessary in severe outbreaks.

# 16 Diseases of the Young Lamb

The results of many investigations have shown that the greatest loss of lambs occurs in the neonatal period and, as already described, many of these are the result of management or environmental factors. After the neonatal period, overall loss of lambs is small, although isolated incidents can still result in heavy losses on individual farms. The situation at the present time is very different from that of 40 or 50 years ago when such lamb losses were enormous. This difference is due to the very considerable and successful research effort which led to the production of a number of sheep vaccines, etc. which are now widely used throughout the world.

The most common diseases seen are the enteric diseases ranging from the readily recognised and specific lamb dysentery to the more complex conditions such as colibacillosis and viral infections. Enterotoxemia or pulpy kidney disease can also occur in young lambs.

Other infections obtain entry into the bloodstream by the navel and then localise in the joints, liver, spinal cord and other internal organs causing local abscesses – hence the names joint ill, navel ill, liver abscess, etc. In areas where ticks are present, infection enters the lamb by tick bites giving rise to tick pyaemia, tick-borne fever, and louping ill. Similarly, tetanus is caused by contamination of the navel or the wounds left after castration and docking. Most of these infections are caused by microbes which are ubiquitous in nature so that general preventive measures depend upon attention to hygiene, clean lambing sheds and pens and the dressing of the navel with iodine or antibiotic. However, more details of prevention and treatment are given under the descriptions of individual diseases.

## Enteric diseases

### Lamb dysentery

Lamb dysentery is an acute and fatal illness of young lambs under 10 days old, caused by a bacterium called *Clostridium perfringens* (*welchii*) type B which may invade the bowel of the young lamb very soon after it is born. This organism is known to be present in the ground or in buildings and pens on many farms, especially those in the Border counties of England and Scotland, in Wales, and in Northern England, where the disease is endemic. In addition, the organism may be carried to clean farms by the purchase of sheep, especially ewes with lambs at foot, or by carrion-eating birds and vermin and when this does occur the lamb dysentery bacteria are very apt to remain on the farm and produce the disease year after year. Although these facts regarding the prolonged resistance of the microbe and the method by which it is spread are well known, exceptions do seem to occur, and it is occasionally difficult to explain the origin of infection on a clean farm along with the fact that it may disappear just as quickly as it appeared. On the whole, the microbe must be considered as being resistant to normal climatic factors and that, once introduced to a farm, it remains potentially dangerous for years to come.

The fatal illness of the lamb is caused by the multiplication of the specific microbes in the intestine, which, at the same time as they damage the lining of the bowel, produce a most potent toxin or poison which is absorbed into the body of the lamb, so upsetting vital centres and causing death. The organism *Cl. perfringens* is divided into a number of types by the different toxins which it secretes and, strangely enough, each type gives rise to a different disease not only of animals, but one type causes illness in man as well. This is the reason why it is always stated that lamb dysentery is caused by *Cl. perfringens* type B, since *Cl. perfringens* type A, secreting slightly different toxins, rarely causes illness in sheep, while *Cl. perfringens* type D, is the cause of a different disease of sheep known as 'pulpy kidney'.

The symptoms of lamb dysentery vary from farm to farm and from year to year on the same farm. As many as a quarter of the lambs born may die, although such severity of attack is now unusual. While the disease may be very acute, killing lambs in considerable numbers without causing observable symptoms, it may also cause severe diarrhoea (often bloody in character) for a day or so before death occurs. Exceptionally the illness may be more chronic, in which case the diarrhoea continues for a few days. Once the disease appears on a farm in any given year the incidence always increases as the lambing season progresses.

Confirmation of the presence of the disease depends on sending the lamb carcass to a diagnostic laboratory where the toxin of the lamb dysentery bacillus, if present, can be demonstrated in the bowel. Considering the serious nature of the disease and the expense of the methods of prevention such confirmation of diagnosis should always be carried out in cases of doubt.

Once the disease occurs in a flock, prevention of further cases depends essentially on the use of lamb dysentery serum and no other method of prevention or treatment is of real value. Lamb dysentery serum, containing the antibodies which react with the toxin, should be given to all lambs immediately after birth, and in this way each lamb is immediately protected for a period sufficiently long to enable the lamb to develop its own age-immunity. The most effective way of preventing the occurrence of lamb dysentery is by vaccination. The vaccine, when first used on a farm, must be inoculated twice into each breeding sheep, once in the autumn and again 2–4 weeks or so before lambing, but in succeeding years only the spring vaccination is necessary. Protection is dependent on the presence of antibody in the colostrum of the ewe (manufactured by the ewe in response to the vaccine), which when sucked in by the lamb is absorbed into the body, and so affords protection against the infection.

## Pulpy kidney

Pulpy kidney (or enterotoxaemia) is an acute fatal disease of sheep of all ages but it is most frequent in lambs of 3–12 weeks but it can occur in newly-born lambs. Enterotoxaemia also occurs in 6–12-month-old sheep particularly when they are being well-fed. It is the commonest acute killing disease of sheep in Britain at the present time and all sheep farmers would be well advised to use preventive methods against it.

The condition is called enterotoxaemia because it is due to the multiplication in the intestines of the specific toxin-producing organism (*Cl. perfringens* type D). As in lamb dysentery, the organism exerts its lethal effect by producing a very powerful poison in the intestines and this, when absorbed into the bloodstream of the sheep, causes a fatal paralysis of vital centres. Death occurs either suddenly or after a few hours of acute illness.

Although death is directly caused by the specific organism, there can be no better example than enterotoxaemia of the overriding importance of predisposing causes or management factors. *Clostridium perfringens* is in the intestines of most healthy sheep where it causes no harm whatsoever – only when stimulated does it cause trouble – and it can be regarded as one of the many kinds of microbe which normally inhabit the gut of sheep. Like so many of these organisms it may perform a useful purpose in the normal sheep but under certain conditions the fatal toxin is absorbed from the gut into the body tissues, apparently due to a disturbance of stomach and intestinal movements. There is good evidence that the disease is nearly always associated with a thriving condition; for example its classical occurrence is in lambs in the best and most forward condition due to an abundant milk supply of the mother.

On post-mortem examination the pericardial sac which surrounds the heart is often filled with a large quantity of straw-coloured or blood-tinged fluid containing soft, gelatinous material. Blood splashes on the heart muscle both outside and inside the heart are also present. The intestine may be congested with blood. The pulpy appearance of the kidney, after which the disease in lambs is named, is the most difficult change to assess even by those experienced in the disease for pulpy kidneys occur at post-mortem examination in a number of circumstances. Confusion also arises when wool-ball or milk-curds occur in the stomach, for shepherds frequently consider these to be the cause of death rather than the enterotoxaemia organism, as is generally the case.

Diagnosis in the laboratory depends on the demonstration of the specific toxin in the intestine of the dead animal and since this rapidly disappears after death, examination must be carried out reasonably quickly. It is stressed that only by examination of the gut contents can the disease be diagnosed so that submission of the suspect pulpy kidney for examination is valueless.

Specific preventive methods depend on the use of pulpy kidney antiserum or vaccine. The latter product cannot be used to control an outbreak once it has commenced. In these circumstances pulpy kidney antiserum gives immediate protection. Many farmers, however, do not allow pulpy kidney disease to occur since continuous protection during periods of risk can be maintained by using pulpy kidney vaccine. Breeding stock can be immunised by vaccination and the resulting immunity is passed on via the colostrum to the lambs which remain immune for a period of several weeks. In order to prolong this immunity for a number of months the lambs themselves must be vaccinated twice at appropriate times which depend upon the vaccine used and the type of farm. If the breeding stock is not vaccinated, then the lambs must receive pulpy kidney serum or combined lamb dysentery pulpy kidney serum at birth followed by a double vaccination with vaccine during the first few weeks of life.

*Tetanus*

Tetanus, sometimes known as 'lockjaw', is an infectious condition of many animals including sheep and especially lambs.

The disease is caused by the bacterium *Clostridium tetani* which belongs to the same group of microbes that cause lamb dysentery, pulpy kidney, braxy, etc. and, as in these diseases, the clinical symptoms and death of the animal is brought about by the powerful toxin which the organisms secretes.

*Clostridium tetani* occurs widely throughout the world in soil, especially when it is rich in animal manure, and in its spore form it can survive for long periods of time. The organism is also commonly found in the intestinal contents and faeces of normal healthy animals, particularly horses, cattle and sheep, where it does no harm until favourable circumstances arise for its rapid multiplication.

Tetanus generally occurs following infection of the navel with the causal organism soon after birth or following infection of castration or docking wounds. The use of rubber rings appears to increase the incidence. The microbe multiplies readily, especially when the wound is deep and dirty, since absence of air favours the organism. No invasion of the body takes place but the very powerful toxins manufactured in the wound quickly reach the brain and spinal cord causing symptoms and death.

Symptoms generally develop 6–10 days after the wound is made, the first sign being that the lamb begins to walk stiffly with a stilted gait. In mild cases, no further symptoms follow and the lamb begins to recover. More usually, the animal quickly loses its balance and falls over to remain lying with legs outstretched and head thrown back, relaxing only to go into spasm again. Death quickly follows. Vaccination with tetanus vaccine of both the mother and lamb is very efficient while tetanus antiserum can be used in an emergency. No characteristic post-mortem changes can be seen except for the original wound where infection entered. Diagnosis is mainly dependent on recognition of the clinical symptoms.

Since treatment is nearly always useless once severe symptoms have developed, destruction of the animal on humane grounds is indicated. The prevention of tetanus in the sheep is also a matter of cleanliness, not only of personnel but also of instruments and sheep pens particularly during lambing, castration and docking.

A vaccination programme against the clostridial diseases is an essential component of any sheep disease control programme. There are a number of multicomponent vaccines available, some being made up in oils or adjuvants and many flockmasters are now using these since they protect against one or all of the common clostridial infections including those described above and others to be described later such as blackleg, black disease and gas gangrene of the womb. In flocks not having been previously vaccinated, breeding ewes and rams should be given a sensitizing dose of one of the multicomponent vaccines in September followed by an immunising dose a few weeks later prior to tupping. Gimmers and ewes are again vaccinated just before lambing and by this the colostrum will then provide sufficient antibody to protect the sucking lambs with passive protection for 12 weeks or so. Lambs which are to be kept or fattened should then be vaccinated.

## Blood infections

There are a number of diseases of sheep and particularly of lambs which are caused by microbes reaching the bloodstream by one route or another. This generally happens when the microbes overcome the usual defence of the body at the site of invasion and enter the blood either by the lymphatic system or by direct invasion of blood vessels. When this happens the condition is known as septicaemia or bacteraemia depending upon whether the organisms multiply in the bloodstream or not. Death of the animal often follows quickly but if not, the organisms circulating in the bloodstream settle out in the joints, brain, liver, lungs or other tissues. If the microbes cause pus to be produced the condition is termed 'pyaemia'. The site where the microbes first enter the body may be the navel or into the wounds following castration and docking. The bites where ticks attack may also be the site of entry but with tick-borne fever the causal organism is carried within the tick itself. In pasteurellosis the causal organism is a normal inhabitant of the nose or throat of the sheep and the reason why on occasion it enters the bloodstream is not known.

The following diseases are common examples of this type of infection: (1) Pasteurellosis; (2) Joint-ill; (3) Tick-borne fever; (4) Tick pyaemia; (5) Necrobacillosis; (6) Louping-ill; (7) Infectious diarrhoea; and (8) Brain disease.

### *Pasteurellosis*

Pasteurellosis of young lambs is caused by the organism *Pasteurella haemolytica* type A of which there are a number of different strains all found in the nasal passages of healthy lambs.

Usually no symptoms are noticed, the first indication of trouble being the sudden death of a few lambs. Other lambs, however, are often seen to have laboured breathing, a very high temperature and perhaps a slight frothing at the mouth or nose. Death then follows quickly within a few hours.

The post-mortem findings are characterised by general signs of septicaemia, the causal microbe having multiplied very rapidly in the bloodstream and other organs, particularly the lungs and liver. Blood splashes occur over the heart muscle and there is considerable congestion of blood in the abdominal and chest cavities. The air passages of the lungs are severely congested and contain a blood-stained frothy fluid while the lungs do not collapse when the chest is opened and the surface of the lung has a characteristic slate blue colour over which are scattered numerous darker areas. An abundance of fluid flows from the lung when it is dissected and the chest cavity may contain a quantity of blood-stained fluid. On occasions the liver may have white flecks over its surface.

Diagnosis is based on the sudden death of the lamb, the post-mortem findings and the demonstration of the causal organism by the laboratory.

Unfortunately this is not a disease which can be readily controlled although antibiotic therapy is indicated if death is not too sudden. Much research is presently being carried out to determine more exactly the predisposing causes and also into the development of numerous vaccines.

## Joint ill

Joint-ill is the common name for osteo-arthritis or inflammation of the bones and joints and is seen usually in lambs. The condition is essentially a bacterial infection in that one of a number of organisms normally present on the skin or in the environment of animals is introduced into the tissues of the lamb. The main routes of infection are either through the lamb's navel or by various wounds such as are caused by castration or docking. After gaining entrance to the body, the microbes invade the bloodstream and from there establish themselves in the bones, joints or tendon sheaths. The usual bacteria involved are those which produce pus, and are called 'streptococci' and 'staphylococci', but other organisms which may be involved include *Corynebacterium ovis*, *Escherichia coli*, *Pasteurella haemolytica*, *Fusobacterum necrophorus* and *Erysipelothix rhusiopathiae*.

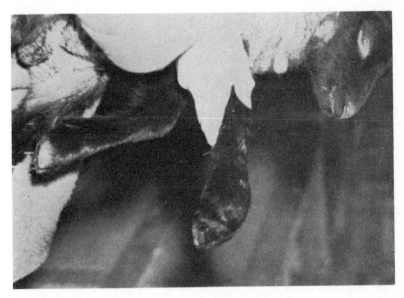

*Plate 16.1* Joint-ill in a lamb.

Joint-ill frequently occurs in lambs born in lambing pens, lambing fields, or other confined areas and this is due to the fact that such places may become heavily contaminated with the germs which cause the disease. Where slightly older lambs are affected the disease is caused by the use of dirty instruments for castration and docking or by the confinement of such lambs in dirty pens, etc. Lack of colostrum is also an important predisposing cause. (Plate 16.1.)

The symptoms of osteo-arthritis vary to some extent depending on the type of infection present. In the early and septicaemic phase of the disease the lamb will be dull, fevered and unwilling to suck. This is followed by sudden lameness or stiffness and affected joints become hot, swollen and painful. In the purulent forms the joints (particularly the knee, hock, elbow, and stifle) become swollen and painful, while the animal may show loss of appetite, diarrhoea, and extreme lameness, in some cases leading to rapid death. If the animal survives, as many do, the joints begin to be filled with pus and this may cause rupture of the skin through which the pus discharges. Recovery from the acute lameness is common but the damage to the joint is such that the animal may remain crippled and unthrifty. As already explained the infecting organisms reach the joints by the bloodstream so it is only to be expected that absesses may occur at other sites than the joints of the limbs as, for instance, in the joints of the backbone, in the internal organs and even in the brain substance. Spinal lesions often give rise to paralysis.

In lambs suffering from the purulent forms of arthritis, examined after death or slaughter, the joints are found filled with pus and the joint surfaces often show considerable destruction. Abscesses may also be found in the joints of the spinal column pressing on to the spinal cord or ruptured into the nervous tissues. Single or multiple abscess formation may be found in the kidneys, heart, liver, spleen, etc. In non-purulent cases the joint surfaces are ulcerated and the tendon sheaths around the joint very thickened.

Prevention of joint-ill following navel infection depends essentially on good lambing conditions. When the disease appears in newly-born lambs the use of the existing lambing pens or fields should be discontinued and the navels of all lambs be treated with antiseptic solution. Instruments used for castration and docking must be surgically clean.

Antibiotic therapy is the most satisfactory form of treatment but large doses are necessary. Lambs which show little or no improvement are best killed since they are very unlikely to grow properly.

*Tick-borne fever*

This has been described in Britain, Norway, Africa and Asia. In tick-infested areas of Britain lambs become infected within the first few weeks of birth. The causal agent is a small microbe called *Cytoecetes phagocytophilia* and the sheep tick acts as a true vector. The organism gets its name because after gaining entrance to the bloodstream of lambs it causes a reduction in the number of phagocytic white cells which are normal constituents of the blood.

The symptoms are those of dullness and a high degree of fever which quickly subside after 2–3 days so that they often go unnoticed. The reduction in number of phagocytes which tick-borne fever causes is thought to be the reason why the staphylococci which cause tick pyaemia become invasive from the skin. Sheep also become more susceptible to other infections such as louping-ill, the parainfluenza viruses and pasteurella organisms. Pregnant ewes infected with tick-borne fever for the first time will often abort.

It is probably not necessary to treat lambs affected with tick-borne fever but both sulphonomides and antibiotic therapy are reasonably efficient at curing the symptoms but not at eliminating the organism. Attempted tick control is of little or no value for the control of tick-borne fever.

### Tick pyaemia (cripples)

*Tick pyaemia* is similar to joint-ill except that the causal microbe is always *Staphylococcus aureas* which enters the bloodstream following the bites of ticks. The tick, however, is not a true vector but because its bite damages the skin of the lamb, staphylococci which are always present on the skin penetrate into the lamb's body and into the bloodstream. The disease tick-borne fever which is very common in tick areas enhances the invasion of the staphylococci. So far as is known tick pyaemia only occurs in Britain.

*Plate 16.2* A lamb badly affected with tick pyaemia.

The clinical signs are in most ways similar to those described for joint-ill except that abscess formation always occurs and particularly so in the brain and spinal cord, so that as well as marked lameness, blindness, incoordination of movement and paralysis are commonly seen. In later stages a chronic arthritis develops and lameness becomes permanent and growth rate is poor. It is a characteristic feature of tick pyaemia that the brain and spinal abscesses on occasion remain occult until they break down in growing or even adult sheep. (Plate 16.2.)

The obvious way to prevent tick pyaemia is to prevent ticks attaching to the skin of the lamb and a number of farmers attempt this by dipping new-born lambs but since there is no correlation between the number of ticks on a lamb and the occurrence of tick pyaemia this is not a very satisfactory method of prevention. It is well recognised that even present-day tick dips have a poor persistence on the fleece (up to 1 week) so that to be a reasonably efficient method of preventing tick pyaemia several dippings at frequent intervals are necessary. Injection of new-born lambs with long-acting antibiotic preparations gives some protection. Vaccination with experimental staphylococci vaccines has so far been of little value.

Antibiotic injections over a period of several days or the one injection of a long-acting antibiotic preparation are of value in treating early cases but once joint lesions are present, little can be done.

The problem remains a serious one in many areas of Britain where ticks occur.

## Necrobacillosis

The cause of necrobacillosis is an organism called *Fusobacterium* which gains entry to the bloodstream of lambs through the umbilical cord following birth. The source of infection is believed to be lambing ewes which have foot rot and foot abscess, diseases associated with *F. necrophorum* infection. The causal organisms settle out in the livers of lambs from the bloodstream and small circular areas of liver necrosis (death) follow, hence the name of the disease.

Affected lambs appear healthy for a few days after birth but then quickly become dull and depressed and refuse to suck. The walk of the lamb is stilted and because of pain the abdomen becomes 'tucked up'. The disease can readily be confused with the early stages of joint-ill.

Treatment is of little value and prevention is dependent upon good hygiene at lambing time, treatment of the navel and the elimination of foot rot, etc. from the flock.

## Louping-ill

Louping-ill is a virus disease of sheep whose vector is the sheep tick. The louping-ill virus is injected into the bloodstream by the tick and after mutliplying in the blood it often passes to the brain causing severe inflammation which is called encephalitis.

On those farms where louping-ill occurs in young lambs it can readily be prevented by vaccination of the ewes.

The disease will be described fully under diseases of sheep of most ages.

## Infectious diarrhoea

Diarrhoea is a common infectious condition affecting young lambs which occurs mainly where intensive lambing systems which use paddocks, pens and indoor lambing sheds are employed. Such systems, unless very carefully managed, encourage the progressive build-up of infection.

The cause of neonatal diarrhoea is not very well understood. For many years the name colibacillosis has been given to all lamb scours and often also to the condition called watery mouth. As in the calf, it was considered that the organism *Bacillus coli* (now termed *Escherichia coli*) was the cause. Undoubtedly this organism is the one commonly isolated from the intestines and often from the bloodstream of scouring lambs at post-mortem examination. The fact that *E. coli* can also be found in healthy lambs and that with only a few exceptions the organisms from healthy lambs are in no way different than those from sick lambs is clearly unsatisfactory and calls into question the exact role of *E. coli*. Recently several viruses have been demonstrated in the droppings of lambs and in particular one virus called rotavirus has been reported as being present in diarrhoeic lambs in Scotland and Northern Ireland. Although the significance of the enteric viruses of sheep in general is not known it would appear that this rotavirus is capable of causing diarrhoea and even death when given experimentally to newborn lambs which have not sucked colostrum. It could well be that a rotavirus infection is the cause of at least some cases of diarrhoea and scour of lambs.

The symptoms can readily be mistaken for those of lamb dysentery and since confusion in diagnosis occurs it is essential to carry out laboratory diagnosis as soon as trouble appears. Affected lambs are dull, apathetic and stand around with their backs arched, whilst regurgitation of watery stomach contents and a yellow fluid scour are characteristic signs in lambs that survive for a few days. Very young lambs often die before the scour develops. The disease usually increases in severity and incidence as lambing progresses.

Treatment with antibiotics or sulphonamides is disappointing and unsatisfactory.

At the present time prevention is dependent upon hygienic measures aimed at preventing a build-up of infection but since disinfection of buildings, pens and lambing sheds is difficult the removal of all unlambed ewes to a new and clean environment is by far the most satisfactory method of preventing further disease following the beginning of an outbreak. There is no doubt that colostrum is a very important factor so that it is necessary to ensure that all lambs be encouraged to suck their mothers as soon as possible after birth. It is also essential to make sure that there is adequate udder development in the ewes at lambing time so that colostral content is plentiful. A number of vaccines have been tried which contain antigens of *E. coli* with little success, but very recently an experimental vaccine containing inactivated lamb rotavirus has been formulated. When given to ewes the vaccine stimulates a very high colostral and milk antibody titre which protects against rotavirus infection of the lamb. This is a very promising development.

## Brain disease

Brain disease is a disease of young lambs in which characteristic nervous symptoms occur – swayback, border disease, and daft lamb disease. These have already been described in Chapter 14, Reproduction and Genital Diseases, but occasionally the signs and symptoms of all three diseases may be very slight at birth and readily overlooked or they may be 'delayed' for several weeks after birth. As a result, confusion in differential diagnosis sometimes occurs.

# 17 Diseases of Older Lambs and Young Sheep

**Pulpy kidney disease (enterotoxaemia)**

This disease has already been described in Chapter 16, Diseases of the Young Lamb, but it is also a very common fatal disease of older lambs and young sheep. As with young lambs, sheep that are very thriving are the ones affected so that disease is associated with a plentiful supply of lush grass and in the autumn with catch crops, excessive feeding with concentrates or by sudden introduction and lengthy feeding on the turnip break. The period of greatest danger is always a few days after sheep are introduced to a new feed.

Although enterotoxaemia commonly appears with good feeding, deaths are also relatively frequent in other circumstances as, for example, in sheep in poor or hard condition when they begin to thrive. For instance, it is not uncommon for a few losses to occur among wormy sheep after treatment with anthelmintics. Store sheep being moved from a poor to a better pasture are very susceptible.

Unlike in young lambs, the post-mortem findings are usually very insignificant consisting only of a slight increase in the fluid in the chest and abdominal cavities. The bowel may be slightly congested with blood. These changes are not specific and are of very little help in diagnosis.

The obvious method of control is to eliminate the management causes which means that sheep should be prevented from thriving too rapidly. This may be achieved by putting them on a poorer grazing and by introducing sheep slowly to any change of feed. Unfortunately, this generally defeats the purpose of the flockmaster which is to have thriving sheep so that more expensive specific preventive measures using enterotoxaemia serum or vaccine are more commonly used. The serum is used to prevent further deaths when an outbreak is occurring and the vaccine is used to prevent the disease during the danger periods. Unfortunately for lifetime protection it is necessary to vaccinate every 6 months or so. It is advisable to consult a veterinary surgeon for individual requirements.

## Coccidiosis

Coccidiosis is a condition which has caused considerable controversy amongst veterinarians and some difficulties in diagnosis. The cause is due to organisms of the genus *Eimeria* of which there are numerous species which differ considerably in their capacity to cause disease and which are difficult to differentiate. It is well recognised that large numbers of coccidial oöcysts occur in the faeces of lambs in the months of June and July in Britain. However, these oöcysts can be found in both healthy and sick lambs so that it is often difficult to assess their significance. As is well known, grazing lambs are not infrequently affected by a persistent diarrhoea during the early summer months and both high worm egg counts and coccidia can readily be found in their scour. Anthelmintics often fail to do any good in such cases and the coccidia, unlike the worm eggs, are not cleared so that it is often assumed that the coccidia are the cause of the diarrhoea. Whether this is so or not is unsure since more often than not the treated lambs are returned to pastures infected with worm parasites and particularly with heavy infestations of nematodirus worm larvae and these may be the cause of the continuing diarrhoea. However, it is probably correct to assume that coccidios can be a cause of fatal disease or more commonly be a contributory cause of diarrhoea and ill thriving in older lambs. The fact that a similar condition sometimes occurs in housed lambs free or virtually free of worms suggest that coccidia can cause ill health on their own.

The symptoms are not specific but consist of intense scouring accompanied by considerable straining.

Diagnosis for the reasons given is often difficult and a thorough investigation is necessary before beginning treatment.

Treatment is by specific drug therapy and a veterinarian should be consulted about both diagnosis and treatment.

Preventive measures are based on routine giving of drugs but with the present lack of knowledge it is of doubtful value and again a veterinarian's advice should be sought. Since studies suggest that first infection occurs when the lamb is very young hygiene in the lambing fields or sheds is important in prevention.

## Braxy

Braxy is a very acute disease of young sheep which occurs mainly in hill flocks during the autumn and winter months. Braxy at one time was one of the most feared diseases especially on the upland sheep farms of the English

– Scottish Borders. For unknown reasons the incidence is now very low and the disease is rarely encountered, at least in Scotland. It is caused by the organism *Clostridium septique* which is of the same group of microbes as those causing lamb dysentery and enterotoxaemia and, as already described in these diseases, death is caused by the toxin or poison secreted by the specific organism. Predisposing factors are again of considerable importance in the production of the disease, since the causal organism is ubiquitous in its occurrence and is generally thought to be a normal inhabitant of many animals, including sheep, and is commonly found in sheep dying of a number of conditions not directly caused by this microbe. As in the case of enterotoxaemia, sheep in thriving condition are the ones usually affected but it is generally thought that the ingestion of frosted food stimulates the disease. There is considerable doubt as to the real predisposing factors. Whatever these may be, they allow the specific organism to become active and multiply in the walls of the abomasum or fourth stomach and there secrete its powerful killing toxin. The organism also breaks through the stomach wall and invades the abdominal cavity where it may be found in considerable numbers immediately after death.

Symptoms are rarely seen, the affected animal being found dead, even though it was in the best of health and condition when seen previously. On the rare occasion when symptoms have been seen they are those of abdominal pain and swelling followed by collapse, symptoms characteristic of a number of diseases of sheep.

Diagnosis of the disease depends on the post-mortem findings which consist of an inflamed and possibly ulcerated patch on the lining membrane of the fourth stomach or, less frequently, in the first part of the intestine. In addition blood splashes may be seen over the heart muscle, while after death decomposition is rapid. These latter findings are common to a number of conditions, so that accurate diagnosis depends on seeing the stomach lesion. For practical reasons it is unfortunate that the stomach lesion, although very obvious in the newly dead animal, quickly becomes unrecognisable if the carcass is not examined within several hours of death, and in a number of cases the author has failed to recognise it even in as short a time as 2 hours post mortem. Laboratory diagnosis is dependent on demonstrating the causal organism in the stomach lesion or in other tissues, but this is again very difficult or confusing unless the examination is quickly carried out. For accurate and proper diagnosis, the animal must be submitted to the laboratory immediately after death; it is useless to expect a proper diagnosis if examination has been delayed.

Prevention of the disease obviously cannot depend on controlling the predisposing factors since, as already mentioned, these are really unknown, nor is it possible to eliminate the causal microbe from a flock. Fortunately, a specific vaccine has been developed and over many years braxy vaccine has been shown to be very efficient. Vaccination is carried out in the early autumn so that at least a month is allowed for the sheep to develop immunity before the onset of the disease is expected. On some farms a double vaccination within 14 days is practised so that immunity will be very strong. The belief that braxy vaccine in full dosage causes a check in the thriving of the vaccinated sheep is widespread but considerable experimental evidence has failed to confirm the impression. Braxy vaccine forms part of a number of multivalent clostridial vaccines which are on the market to protect sheep against lamb dysentery, pulpy kidney, braxy, etc.


## Parasitic gastroenteritis

Round worms of the stomach and intestine of sheep are primarily infections of lambs and sheep characterised by weight loss and diarrhoea. But round worms in sheep not only cause a slow drain on health but also account for outbreaks of acute disease in which sheep may be very sick or even die before wasting is seen. A number of different worms can be involved but the life-cycles of all of them are very similar albeit with important differences between them.

When fully grown, the worms measure from 4 to 12 mm long and are of the thickness of a fine hair; only the larger kinds can be seen directly when an infected gut or stomach is opened and each kind of worm has its own favoured station. The way that a sheep becomes infected and passes the infection to other sheep can be understood by following the life-cycle which can be divided into four phases, egg contamination of the pasture, growth of infective larvae on the pasture, infection of the sheep and development of the worms to maturity in the sheep.

The adult worms in the stomach or gut fertilise themselves and the female worms lay eggs which pass down the gut with the partly digested food so that 24–36 hours later they are dropped out on to the pasture with the dung. In the second phase, which takes place on the pasture, each egg develops into a larva but the details of this development in different kinds of worm, and the time required, vary considerably with the climatic conditions. The only way in which any sheep can become infected for the first time or can acquire more worms is by taking in these infective larvae. This is the third or infection phase of the cycle. Worms do not multiply inside sheep. After they have been swallowed the infective larvae stick to the lining of the gut in their favoured place; some worms actually burrow into the tissue. There they feed, grow, and develop into sexually-mature worms capable of repeating the cycle. This fourth, or maturation, phase takes upwards of a fortnight. With some worms it is these maturing stages which are most damaging to sheep, whereas in other cases it is the full-grown adult worm which causes trouble.

Lambs when they are born are free of worms and only become infested when they eat pasture contaminated with worm larvae so that it is important to understand how pasture can become contaminated.

In most areas of the world which are temperate with summer and winter periods, or in semi-arid or arid zones with alternating periods of rainfall and dry conditions, there are definite and constant fluctuations in the number of worm larvae on pasture according to the season of the year. Fluctuation of this type is not so definite in tropical countries where rainfall and temperature are high throughout the year.

There are two factors which are important in determining the number of worm larvae on pasture. The first factor is the length of time that the larvae on the pasture remain alive and potentially infective. The second factor is the rate at which new populations of larvae can build up on pasture. Both are complicated and not very easy to understand and it is only in recent years that a beginning has been made in understanding the epidemiology of parasitism.

It is accepted that in temperate climates such as encountered in Britain a large proportion of the round worms of sheep can overwinter on pasture either as eggs or as ensheathed larvae. This means that any pasture which has carried sheep the previous summer or autumn will be to different degrees contaminated during the following spring and early summer. Development of these eggs or larvae can of course occur during mild winters although this is not so very usual, but if they do, susceptible sheep will be continuously infected and will shed even more eggs on to the winter pasture. Sheep housed during the winter will be infected when they are turned out on to grass carrying over-wintered larvae. However, if the weather in the spring is dry and warm many of the over-wintered eggs and larvae will die off as they begin to be active, with the exception of the larval stages of the worm *Nematodirus battus* which is protected by remaining within the egg.

For these reasons, Nature ensures that the worm parasites of sheep survive adverse conditions and are ready to infect the next batch of lambs usually born when climatic conditions are suitable for lamb survival, pasture growth and the development of the free living infective stages of sheep stomach and intestinal worms.

Nature, however, has other ways of ensuring the survival of these parasites over the winter period and also for ensuring that the number of infective larvae on the pasture in the spring is sufficient for the purpose. It is known that the larvae of a number of species of round worms when they are ingested by sheep in the autumn or early winter do not develop in the normal way to adult worms but remain dormant for several months. In the spring they resume their development to sexual maturity ensuring that their eggs encounter an environment suitable for them to mature to the infective stage. This arrested larval development within the sheep is known as hypobiosis. The precise biological reason for how it occurs is not understood. It is also known that during the spring lambing time the female worms present in the stomachs and intestines of pregnant sheep become much more prolific egg-layers so that many more eggs are passed on to the pastures. Again, the precise reason for this is not known but it is thought that it might be due to hormonal changes in the ewe at the time of parturition. At this time, sheep and particularly breeding sheep become much more susceptible to reinfection with worm larvae which have survived the winter and these quickly become adult and fecund. Again the reasons for this are unclear but a lowering of the sheep's resistance, changes in hormone levels, and dietary deficiency may all play important roles.

Whatever the reasons for all these phenomena it has been recognised for many years that in spring time there is a considerable rise in the faecal worm egg counts of ewes and this has been termed the 'spring rise', or more recently and probably correctly called the 'periparturient rise', since it is associated with the birth and early lactation period.

It is therefore very clear that when lambs begin to nibble the grass on pastures which have carried sheep during the previous summer, autumn or winter they will ingest varying numbers of infective worm larvae, many of which will develop into adult egg-laying worms. However, with the exception of *N. battus* infestations the number of worms will be insufficient to produce disease. This first generation of worms in the lamb will in due course add to the larval worm numbers on the pasture and these after a period will produce a second generation of adult worms in the grazing lambs and similarly third and fourth generations will be produced. The development period between generations shortens as the summer weather becomes warmer and this results in a considerable build-up of larvae from midsummer onwards. Because of this and because the growing lamb begins to eat more and more grass and therefore more larvae, outbreaks of acute parasitic gastroenteritis begin to occur. There are differences in the generation times of various species of worms so that ostertagia and haemonchus occur earlier in the summer than do trichostronglyus, etc.

The clinical signs of parasitism vary a great deal depending upon a number of variables which include the rate of larval intake, the species of parasite involved and the breed, age and nutritional status of the infested sheep. A rapid intake of many thousands of larvae of most species of worm can result in sudden death or in acute disease and scour, and the best example of this is seen in lambs infested with *N. battus*. At the other end of the spectrum small or gradually increasing larval intakes will permit the sheep to adapt to the infection with only minor metabolic changes; good nutrition, increasing age and previous worm intakes all help to overcome the effects of parasitic invasion. Between these two extremes there can be seen in the field all levels of larval intake which often cause severe loss of performance without the sheep showing any specific symptoms. The reason for the loss of performance and clinical signs when they are present differs between the species of worm involved, the region of the gut which is parasitised and the development stage of the parasite. Some worms cause blood loss and anaemia while others cause anaemia by impairing the nutrition of the sheep. Others interfere with the digestive functions of the gut by impairing the gut secretions so bringing about a reduction in digestibility of the food, particularly protein, minerals, and energy. In some instances there is a complete stop to skeletal mineral deposition and hence reduction of bone growth. Not only is food utilisation affected but there may be a reduction in food intake. Wool production may also be affected due to a loss of essential proteins leaking into the gut. It has been found that probably the biggest economic loss in parasitism is that moderately or even lightly infested sheep have to eat more food than normal sheep in order to achieve similar weight gains.

Outbreaks of real clinical parasitic disease are predominantly seen in lambs during July, August and September. The symptoms arise suddenly in several lambs at the same time, the main sign being a profuse, watery diarrhoea. The sheep is dull and lacks bloom, growth rate falls and there follows loss of weight. Finally, if treatment is not carried out the sheep becomes dehydrated, collapses and dies. However, sub-clinical parasitism, although less spectacular, undoubtedly is much more serious from the economic viewpoint.

There are two forms of parasitic gastroenteritis which although they fit into the general pattern described above are sufficiently different to require separate mention.

Nematodiriasis has the most definite and distinct pattern of all worm disease. It makes a sudden and dramatic appearance in the lambs of low ground flocks at some time between the end of May and the second week of July, the date varying with the locality and the year. Although adult sheep only carry small numbers of *N. battus* worms and do not suffer any detectable ill-effects, they do contaminate pastures lightly with worm eggs and are often responsible for carrying the parasite to clean pastures. Lambs also often carry adult worms without harm and in greater numbers but on occasion lambs unlike ewes become extremely ill as a result of extensive damage to the intestine caused by taking in large numbers of infective larvae which proceed to mature in the lining of the gut.

Symptoms appear suddenly in a flock, a number of lambs in a field being affected simultaneously. The lambs begin to scour violently, but soon only quantities of mucus are passed which cakes on the hindquarters and legs. The wool quickly loses its bloom while the face darkens and the eyes sink and are dirty with a muco-purulent discharge. The lamb becomes 'tucked-up', the ears droop, the animal is reluctant to move and when forced to do so assumes a stilted gait. All affected animals seek avidly for water. In fatal cases illness lasts 4–5 days and the animal dies from the severe loss of body fluid. Animals which do not die may take months to recover and often remain stunted for the rest of their lives.

The post-mortem findings in a lamb which has died from *N. battus* infestation are not unlike those seen in pulpy kidney disease. The carcass is extremely dehydrated, the eyes sunken and the fleece 'staring'. The intestine may or may not be inflamed and congested with blood, the contents being generally very fluid and often blood-stained. Blood splashes may be present on the heart muscle and on the kidneys, the latter being soft and rather mushy. Worms may or may not be present in large numbers. A dirty white in colour, they are readily seen in sieved and washed preparations of the dead lamb's intestinal contents when they look like a mass of wet cotton wool. Individual worms are about 15 mm long and are easily observed when freed from the intestinal debris but they are not difficult to distinguish even in the freshly opened gut. The developing larvae of the parasite (the stage which causes the disease) are partly buried in the lining of the gut and can only be seen by microscopic examination.

The life-cycle of *N. battus* takes approximately a year to complete. First-year grass on which ewes and lambs are running is lightly contaminated with eggs from the ewes and when such grass is again grazed in April, May, or June in the following year the lambs pick up the infective larvae which have now developed from these eggs. The worms which develop from the larvae in the intestines of the lambs then proceed to lay further eggs in large numbers and the cycle is built up and repeated in the following year, but this time the numbers of infective larvae may be sufficiently large for disease to result. If lambs have even a relatively short time on old grass before going on to first-year grass the sequences described may be speeded up so that second-year grass pasture becomes sufficiently contaminated to cause illness. Unlike other worm parasites the larvae of *N. battus* are protected during the winter and spring by being encapsulated in the egg membrane. When given the right conditions the larvae hatch out and become infective in large numbers all at the same time.

Black scour is a type of acute enteritis often caused by Trichostrongylis worms. It occurs in all ages of sheep during the winter months and is often found among sheep that are herded in a confined space on the turnip break or inbye hill grazing. Severe clinical symptoms occur more commonly in sheep whose physical resistance has been reduced by low levels of nutrition or by debility in late pregnancy. It is thought that the large number of worms present in the gut originate from the development of larvae 'arrested' since the previous late summer and autumn.

Diagnosis of parasitism is based on clinical signs or lack of well doing, the season of the year, the age of the sheep and a knowledge of the grazing history. In order to establish a clear diagnosis a post-mortem examination is necessary even if an ailing sheep has to be sacrificed. Examination of faecal samples for worm eggs is a useful aid but can be very misleading particularly for *N. battus* infestations. A good response to anthelmintic drug treatment is often as good a method of diagnosis as any.

Treatment of most forms of parasitic gastroenteritis and also subclinical infestations is by the use of one or more of the modern anthelmintics which are very efficient. However, they should be used not only for treatment but also as a form of proplylaxis by helping to keep pastures relatively clean of worm eggs. To do this ewes should be dosed during the fourth month of pregnancy to eliminate the worm burdens at the time of the 'spring rise'. A further treatment of the ewes can be given about 4–6 weeks after lambing. Energy blocks containing anthelmintics can be used at these times instead of dosing but this is only recommended when the ewes are on confined grazings. Lambs should be treated at weaning and then moved to clean pasture; if this is not possible frequent treatment until late in the autumn especially on low ground farms is necessary. One such treatment is often sufficient for hill lambs, usually about September. For *N. battus*, treatment must be given every 3 weeks to the lamb crop during May and June, and on susceptible farms the flockmaster should not wait until symptoms occur in the lambs. Tups and hoggs should be treated at the same time as the ewes. There is some evidence that frequent dosing with the same group anthelmintic may induce resistance in worm populations but before accepting this and before changing to an anthelmintic of a different group expert advice should be sought.

Prevention of helminthiasis has been practised with varying success for many years. All prophylactic measures are based upon providing 'clean' pastures for lambs which is perhaps not so easy as it may sound. The oldest method was based upon rotational grazing methods including creep grazing through fields or paddocks but it generally failed because on most farms it involved the eventual return of the sheep to previously grazed ground, generally because sufficient grass was not available. More recently, clean grazing has been provided in one way or another by dividing a farm into two areas which are grazed by sheep and cattle in alternative years and by controlling heavy pasture contamination by anthelmintic dosing of the adult sheep particularly at the time of the periparturient rise in egg counts. Modern anthelmintic drugs have an enormous advantage over the older drugs such as phenothiazine in that they are very effective even against worms that are in the 'arrested' stage. The part of the farm not being used for sheep can be safely grazed by cattle since cross-infections between cattle and sheep do not occur. *Nematodirus battus* disease can be prevented by following the maxim 'never' put the lambs on the same pasture 2 years running.

A number of attempts to formulate vaccines against alimentary paratism have been attempted but up to now none have been successful.

## Acidosis

This is an acute metabolic disease mainly seen in older lambs at the end of the summer when the lambs are being finished for sale on grain feed and particularly on barley feed. It also occurs in store lambs that have been turned on to barley stubble rich in blown grain on the ground. Sheep which gain access to a grain store can be affected at any time of the year.

Sudden deaths in healthy lambs is characteristic while other lambs will stop eating and become depressed and weak and have intense abdominal pain. Later a profuse mucoid diarrhoea develops if the lambs survive.

Post-mortem examination will show the rumen full of whole grain in a very watery fluid which is acidic when tested for this. The lining of both the rumen and the abdomen may be haemorrhagic, particularly in patches, and the surface will be peeling off.

Differential diagnosis between this condition and pulpy kidney disease is necessary.

There is no successful method of treatment. Prevention depends upon controlling the grain intake of the lamb.

## Pneumonias

Pneumonias commonly occur in sheep of all ages, and since the term 'pneumonia' is the medical name given to inflammation of the lungs it is expected that there are many causes – microbial, viral or mechanical (as, for example, when drenches are wrongly administered). It is therefore not surprising that the symptoms and post-mortem findings can differ greatly from sheep to sheep when the pneumonia may be acute, subacute or chronic.

### *Pasteurella pneumonia*

Pasteurella pneumonia is caused by *Pasteurella haemolytica* type A but there is good evidence that parainfluenza virus type 3 is also an important factor. All ages of sheep are affected but the disease is seen more commonly in lambs less than a year old. It can be either sporadic or enzootic and, on occasion, losses are heavy among lambs and young sheep. The onset of the disease often follows the changing of sheep from poorer pastures to aftermath or foggage in late summer. Since the incidence varies considerably from year to year climatic factors may also be important. Commonly, no symptoms are noticed, the first indication of trouble being the sudden death of a few animals while others are then seen to have laboured breathing, a very high temperature and a slight frothing at the mouth or nose. Death often quickly follows within a few hours.

The post-mortem findings in the very acute form are characterised by general signs of septicaemia, for the causal microbe *P. haemolytica* multiplies very rapidly both in the lungs and in the blood, thus killing the sheep. As would be expected on post-mortem examination, blood splashes occur over the heart muscle and there is considerable congestion of blood in the abdominal and chest cavities. The air passages are severely congested and contain blood-stained, frothy fluid while the lungs are large and do not collapse because of severe congestion and waterlogging, the lung surfaces exhibiting a characteristic slate-blue colour over which are scattered numerous darker areas. An abundance of fluid flows from the cut surface when the lung is dissected and the chest cavity may contain a quantity of blood-stained fluid. In the less acute cases, real consolidation is present in the lungs, especially in the fore portions, the disease tissue being dense and liver-like in consistency and either dark red or grey in colour. There may be a quantity of discharge over the surface of the lung and chest wall.

Diagnosis of the condition depends on the clinical history and symptoms of the outbreak and the demonstration of large numbers of the causal organism in the lung, liver and spleen.

The most satisfactory method of preventing the disease once an outbreak has arisen is to remove the predisposing cause if this can be determined. Outbreaks of acute pneumonia amongst feeder lambs can often be stopped by a complete change of pasture, for example by moving them off foggage to a poor grass field. Careful transport of sheep with appropriate watering and feeding will prevent losses which might occur after sheep have been transported. Specific preventive vaccines are available against *P. haemolytica* infection and these are used in certain areas of Britain as a routine method of preventing the disease; these vaccines do not appear, however, to be very efficient. However, a recent adjuvant vaccine containing a number of serotypes of *P. haemolytica* type A has been produced at Moredun Institute and is now commercially available and appears to be more satisfactory.

Pneumonic pasteurellosis can be treated by giving penicillin, etc. but some types of the organism are not susceptible to antibiotics.

### Atypical pneumonia

This disease is so-named because it is very different from *P. haemolytica* pneumonia. A considerable number of lambs in a flock can be affected but the illness can be overlooked by the flockmaster because the signs and symptoms are not very evident. The main clinical sign is coughing and odd animals may have a raised temperature and breathe rapidly. Growth rate is very much reduced. Outbreaks are, however, often complicated in that *P. haemolytica* infection also occurs so that both forms of pneumonia may be found in the same animal. This makes differential diagnosis difficult. (Plate 17.1.)

The cause of atypical pneumonia is probably an organism called *Mycoplasma ovipneumoniae* but other organisms including parainfluenza type 3 can be involved.

The disease was originally defined by post-mortem findings which distinguish it very clearly from *P. haemolytica* pneumonia. The anterior portions of the lung are fleshy and grey to brown-red in colour and the microscopic lesion is a productive one rather than exudative.

Transmission of the disease is almost certainly by contact so that it is much more prevalent on intensively stocked sheep farms and in lambs kept indoors.

Control and prevention depend mainly on good husbandry, since up to the present there is no commercial vaccine except that *Pasteurella* vaccines can be used to control the worst forms of the disease. Various antibiotics are useful for treatment.

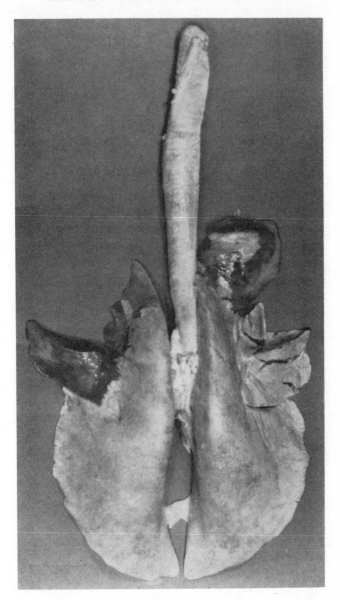

*Plate 17.1* Lung of a sheep affected with Pasteurella Haemolytica pneumonia.

## *Parasitic pneumonia (hoose or husk)*

This is a parasitic disease of the lungs of young sheep which occurs in the months of August and September but it is now of little economic importance in Britain. For no known reason lungworm disease appears to be much more of a problem in Eastern Europe, the Middle East and particularly in the Soviet Union. The causal worm is called *Dictyocaulus filaria* and sheep become infected in exactly the same way as they do with stomach worms by eating the infective larvae with grass. When these infective larvae reach the gut they make their way through the lining and are then carried to the lung where they first force their way into an air sac and then, as they grow larger, make their way into wider air tubes. The adult worms lie in the main airways, the females laying eggs coughed up by the sheep and then swallowed to pass out in the dung and thus contaminate the pasture. The larvae can overwinter on pasture or in carrier sheep and these are the main ways in which infestations occur year after year.

The worms cause irritation and inflammation of the air passages and of the lung tissue which allows bacterial infection to take place with the consequent development of pneumonia which can be difficult to distinguish from atypical pneumonia except by laboratory examination. The degree of infestation and damage is to a large extent dependent on the nutritional status of the sheep although this is not always the case, since if the pasture is very heavily contaminated with infective larvae these may be able to overcome the resistance of even the best-fed sheep. The most obvious symptom is a characteristic irritating cough along with progressive loss of bodily condition. In some instances death is due to poverty and exhaustion but in other cases acute pneumonia may be the cause. Conditions which favour the infective larvae of lungworms will also favour other kinds of roundworm so that lungworm disease will often be accompanied by scouring due to intestinal worms.

Prevention depends on maintaining adequate nutrition and proper stocking rates. There are several effective drugs against lungworms, and treatment to remove all other kinds of worms will help the sheep to overcome the lung infestation. Lungworm vaccine containing irradiated larvae has given good results in the Middle East, Yugoslavia and the Soviet Union and, as with gastroenteritis, alternative grazing can be helpful. In Britain, however, the disease does not warrant the use of vaccine nor the trouble of grazing systems.

Another lungworm of sheep is *Muellerius capillaris* which occurs in the majority of sheep but although it causes lung lesions, these are not important as a cause of illness nor even of ill-thriving. The life-cycle is similar to *D. filaria* except that the free-living larvae develop in snails.

## Cerebrocortical necrosis

This is a degenerative disease of the brain of sheep of all ages but the majority of cases occur in older lambs.

It is thought that a deficiency of thiamine is an important factor but since thiamine is part of the vitamin B complex which is manufactured by the microbes of rumenal digestion it is unlikely to be a dietary deficiency. Thiamine can, however, be destroyed by drugs such as amprolium which when given experimentally results in the development of cerebrocortical necrosis (CCN). Similarly, certain bacteria multiplying in the gut of sheep may destroy thiamine before absorption and so bring about the disease but since CCN does not appear to be associated with any particular environment or dietary regime the cause is not really established.

Initially affected lambs move about aimlessly and then develop a stagger or swaying walk and appear blind; finally they become incapable of standing. Convulsions follow before death occurs. In some cases death occurs without symptoms.

As the name suggests, the primary lesion of CCN is death of brain cells followed by small yellow foci of necrosis (death) throughout the frontal area of the brain which later extend to the whole of the main part of the brain. These can be seen by the naked eye but microscopical examination is often necessary to confirm the diagnosis.

Treatment of affected sheep with thiamine as multivitamin B preparations has been very disappointing so that little can be done. This is in contrast to treatment of a similar disease in cattle.

## Erysipelothrix arthritis

This is a form of joint-ill but it occurs in older lambs usually when they are several months old. The causal organism is *Erysipelothrix rhusiopathiae* which is a normal inhabitant of many soils. This organism also causes erysipelas of pigs but since it is a soil organism the disease in lambs has no association with the pig disease. This form of arthritis follows soil contamination of docking and castration wounds and was at one time common in lambs that had just been dipped due to contamination of the dip bath. Most dips, however, now contain bacteriostats which kill off erysipelas organisms so that post dipping lameness now only occurs as a result of poor management methods which allow sheep dip batches to become very soil contaminated and dirty. As many as a quarter of the lamb flock may be affected.

After infection the organism invades the bloodstream and may cause an acute phase of the disease. Affected lambs are dull and unwilling to feed but the disease is mainly seen in the sub-acute or chronic form after the erysipelas bacteria leave the blood and enter the joints. There is very obvious lameness but the joints are rarely swollen but when touched are hot and painful. Later the joints, particularly those of the limbs, become permanently deformed and twisted. Death may occur in the acute phase and occasionally in the chronic phase due to vegetations growing on the heart valves as a result of infection of the valves. Economic loss is mainly due to the prolonged unthriftiness which can be very marked. In less severe cases the tissue around the cornory band of the foot is infected and becomes inflamed, hot and swollen.

The erysipelas organism is sensitive to antibiotics and treatment is very successful if begun immediately. Such treatment is much less effective if delayed until lameness is well established.

Control is entirely dependent upon good hygiene and good management at docking and castration and during the dipping operation.

## Orf or contagious pustular dermatitis (CPD) or contagious ecthyma

Orf is a virus disease which mainly affects older lambs although sucking lambs and ewes can also be affected. The virus has been classified as one of the parapoxvirus group of pox viruses.

The disease is a very common one both in Britain and throughout the world. In older lambs it occurs in the late summer and autumn but in springtime it can occur in suckling lambs and on the teats and udders of their mothers. The sites most frequently affected are the lips and muzzles and even on the nose, tongue and inside the mouth. In some outbreaks lesions are confined to the skin in the region of the coronet of the hoof and such affected lambs are very lame. The early lesion consists of small vesicles but these are rarely noticed and the disease is only observed when thick scabs form on the lips and feet. When these scabs are removed a red ulcerated wound remains. As the disease progresses wart-like growths develop on the raw surface of the skin and these bleed very easily when damaged and then become infected with a variety of pus-producing bacteria. Feeding is very seriously interrupted and the lambs begin to lose condition. In affected ewes very acute mastitis may develop as a result of secondary infection with staphylococci when death is frequent.

The disease is spread by contact and spreads rapidly. Since orf tends to appear year after year on some farms the infection must be carried over possibly in dried scabs but the method of carry-over is not very clearly understood.

Treatment is not very successful unless severe secondary bacterial infection is present when antibiotics can be used. Fortunately, most cases of orf get better on their own in due course.

A vaccine which contains live orf virus is used on farms where the disease occurs year after year. The vaccine is applied to a scarified area on the inside of the thigh or fore leg about 2–3 weeks before the time the disease is expected to appear. The vaccine has given very variable results but many farmers use it, believing it prevents severe disease. Since the vaccine is a living orf virus, it should not be used on closed farms where the disease has not been seen, and when it is used all lambs must be vaccinated. Immunity is not passed on in the colostrum. A venereal form of the disease was thought to occur in ewes and rams but this will be described later in the section on ulcerative vulvitis and balantitis in Chapter 19.

Both the scabs of the natural disease and the vaccine itself can infect human beings giving rise to a nasty and persisting infection of the skin and local tissues.

### Strawberry footrot

This occurs on the coronet of older lambs and is very similar to orf and indeed many think it is the same disease. Some, however, consider it to be a separate disease caused by *Dermatophilus congolensis* or by a mixture of orf virus with this organism.

### Pyodermas

Pyodermas is a further disease which can be confused with orf. It occurs in lambs and on the udders of nursing mothers. Pustules develop in the skin around the lips or under the tail of lambs and the udders of ewes. The cause is a staphylococcus which causes the pustules. The condition is not a very serious one and clears up within a week or two. More serious is facial or periorbital eczema of older sheep which is caused by a similar organism. Badly affected animals suffer severe symptoms which clear within 6–8 weeks. Antibiotic therapy either by injection or topically is indicated in severely affected animals.

## Yellowsis (facial eczema, head grit, photosensitisation)

The symptoms of sensitisation to light are caused by exposure of sheep to sunlight after they have eaten certain plants or after they have been dosed with one of several drugs. The reason why the ingestion of such substances should make the skin sensitive to sunlight is complex, but in some cases it is due to the fact that even in a small quantity the ingested substance is itself a photosensitising agent or is broken down into a photosensitising substance in the animal's body. In other instances the photosensitising agent is a substance which is normally present in the body but only builds up in sufficient quantity to cause harm when the liver fails to excrete it; this type is represented by those photosensitivities in which the substance phyloery-thrin, a product of the green substance of plants, accumulates in the blood as a result of liver dysfunction. In the majority of such diseases the liver dysfunction is due to a plant or chemical poison which may also cause sufficient liver damage to upset excretion of bile, so bringing about jaundice of the animal as well as photosensitisation; an example is the important sheep disease of New Zealand called facial eczema, which is characterised by liver damage and jaundice caused by a fungus present on pasture. Finally, there are various types of photosensitivity about which little definite information as to their origin is available.

The following examples of the disease are most frequently seen in Britain: (1) yellowses; (2) rape scald; (3) photosensitisation of lambs.

(1) *Yellowses, or 'head grit'* is a disease of lambs grazing rough permanent or hill pasture in certain areas of Scotland which occurs during the months of June and July. Its incidence is never high but it varies from year to year on the same farm. The photosensitising agent is probably phyloerythrin but it is uncertain whether liver damage is present. The cause is unknown although the bog asphodel plant is considered to be responsible for a similar condition in Norway and now possibly in Britain as is also St John's Wort.

(2) *Rape scald* is common in sheep running on rape. In this case no liver damage has been found nor is there an excess of phyloerythrin in the blood. As no unusual pigments have been found in rape, the nature of the photosensitising reaction is unknown.

(3) *Photosensitisation of lambs* occasionally occurs as a result of normal dosing with phenothiazine if this takes place on a very hot, sunny day. It is considered that the substance which causes the disease is a normal breakdown product of phenothiazine. This condition does not appear to be serious.

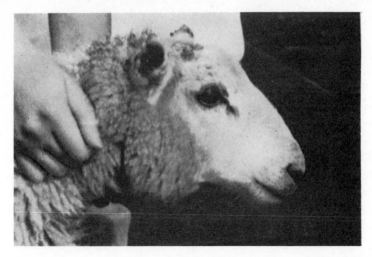

*Plate 17.2* Necrosis and sloughing of the ears associated with photosensitisation. Yellowses or head grit is one form of this condition.

The symptoms of photosensitisation seen mainly in lambs consist of swelling of the soft white parts of the head and ears, the latter hanging downwards with the weight of the fluid. The eyelids, face, and lips also become swollen. Later, exuded fluid coagulates and forms crusts or scabs which dry up and fall off leaving fresh normal skin beneath. The irritation set up by the oedema of the skin causes the animal to rub its head against fences and the ground. Finally, the ears may become crumpled and withered. Similar lesions may occur in other white areas of the skin unprotected by wool. In some instances the skin and lining membranes of the eyes and mouth become yellow due to jaundice. Death is frequent only when the trouble occurs among lambs on free range where they often get lost from their mothers and lie in the bracken or in drains. In other cases recovery is usual, although affected animals may prove uneconomic owing to a considerable loss in physical condition. (Plate 17.2.)

### Bone diseases (osteodystrophy)

Osteodystrophy is particularly prevalent in young growing sheep from 5–12 months of age due to the fact that their immature skeletons are particularly susceptible to dietetic disturbances such as partial starvation, simple or conditioned deficiency of one or more of the nutrients required for bone formation, or an imbalance of these substances. It has also been established that intestinal worms, even when present in comparatively small numbers, are an important cause of poor calcium and phosphorus retention, which seriously interferes with skeletal mineral deposition. In young sheep these factors cause retarded or arrested growth and also make the animals more susceptible to intercurrent disease. Although a number of conditions exist only three will be defined: (1) double scalp; (2) rickets; and (3) open-mouth.

(1) *Double scalp (cappie; scappie)*, at one time common but much less so now, occurs in autumn and winter on both hill and lowland pasture in many parts of Britain (and elsewhere), but used to be particularly prevalent on the poor hill pastures of Scotland. The clinical features are poor condition and ill-thriving with no scour unless heavy worm burdens are also present. Characteristically the frontal bones of the skull, that is those outer plates of bone which overlie the frontal sinus, become very thin and flexible and yield easily to pressure with the thumb. In this way it is possible to feel the inner layer of bone, hence the name 'double scalp'. This lesion is only a local manifestation of a generalised thinning of all the bones of the body although strangely, bone fractures are not particularly common in affected lambs. (Plate 17.3.)

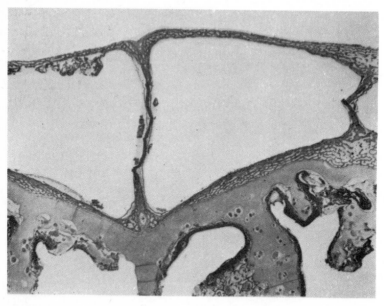

*Plate 17.3* Cross section of the skull of a young sheep suffering from double scalp (cappie). Note the very thin sheet of bone over the sinus.

Post-mortem signs are those of emaciation and marked reduction in the thickness of the bone walls, especially the frontal bones which may be paper thin.

The cause of cappie has not been defined as a simple mineral deficiency but the disease is the result of an overall deficiency of mineral, protein and energy foods. Treatment and prevention therefore depend upon improvement of pasture and feed and affected animals begin to recover as soon as summer grazing becomes available. Supplements of minerals and trace elements are of little value, but anthelmintics are of help if intestinal worms are numerous.

(2) *Rickets (bent-leg)* is essentially a disease of young animals and the severity of clinical signs increases with the rate of growth of the animal. For this reason rickets occurs more commonly in very well-fed animals, for example young rams which are being brought out for sale, and when climatic conditions produce a plentiful supply of herbage in the spring on which young sheep are put to graze. (Plate 17.4.)

*Plate 17.4* Sheep suffering from rickets.

The clinical manifestations vary from unthriftiness to lameness, but rarely include total inability to walk. The legs are bent at the knee, hock, or fetlock, or the leg bones themselves may be bent and this causes the feet to be splayed outwards. Sometimes the joints are very swollen and painful.

Apart from rapid growth, the factor concerned in causing rickets is a deficiency of phosphorus possibly coupled with a vitamin D deficiency, the latter caused by lack of sunshine during winter months.

Prevention and treatment consists of administering vitamin D along with adequate calcium and phosphorus supplements during the period of rapid growth.

(3) *Open-mouth* of young sheep has been reported in Britain and New Zealand and probably occurs elsewhere. The most obvious clinical sign, as the name suggests, is an inability in affected sheep to close their mouths so that the incisor teeth will not reach the dental pad. Other less obvious features are those described for rickets and for double scalp, a thinning of the frontal bones being very obvious. As grazing and eating are difficult or impossible the sheep are often in poor condition or emaciated.

Post-mortem examination will show that the lower jawbone is spongy and easily bent and for this reason the mouth will not shut. Other bones of the body are also thinner and lighter than normal.

Open-mouth appears to be a combination of rickets and double scalp, the causation probably being a complex of both these diseases.

Treatment of badly affected animals is useless and prevention depends on those methods advised for double scalp and rickets.

## Trace element deficiencies

### Cobalt deficiency

Cobalt is a mineral found in metallic ores and it belongs to a group of elements termed by animal nutritionists 'trace elements', since only very small quantities seem to be necessary to keep an animal in optimum growth and health. It has for long been known that in various parts of the world including Britain, many fields and hill pastures are unable to support sheep, especially lambs and young growing sheep, in good bodily condition owing to the fact that the soil is deficient in even the very small quantities of cobalt needed.

Dietary cobalt is essential in sheep and cattle since when it is absent or deficient the microbes in the rumen are unable to manufacture vitamin $B_{12}$ which is necessary for ruminant digestion. Without vitamin $B_{12}$ or when it is deficient, sheep will lose their appetite and eventually die from starvation. (Plate 17.5.)

Since the severity of cobalt-deficiency disease is determined by the amount of cobalt taken in, factors which control the availability of soil cobalt to the growing plant will influence the amount of unthriftiness in the grazing animal. For example, the cobalt content of pasture decreases in the spring and summer and, hence, the disease is worst at this time of the year. In addition, it is known that the cobalt content of a pasture may vary from

*Plate 17.5* Cobalt deficiency – compare carcass of affected animal with those of two healthy sheep of the same age.

year to year; therefore in some years sheep are worse affected than in other years. Finally, considerable differences are found in different plant species so that different pastures alter the cobalt picture of any grazing both from time to time or from place to place.

The symptoms of cobalt deficiency are non-specific since they are common to a number of diseases. Loss of appetite, poor growth and eventual loss of condition are typical of the disease. Lambs instead of being bright and alert are dull and listless and become anaemic and most animals in the flock are similarly affected. Ewes have a low fertility rate and lambing is often prolonged.

Diagnosis is often difficult because of the non-specific features of the clinical signs. Blood levels of cobalt or vitamin $B_{12}$ are useful indicators of deficiency providing the levels are sufficiently low but often they are not so in field situations. Liver vitamin $B_{12}$ concentration is more sensitive and reliable but obviously samples are more difficult to obtain for analysis. Soil and pasture sampling is of little value unless large numbers of samples are taken and all found to be deficient. Without doubt the most sure and easy way to carry out a differential diagnosis is to observe the response or otherwise of affected sheep to cobalt or vitamin $B_{12}$ therapy.

Treatment depends upon giving cobalt or vitamin $B_{12}$; the latter can be given by injection but is expensive. Cobalt must, however, be given by mouth. In both cases a number of doses have to be given over a prolonged period of time.

Prevention of cobalt deficiency is dependent upon giving cobalt continuously to grazing lambs. Pasture dressing with cobalt-rich fertiliser is one way to do this but often this is impossible on hill ground even though it only needs to be done once in every 4 years. The present-day price of cobalt also makes this an impractical method even when the treatment is confined to small areas for special grazing. The feeding of mineral mixtures supplemented with cobalt is a satisfactory method of prevention on many hill farms where the shepherding is good. Cobalt bullets made of glass, etc. given by mouth appear to be reasonably efficient but again is a rather expensive method. The bullet remains in the rumen for many months constantly supplying small amounts of cobalt to the rumenal microbes. Individual dosing of lambs with cobalt salts often contained in anthelmintic drenches is probably the practice most commonly used but the infrequency of dosing makes such methods of little use, weekly dosing is necessary for the method to be of any value.

### Muscular dystrophy (white muscle disease)

White muscle disease, or degeneration of skeletal and heart muscle, occurs in both cattle and sheep and is caused by a deficiency of vitamin E in the diet. In some instances a deficiency of the mineral selenium may also be involved in white muscle disease since it is a selenium-responsive disease. The ingestion of polyunsaturated fatty acids accentuate this clinical condition.

Muscular dystrophy occurs in lambs of all ages up to about 9–12 months of age. There is also a form of the disease in new-born lambs which are either born dead or die within the first few hours.

As one would expect, this deficiency occurs in lambs whose mothers have had a poor-quality diet but it is accentuated in those areas of the world where the soil and pasture are deficient in selenium.

Affected animals walk with a stiff movement, tremble and have great difficulty in regaining their feet when down. Very badly affected lambs have bodies which 'hang down' between the shoulders. If the heart muscle is affected the pulse is fast and often irregular and the respiratory movement is quickened and this can cause confusion in diagnosis since pneumonia may be mistakenly suspected. On many farms where these clinical signs are never seen there nevertheless can be a growth response when the lambs are dosed with selenium.

Confirmation of clinical diagnosis is by demonstration of affected muscle or parts of muscles including the heart at post-mortem examination. As the name of the disease suggests the muscles are pale or even white in colour.

Both treatment and prevention of muscular dystrophy is by giving either vitamin E or selenium or both, by oral drenches, injections or supplements to the diet.

## Poisons

### Kale poisoning

Kale poisoning occurs in fattening lambs grazing on kale for considerable periods of time. Its incidence varies from year to year and even during different stages of growth of the kale in any one year. The active principles in the kale which cause the condition are complex but are well defined by work done at the Rowett Research Institute. Less commonly the disease may occur when lambs are grazing rape or yellow turnips.

The sheep are usually in good condition. The worst affected hang their heads in a dejected manner and stand with arched back and sunken eyes, while respiration and pulse rates are very fast. The visible membranes of the eyes, mouth, and nose are dark yellow-brown in colour and when the wool is parted it is seen that the skin is yellow. There is considerable diarrhoea and the urine is dark-brown. The temperature is raised to 104–105°F. Death occurs rapidly within a day or two after a short period of coma. Abortion may occur in the pregnant ewe grazing kale.

On post-mortem examination the most striking feature is the deep yellow pigmentation of all the tissues and the dirty brown coloration of the blood. The liver is easily broken and is a bright yellow in colour – although it may be yellow-brown – and the surface looks rather like a nutmeg. Both kidneys are embedded in very yellow fat and are black in colour, but the tissue is firm and of normal consistence, unlike the kidney of enterotoxaemia. The bladder is generally very extended and filled with clear, dark-brown urine.

Treatment is by removal of the sheep from the crop and by feeding them carefully on good food.

### Other poisons

These are not so common. Rhododendron leaves can cause a high death rate when eaten but it has only infrequently been recorded. All species of yew trees are very poisonous and when eaten by sheep cause sudden death.

The illness of the sheep suffering from kale poisoning is due to severe anaemia caused by poisons in the kale, etc. which cause a very severe breakdown in the blood cells. Damage to the liver follows.

# 18 Diseases of Pregnant and Lambed Ewes

## Metabolic diseases

*Pregnancy toxaemia* (twin lamb disease or snow-blindness)

Pregnancy toxaemia occurs late in pregnancy, generally in ewes which are carrying more than one lamb, although in hill sheep it is not infrequent in ewes with only a single lamb. It is very common in all sheep-rearing areas of Britain, and is, indeed, of worldwide distribution. However, with the present-day improvement in the management and feeding of sheep, and particularly of pregnant sheep, the incidence of pregnancy toxaemia is very much lower than it was 30 or so years ago.

The disease occurs under a variety of conditions, so that there is considerable doubt as to whether or not the clinical symptoms are common to several diseases.

All of the predisposing causes of pregnancy toxaemia, with one exception, are associated with some form of nutritional deficiency. The exception is the form of the disease associated with overfat ewes and obviously in this case there is adequate food available. Ewes of this kind stop feeding and this is particularly so during the last 6 weeks of pregnancy. If the quality of the food offered is poor, loss of weight before the onset of symptoms is usual. Snow storms often predispose to the disease.

During late pregnancy there is a very considerable increase in the size and weight of the major foetal tissues and organs and this process is a very expensive one in terms of nutrient intake since the efficiency of conversion is low and this is particularly so for energy. This means that by the end of pregnancy a ewe carrying a single foetus requires about twice as much food as a non-pregnant ewe, while a ewe carrying twins requires almost three times as much. As pointed out earlier in the book it is not economic to attempt to meet in full these high-energy inputs but it is essential to go part way to meet them if pregnancy toxaemia is to be avoided.

The nature of the disease process is very complex and even now not too well understood but the clinical symptoms are primarily due to a profound change in the carbohydrate metabolism of the affected animal. Associated with this are low levels of glucose in the blood and in the brain and an increase in the breakdown of body fat with excessive amounts of ketone bodies and free fatty acids in the blood. There is also an upset in various hormone and enzymic systems.

The earliest symptoms of the disease is that one or more ewes, heavy in-lamb, are found standing apart from the rest of the flock. Such ewes appear to be completely lost to their surroundings and make no effort to move even when approached, but if forced to do so they walk with a staggering and swaying gait in an aimless way as though blind. The ewe often trembles and this is especially noticeable around the head and neck. The appetite is completely lost while constipation is a feature of the disease. In the course of a day or two the ewe collapses, lies quietly, often with the head round to the flank or stretched out in front until coma and death follow in a matter of a few hours or a day or so. At no time is the temperature raised but there is generally a strong sweetish smell of acetone from the breath. The exceptional case may survive for a number of days. In some outbreaks of pregnancy toxaemia abortion of dead, although well-developed, foetuses occurs in both affected and apparently healthy animals. When this occurs in ewes showing symptoms of the disease recovery often takes place.

The post-mortem findings are negative, no diagnostic lesions being present. The liver is often, but not always, yellow in colour due to fat infiltration but this is not unusual even in normal pregnant animals. The carcass is usually in poor condition.

Diagnosis is dependent upon recognising and differentiating the various clinical signs. The recognition of the causal factors operating on any particular farm goes far towards determining the nature of the trouble.

In the differential diagnosis of pregnancy toxaemia it must be remembered that hypocalcaemia (milk fever), hypomagnesaemia, gid, cerebrocortical necrosis and louping-ill can be readily confused on clinical signs. Laboratory examination of a blood sample can be helpful in confirming a diagnosis when glucose would be low and ketone bodies high.

Treatment of established cases of the disease is often difficult and unsatisfactory. As recovery usually occurs if the lamb and placenta are expelled a number of cases of the disease occurring very late in pregnancy recover spontaneously after lambing has taken place. Abortion, if it can be induced, also leads to recovery but the most satisfactory way to remove the foetus is by caesarian section and a number of veterinary surgeons are successfully carrying out this type of surgical treatment. This should, however, only be attempted in the earlier stages of the disease or when all else has failed. In the pregnancy toxaemia of overfat ewes forced exercise has a beneficial effect in early cases and also acts as a temporary preventative measure until nutritional changes have been made. Rational treatment of the disease is dependent on replacement of the deficient glucose. Some success has been obtained by giving repeated injections of glucose or glycerine or propylene glycol but inducing abortion or carrying out caesarian birth is much more satisfactory.

Since pregnancy toxaemia is rarely seen on well-managed farms, with good shepherding prevention is clearly a management problem. In this section of the book it is not necessary to try to make general recommendations regarding the nutrition of the pregnant ewe since this varies considerably from farm to farm depending upon the environment, the body condition of the ewes and whether the farming system is intensive with high lamb production or extensive as in the hill situation. Reference should be made to the previous sections which deal with the nutrition of the pregnant ewe.

*Hypocalcaemia* (lambing sickness, milk fever)

Hypocalcaemia occurs in late pregnancy or during the first days of lactation, hence the name 'lambing sickness', which is often used to describe the condition. (This name is rather confusing since the same name is used for different diseases in different countries and even in different areas of Britain.) It does occasionally occur at other times of stress. A number of ewes are often affected at the same time. The disease, mistakenly called a fever, is always associated with a lowered quantity of calcium in the bloodstream to which the symptoms may be ascribed. Unfortunately the reason for the lowered blood calcium is not understood. The disease in sheep is thought to be very similar in its nature to milk fever in the cow and if this is so a number of observations as to its possible cause can be made. For instance, it is considered that the blood changes that occur in milk fever are probably an exaggeration of the changes which occur at parturition in perfectly healthy sheep and cattle, and that the lowered blood calcium which occurs at this time is due to a large extent to the drain of calcium into the colostrum.

It has been observed in some instances that the fall in blood calcium which occurs in normal, healthy animals at parturition or just before may be as great as that recorded in cases of milk fever but, by and large the average fall in blood calcium in milk fever is greater than the maximum decrease in normal animals. Since milk fever is associated with excretion of calcium into the colostrum it could be suggested that milk fever occurs as a result of excess secretion of colostrum or of colostrum containing excess calcium, but there is no evidence that this is the case.

Further, milk fever commonly occurs before lambing and therefore before lactation begins. It would appear that animals which become affected are probably less able to mobilise their calcium reserves at times of stress or they have less reserves to mobilise. For instance, it has long been considered that milk fever is due to some deficiency in the parathyroid glands which secrete a hormone which controls the level of calcium in the blood. The question of a deficiency in calcium intake has also been a subject of considerable discussion. Both these aspects of the disease lack experimental proof and indeed attempts to treat milk fever cases with parathyroid extracts have not been encouraging. On the other hand, it is difficult to correlate a calcium deficiency with the rapid recovery of treated animals or indeed with the number of untreated cases which recover spontaneously. It would seem, therefore, that in milk fever the basic cause of the disease is an inability of the individual animal to mobilise calcium sufficiently quickly from its bone reserves at periods of calcium stress.

Post-mortem examination of a sheep dead from milk fever shows no characteristic lesions so that the condition cannot be diagnosed in this way. The only method of accurate diagnosis is by the laboratory examination of a blood sample from a sick sheep but for obvious reasons curative therapy of the individual animal must be carried out on the strength of clinical diagnosis.

Sheep seen in the early stages of the disease walk with a staggering gait and are weak on their hind legs. Breathing is often accelerated, probably due to the restlessness and excitability of the animal, which often shows considerable trembling. The animal quickly collapses, becomes comatose and dies within 24 hours. Not infrequently the course of the disease is more rapid and affected animals are often found dead, any previous indication of illness not having been noticed.

No method of preventing the disease is known although it is considered that over-exertion and fatigue may be predisposing causes and should be avoided near lambing time.

The treatment by subcutaneous or intravenous injection of calcium salts gives spectacular results and is the only method of treatment.

*Hypomagnesemia* (grass staggers grass tetany)

Hypomagnesemia during recent years appears to have become more prevalent and is now a condition which must always be considered when a sudden death is being investigated. It occurs in both hill and low ground sheep generally in ewes a month or so after lambing and it is especially common in cast hill ewes which have been brought down to low ground and crossed with a halfbred or Down ram. The disease also occurs in non-pregnant adult sheep at other periods of the year.

The term 'hypomagnesemia' means that the sheep has a low content of magnesium in the blood which is the characteristic feature, and it seems that sudden death occurs because there is insufficient magnesium in the blood to keep vital centres functioning. The reason for the drop in blood magnesium is not very clear but there are several possibilities. The first is that the pasture grazed by the sheep contains insufficient magnesium to meet the requirements of the animal but it is difficult to correlate this with the fact that the pasture on which the disease occurs often has a magnesium content as high or even higher than pastures on which it does not occur. Recently, the suggestion has been made that in the spring of the year when it is known that the magnesium content of the herbage is at its lowest, the amount present in the food may be insufficient for the immediate needs of the ewe heavy in lamb or in milk, while the ewe is unable to call upon the magnesium reserves of her body which are stored mainly in the bones. Secondly, it has been suggested that there may be some factor in grass, especially in quickly-growing grass, which upsets the absorption of magnesium from the gut, so giving a low content of magnesium in the blood. What the nature of this factor might be is unknown but the disease does appear to be more frequent on improved grassland where plentiful use has been made of nitrogenous and potash fertilisers, so that it is possible that the hypomagnesemic factor may be linked with a high protein content of the grass along with a derangement of its normal mineral ratios. Thirdly, the fact that the disease occurs commonly in cast hill ewes carrying or feeding heavy types of lambs might suggest that the pasture plays only a small part in the disease and that the low blood magnesium is due to some incapacity of the animal to deal with this situation. In cattle it is well known that the disease is often associated with a sudden spell of wet, cold weather and this may also be a precipitating cause of the disease in sheep.

The mortality rate in an affected flock of sheep may be quite high. As already stated, the usual sign is that of sudden death without the animal having been seen ailing in any way. In some instances initial symptoms of nervousness, restlessness, twitching of muscles, and grinding of the teeth may be seen which give warning of the approaching crisis, while in other cases the animal may become comatose before dying.

Diagnosis of the disease, unlike in most other conditions in the sheep, is not possible by post-mortem examination since there are no characteristic lesions or laboratory tests. Examination of a dead animal may be of some value in that the other possible causes of sudden or quick death may be eliminated leaving only the possibility of hypomagnesemia. Further help in diagnosis consists of the laboratory examination of a number of blood samples from other sheep in the affected flock when it may be possible to show that other animals, although not ill, have a low blood content of magnesium and are therefore potential cases of the disease. Such examinations are often of considerable help in establishing a true diagnosis.

Despite some obscurity as to the real cause of the disease a high measure of success has been achieved in prevention. The cure of established cases by the injection of magnesium and calcium salts is not so satisfactory, relapses being frequent. Prevention lies in ensuring that the animal has a plentiful supply of magnesium in its diet, sufficient either to overcome any potential deficiency or to swamp the action of any interfering factors. This can be done in several ways. The magnesium may be given as a magnesium-rich mineral mixture (the magnesium content must be as high as 15–20 per cent magnesium) or it may be given as a magnesium-rich food supplement as an enriched sheep nut or protein cube. Top-dressing pastures with magnesium limestone or calcined magnesite or kieserite, 110 kg to the hectare, also significantly raises the blood magnesium of animals grazing such pastures and forms a useful method of controlling the disease on arable farms and on certain hill grazings. It is especially worth while on farms where the disease occurs in both cattle and sheep. There is a considerable time-lag between the application of the top-dressing and the appearance of the increased magnesium content in the pasture and subsequently in the blood of animals grazing such land.

Although lambing sickness and grass staggers have been described as specific entities it must be recognised that a number of deaths may be caused by less clearcut metabolic upsets. Some sheep can be deficient in both magnesium and calcium whereas in others such minerals as sodium, potassium, and phosphorus may be too plentiful or too sparse. In consequence, treatment is often unsatisfactory and the good flockmaster tends to rely on a change of pasture as a more satisfactory method of prevention.

*Kale poisoning*

Kale poisoning, as already described, occurs in fattening lambs but ewes which are heavy in lamb are also very susceptible.

**Microbial diseases**

*Metritis* (inflammation of the womb, post-parturient gas gangrene)

Metritis is one of a number of infections of sheep all of which are of a similar nature. It is often a serious consequence of folding ewes in a confined lambing area. The other 'gas gangrene' conditions are described later.

Symptoms of inflammation of the womb never occur in ewes until the lambing season is well advanced. They arise first in one or two ewes which have lambed with difficulty but later appear in ewes lambing normally so that it would appear that the causal microbe increases in virulence as the disease progresses in the flock. Inflammation of the womb usually appears a day or two after lambing, although in severe outbreaks symptoms or death may occur actually during the lambing process. The affected ewe may collapse and die quickly or it may show signs of severe toxaemia, collapse, rapid breathing and a terminal scour for 24 hours or so before death takes place. In these cases there may be severe swelling and darkening of the tissues within and around the vaginal passage and vulva with oozing of a blood-tinged fluid.

Death is caused by infection of tissues with *Clostridium perfringens*, *Cl. septique* or similar organisms (*Cl. chauvoei*, *Cl. oedematiens*) all of which kill quickly due to their powerful toxins. These organisms are very resistant to adverse conditions and so lie latent for long periods of time either in the soil or in the tissue of healthy animals, becoming dangerous only when conditions are favourable for their multiplication. Some of them are normal inhabitants of either soil or healthy sheep and only occasionally are disease-producing. Others, especially *Cl. chauvoei*, have a more sporadic distribution so that infection is erratic in its occurrence and may be associated only with certain fields on particular farms. In short, one or more of these organisms are always present in and around most sheep.

Since, in each case, death is brought about by the toxins of the multiplying bacteria entering the bloodstream, generalised signs of toxaemia are present. There is often a bloodstained froth at the nostrils and if the animal has been dead for a few hours the carcass will be very bloated and swollen due to gas formation. The internal organs will show very little abnormality apart from numerous blood splashes on the heart muscle, a lesion characteristic of all toxaemias. On cutting into the local lesion the tissues will be dark red in colour and infiltrated with a bloodstained fluid and gas.

Post-mortem diagnosis is not difficult providing the dead animal is examined quickly but the rapid onset of post-mortem degenerative changes quickly confuse the picture. Although diagnosis is relatively easy, if specific preventive measures are to be carried out, it is obviously essential to determine which of the several possible causal bacteria is involved and this can only be done by laboratory examination of a newly dead animal.

Treatment with antibiotics can be effective provided the sheep are seen in the early stages of their illness. Serum hyperimmunised against the type of organism involved in the particular outbreak can be used both for treatment and to 'cut short' an outbreak.

General preventive measures are those of good husbandry, such as well chosen lambing fields and disinfected sheds, along with personal cleanliness of the shepherd. Routine preventive vaccination is by far the best method especially now that combined clostridial vaccines are readily available since these give protection against most of the clostridial microbes of the sheep.

## *Mastitis* (udder clap)

Mastitis may be caused by a number of organisms, but the common gangrenous or suppurative form of the disease is caused by the microbes *Staphyloccus aureus*, *Corynebacterium pyogenes* and *Pasteurella haemolytica*, a less spectacular form of the disease is due to various types of streptococci.

The disease occurs frequently and on many farms it may cause considerable economic loss, many ewes being left with only half an udder or no udder at all while sucking lambs are under-nourished and some die from starvation. It is known that the organisms which cause mastitis are common inhabitants of the skin of normal sheep so that predisposing causes must be as important as the microbes in the production of the disease. Slight injuries to the teat and the occurrence of cold, stormy weather may be causal factors but the sudden weaning of lambs from mothers rich in milk may be a more common cause. An outbreak of orf among suckled lambs and ewes also predisposes to mastitis.

In the acute disease, generally seen soon after lambing, symptoms develop very rapidly and the udder or half-udder quickly becomes swollen and intensely inflamed. Affected ewes have lameness or stiffness in the hind legs while the suckling lambs are often weak and hungry. This is followed by the affected parts of the udder becoming dark-red or black and when the teat is drawn a bloodstained, clotted fluid comes away. Finally, the udder becomes cold and may slough away if death of the animal does not take place first.

The less acute forms of the disease are not so readily recognised but are very common and their occurrence may be realised only at culling time when a number of udders or half-udders are found to be hard and swollen or small and fibrosed. If noticed earlier, such udders are enlarged and slightly inflamed and contain pus. (It should be noted that normal udders in the immediate post-weaning period tend to be swollen and contain clotted milk and such changes should not be confused with those of mastitis.) This form of mastitis is a very common reason for the discarding of sheep.

At a post-mortem examination the udder either shows the lesions association with intense inflammation, haemorrhage, and death of tissues or the udder may be tougher than usual and contain a number of variously-sized abscesses particularly so in the chronic form of the disease.

Diagnosis in both the living and dead animal depends on recognition of the abnormal appearance of the udder.

For treatment, antibiotic drugs given intravenously or into the muscle can save the life of the ewe but the disease must be diagnosed early for satisfactory results. Treatment of well-established cases is of very little value. Surgical removal of the affected half of an udder may be tried when the animal is a valuable one.

There are no specific vaccines in Britain but the infusion of antibiotics into the udder at weaning time appears to be worth trying although expensive and time-consuming.

*Contagious pustular dermatitis* (CPD orf)

Contagious pustular dermatitis has previously been described as a disease of older lambs, but it can also occur on the lips of suckling lambs and on the udders of suckling ewes which quickly become infected with secondary bacterial infections. Severe outbreaks of mastitis follow and the losses of both ewes and lambs can be considerable. (Plate 18.1.)

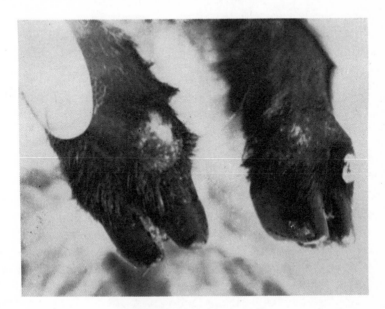

*Plate 18.1* Orf affecting the foot of a lamb. This condition is caused by the virus of contagious pustular dermatitis.

Treatment with both systemic and local antibiotics to control the secondary bacterial infections is often essential.

Since antibodies to the orf virus do not appear to pass in the colostrum, routine vaccination of ewes is only partially helpful and as outbreaks of this type tend to be sporadic, preventive vaccination on clean farms is not indicated since the vaccine contains living orf virus.

## Staphylococcal dermatitis (Periorbital eczema)

Staphylococcal dermatitis has occurred in sheep for many years but the cause has only recently been defined. It occurs mainly in adult sheep and particularly during the lambing period when ewes are being trough-fed. The disease spreads rapidly from ewe to ewe.

The cause of the disease is a virulent staphycloccus aureus which is sensitive to various antibiotics.

The main lesions are suppurative ulcers covered in black scabs which occur on the skin of the head, particularly over the nose and around the upper jaw and around the eyes and ears. The lesions bleed readily and particularly if the scabs are removed. Affected sheep can lose their eyesight for a time and the ears may become necrotic, drooped and hanging. Raised temperature and fever are not usual. Healing of the lesions takes a number of weeks during which time the skin becomes hairless.

Diagnosis is relatively easy based on the obvious lesions but the causal organism can be grown in the laboratory if confusion arises between this condition and orf, etc.

The application of antibiotic lotions hastens healing but is expensive and hardly necessary except in very severe cases. Prevention is mainly by allowing plenty of trough space when feeding ewes.

## Parasitic diseases

### Haemonchosis

Haemonchosis of the fourth stomach of sheep causes very considerable losses in sheep kept in tropical climates but in Britain it causes only sporadic disease, and this mainly in ewes in the immediate post-lambing period when this is in the early spring months. However, when haemonchosis does occur it is both spectacular and economically serious. It can occur in ewes that are at pasture or that have overwintered indoors.

Haemonchus contortus is the largest of the stomach worms and is about 25 mm long and easily visible. The body of the female worm is spirally striped in red and white and is known as the 'barber's pole worm'. *H. contortus* lives on blood which it gets by lacerating the lining of the stomach. This causes heavy bleeding, as much as a pint a day in badly affected sheep which become very anaemic, the main symptom of this disease for Haemonchus does not cause scouring.

When infestation is heavy the death rate may be high even in sheep in good or fair condition, the time from the start of the bleeding to death can be as short as a week. Anaemia will first show in the inability of some animals to keep up with the rest of the flock when running. Later, these sheep become sluggish and hang their heads. Severe anaemia causes very obvious paleness of the membranes of the eyes and mouth and whiteness of the skin which is particularly obvious around the anus and vulva. There is often considerable dropsical swelling below the jaw. The post-mortem findings are those associated with anaemia. Severe inflammation of the fourth stomach is also often present, the lining of the stomach being ulcerated, splashed with blood, and covered with a film of thick mucus. In the newly dead animal the characteristic worms are readily seen moving and squirming about when the fourth stomach is opened, but, because they are large and conspicuous they may be seen even when the numbers present are too small to cause disease. Diagnosis depends on the recognition of the anaemia coupled with the demonstration of large numbers of eggs in the droppings or by observing and then counting the worms in the fourth stomach of a dead sheep.

The reason for the occurrence of the disease in the early spring is that the *H. contortus* larvae ingested during the latter part of the previous grazing season do not immediately develop to the adult stage in the fourth stomach but are totally inhibited in the lining membrane of the stomach. As previously mentioned when describing parasitism in older lambs, the inhibited larvae begin to mature in large numbers in the perinatal spring period for reasons not well understood, and the resultant adult population causes the symptoms to occur. Strangely, although large numbers of worm eggs pass on to the spring pasture these rarely cause any trouble in the lamb crop, at least this is so in Britain. In tropical countries, acute haemonchosis in older lambs is very common and very serious.

Treatments with modern anthelmintics are very good and, if given before lambing, readily prevent the disease occurring. Clean pasture grazing in the autumn also effectively prevents any serious problem.

## Ventral hernia and vaginal prolapse

These are always associated with pregnancy in the ewe when there is a considerable increase in the weight of abdominal contents which may cause damage to the wall of the abdomen and in the pelvis. This leads to herniation of the muscles of the abdomen in the lower flank so that the abdomen sinks to the ground causing difficulty in walking. Prolapse of the vagina is seen at the end of pregnancy and this may involve the bladder so that the passage of urine is prevented and the bladder distends. The vagina may be ruptured and the intestines may be extruded to the ground.

The cause may be overfeeding or it may be due to overcrowding of the ewes and hence mechanical damage to the ewes.

Treatment of ventral hernia is difficult and slaughter is indicated. Treatment of vaginal prolapse is local in that the extruded tissues should be replaced using antiseptic washes and the bladder should be emptied by manual manipulation. Closure of the vulva by suturing is not satisfactory but the vulva can be closed by tying string across it using the surrounding wool as sites for attachment of the string. Caesarian section is useful towards the end of pregnancy. After lambing, suture of the vulva or deep into the vagina if straining of the ewe continues is helpful.

Prevention is by good husbandry including good shepherding and non-overcrowding of ewes and by the feeding of more bulky diets at the end of pregnancy.

# 19 Diseases of Sheep of Most Ages

### Liver fluke (fascioliosis, liver rot)

Liver fluke is a parasitic infestation of sheep and is one of the best-known of all internal parasites, *Fasciola hepatica*. It has a fascinating life-cycle involving the mud snail *Lymnaea truncatula* and in areas where sheep and snail coincide liver-fluke infestations will occur usually in the autumn and winter months. Depending upon the weight of infection, affected sheep may show weight loss, anaemia or even sudden death.

The life-cycle begins with the adult fluke which is flat, leaf-shaped and between 20–50 mm in size. It lives in the bile ducts of the liver where it lays eggs which eventually pass out in the sheep's droppings. These eggs are very resistant to unfavourable conditions and await moisture and warmth before hatching into a form which is termed a 'miracidium' which can move around in water. These miracidia do not survive for more than a few hours unless they are able to penetrate the water snail. In the snail the miracidia undergo further developments and eventually each miracidium multiplies into hundreds of a further stage called metacercariae which are shed from the snail and these again move about in water. These mobile forms quickly attach to firm surfaces such as blades of grass and then become encysted and infective to sheep in which the cyst becomes a small liver fluke. It can take anything from 6 weeks to 3 months or so for the development from egg to cyst. In the infected sheep the young fluke passes through the intestinal wall, over the peritoneum and through the liver capsule to the liver tissue and finally completes its journey to the bile ducts. This journey from gut to bile duct takes about 12 weeks when the cycle begins again by the laying of eggs. Each complete cycle takes at least 18 weeks but it can be much longer. The mud snail, which is an essential part of the life-cycle of the liver fluke, although requiring moisture for normal activity, can resist both freezing and drought; in addition the snail is hermaphrodite and in one breeding season can produce at least 100 000 offspring. Small wonder that the liver fluke is difficult to control. (Plate 19.1.)

307

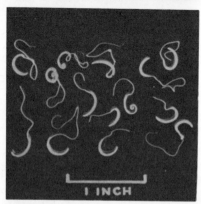

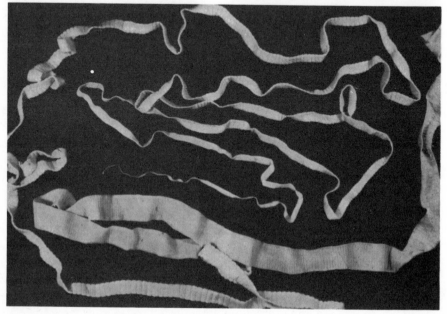

*Plate 19.1* Internal sheep parasites. (Photographs courtesy of Cooper, McDougall and Robertson Ltd)
*Left:* Liver fluke
*Centre left:* Round worm, large intestine
*Centre right:* Twisted wire worm, fourth stomach
*Bottom:* Tape worm, small intestine

Although the epidemiology of the liver fluke in the sheep is dependent upon suitable ground, for the mud snail – which prefers mud to free water – the presence of a moist environment and a temperate temperature allows at least two annual cycles of fluke to occur regularly in the wetter areas of Britain. The first cycle is, however, the more important and is the cause of the heavy fluke infestations which occur in late August and onwards. Surviving metacercariae of this cycle, which are not taken up by sheep, can survive the winter and infect sheep in the following spring.

Clinical fascioliasis in sheep may be acute, sub-acute or chronic. The acute form of the disease, often known as liver rot, is perhaps not so widely appreciated as the chronic form, but nevertheless it is of considerable economic importance and extremely serious to the individual flockmaster. It occurs mainly in late autumn and early winter. However, it can occur from November to March if sheep are put on to dangerous pastures in early winter. Affected sheep may die suddenly but usually they show signs of illness for several days before death occurs. The animals are often in very good condition. Clinical signs are not specific for the disease, the sheep merely hanging back from the flock and showing signs of abdominal pain. The death-rate in a flock may be very considerable.

The post-mortem findings are characteristic and diagnostic. There is an excess of bloodstained fluid in the abdominal cavity due to outpourings of blood from the damaged liver, which is enlarged, dark, and mottled in colour and the surface covered with haemorrhages. When handled, the liver is found to be soft and easily broken hence the term 'liver rot'. The immature flukes, being only about 3 mm in length, are not always easily found and adult flukes may be absent. The large number of immature flukes developing in the liver tissues are the result of the sheep ingesting large numbers of metacercaria over a short period from a very heavily infected grazing.

Sub-acute fascioliasis also occurs during the winter and is simply a slightly less acute form of the disease in which the liver tissue is not so heavily infested and a proportion of the flukes are adult and have reached the bile ducts. The metacercaria ingested, although very numerous, have been eaten over a longer period of time. Affected sheep lose condition rapidly, are very anaemic and a few develop the typical oedema below the jaw known as 'poky jaw'.

Post-mortem examination shows large numbers of flukes in very distended bile ducts. The liver is enlarged but the death of tissue is not so extensive as in 'liver rot'.

The chronic form of fluke infestation is by far the commonest form and causes considerable economic loss especially in the wetter areas of Britain and of other countries. Outbreaks are seen in the latter part of the winter and in the spring. The disease is characterised by a progressive loss of condition and development of severe anaemia resulting in paleness of the membranes of the eyes and mouth and the oedematosis 'poky jaw' typical of chronic infestations. The infection is picked up in the autumn and winter and is due both to first and second cycles of fluke development.

Post-mortem examination shows a very fibrosed liver with swollen bile ducts containing many flukes. The liver is misshapen and often shrunken and is very pale in colour. The carcass is emaciated, pale, and oedematous with fluid in the abdominal cavity.

Diagnosis of all forms of fluke disease is based upon a proper post-mortem examination and on clinical examination. Faecal egg counts are useful for the sub-acute and chronic forms.

Treatment of all forms of the disease is now very effective since the newer drugs are very efficient against young flukes in the liver tissue as well as against the adults in the bile ducts. One or two treatments are probably all that is necessary for the chronic disease but several doses at 3-week intervals are usually needed for the more acute forms. Dosed sheep should be moved to fluke-free pastures after dosing but if this is not possible dosing should continue until deaths cease if one is faced with the acute disease.

The control of liver fluke infestation is mainly achieved by routine dosing of sheep with appropriate anthelmintic drugs which reduces the number of fluke eggs dropping on to the ground. The Agricultural Departments in Britain each year issue forecasts, based on climatic conditions, on the probable severity and timing of fluke infestations of sheep and the frequency and timing of preventive dosing should be based on these forecasts. Recent work at the Hill Farming Research Organisation has shown that 'strategic dosing' can eliminate fluke eggs from sheep droppings so that snails may not be important. Liver fluke can also be controlled by reducing the mud snail populations but this is not readily achieved. Appropriate drainage of infested pastures is probably the best way to do this but if the areas involved are large this can be expensive. Theoretically snail populations can be destroyed by spraying with chemical sprays. Copper sulphate has been used but unfortunately this can be dangerous both to the grazing sheep and to fish in the draining rivers. Other less-toxic chemicals are presently being used under supervision, but are expensive. Advice should be sought as to the optimum time for spraying, as to the chemical to be used and for the areas to be sprayed. If the wet areas are limited in size they should be fenced to keep the sheep out.

Several surveys of the economic loss due to *F. hepatica* infestation of sheep in Britain have shown the losses to be enormous and in bad years millions of pounds have been lost due to loss of production while many thousands of sheep die. There seems to be little doubt that liver fluke remains the worst disease with which a flockmaster has to contend.

### Black disease (infectious necrotic hepatitis)

Black disease is an acute disease of sheep which is characterised by sudden death. The causal microbe is a member of the Clostridium group, its specific name being *Cl. oedematiens* type B. Death is brought about by the potent toxin produced by the organism in its active phase. As with the enterotoxaemia and braxy microbes, the present one is a common inhabitant of both soil and healthy sheep, being often present in the spore form. This means that it is inactive and very resistant to unfavourable conditions and so able to lie dormant for long periods of time awaiting suitable conditions for active multiplication. In Britain the microbe does not appear to be so ubiquitous as the enterotoxaemia organisms and many farms are free of the disease, but its true distribution is not well known. As in some of the diseases previously described the organism by itself appears to cause no harm to the sheep and becomes dangerous only under certain stimulating conditions, the most common one being the invasion by the liver fluke of sheep in good forward condition. When the young fluke burrows into the liver it causes damage to the tissue and this appears to produce a very suitable medium in which the organism multiplies rapidly giving rise to classical 'black disease'.

As would be expected, black disease occurs only during the period when the liver fluke is active, which is during the autumn, winter, and early spring months. It occurs in both hill and low-ground sheep grazing fluke-infested land. Adult sheep are mainly affected but the disease is occasionally seen in sheep under 2 years of age. The disease occurs more commonly in some years than in others depending on suitable climatic conditions favouring the development of the liver fluke.

There are practically no symptoms, death usually occurring during the night or early morning, the animal often being found dead lying on its chest with forelegs outstretched and with no sign of struggling around it.

Post-mortem examination must be carried out within a few hours of death if specific lesions are to be recognised since post-mortem decomposition occurs rapidly, obliterating all lesions. The name 'black disease' is derived from the rapid blackening of the inner surface of the skin due to these post-mortem changes. In newly dead animals there is an excess of fluid around the heart which may contain a gelatinous clot and there are blood splashes on the surface of the heart, findings similar to those seen in enterotoxaemia.

The liver is very congested and often shows haemorrhages on the surface while a characteristic darkening of the edges may be seen. The diagnostic feature of the disease is the occurrence of one or more small greyish-yellow areas in the liver. These are commonly seen from the surface extending into the substance and, when cut, are surrounded by a darker bloodstained rim. Contrary to what might be expected, only a few flukes may be present so that diligent search may have to be made, but the small immature flukes can usually be seen readily on examination with a microscope. More rarely, adult flukes may be plentiful and easily seen.

Diagnosis depends essentially on the demonstration of the typical lesions in the liver, while laboratory confirmation is by demonstrating the causal organism in such lesions. It must again by stressed that it is necessary to make a post-mortem examination quickly after death both for field and laboratory diagnosis. A sheep dead more than several hours is useless for such purposes.

Since true black disease occurs only in the presence of both the specific microbe and the liver fluke, prevention can depend on either the elimination or reduction in numbers of the fluke or by specific vaccination against *Cl. oedematiens*. When outbreaks of black disease occur on a farm the only possible method of overcoming the outbreak is by serum and vaccine therapy. The serum will give immediate protection against further deaths for a period of 10–20 days, but for continuing immunity vaccine must be given at the same time as the serum. Unfortunately, one dose of vaccine is seldom sufficient to control the disease so that it is usually necessary to vaccinate 2 or 3 times at 3-week intervals before complete control is achieved. In such cases it may be necessary to give further doses of serum in the interval.

Routine prevention of black disease depends upon annual vaccination of the flock with black disease vaccine which can be used alone or more commonly as part of a multivalent clostridial vaccine such as is commonly used for pulpy kidney disease, braxy, etc.


**Enterotoxaemia**

Enterotoxaemia is the same acute fatal disease which has already been described and which in lambs is called pulpy kidney disease and to which reference should be made. In adult sheep it is known as enterotoxaemia and, as in lambs, it occurs in thriving sheep or in sheep that are beginning to thrive after a period of poor growth. It is seen most commonly in sheep being fed catch crops, excessive amounts of concentrate or being too long left on the turnip-break. The period of greatest danger is always the few days after sheep are introduced to the particular feed although this is not always so.

In enterotoxaemia the post-mortem findings are usually very insignificant, consisting only of a slight increase in the fluid around the heart along with blood splashes on its walls. A slight congestion of the bowel may also be present. In all cases of sudden death in adult sheep enterotoxaemia should be suspected but it must not be forgotten that anthrax can occur in sheep although not very commonly.

Specific diagnosis can only be carried out in the laboratory and this is essential if it is intended to carry out preventive vaccination or serum inoculation. The laboratory will also confirm the absence of anthrax in such cases. Change to poorer feed is, however, probably as effective and often cheaper since the flockmaster is not generally aiming for rapid growth as he is with lambs.

Vaccination with a multivalent clostridial vaccine can be used as a prophylactic but it is probably more economical to use this on a flock basis for the control of the various clostridial diseases which affect sheep of different ages. Since such programmes vary from farm to farm veterinary advice should be sought.


**Pneumonia**

Pneumonia can, on occasions, occur in most ages of adult sheep although it probably causes most economic loss in cast hill ewes when they are brought down to the lower ground for crossing purposes. Pneumonia can also cause numerous deaths in older hill sheep in certain years. So far as is known the disease is similar to that seen in lambs and young sheep and reference should be made to the appropriate section of the book. The cause is *Pasteurella haemolytica* but environmental factors and food management predispose to the infection. (Plate 17.1.)

Jaagsiekte and maedi diseases also affect the lungs of adult sheep but since they cause chronic wasting are very different from pneumonia. These are described later.


**Louping-ill**

This is a disease principally of sheep, although cattle, pigs, and horses may also be affected, but less frequently. The infection has only been reported as occurring in Britain. It is caused by a virus which first multiplies in the blood of the sheep and from there in a number of instances it invades the brain cells. Multiplication of the virus in these cells brings about the characteristic nervous symptoms after which louping-ill is named. The virus is transmitted from sheep to sheep by the tick *Ixodes ricinus* and so far as is known the disease is never present other than on tick-infested pastures. The tick has a life-cycle which occupies 3 years passing through larval, nymphal, and adult stages. During the spring and early summer and again in the autumn each stage of the tick must feed on a sheep or other warm-blooded animal for a number of days before it can progress to the next phase of its cycle or before the adult female tick can lay its eggs. It is during these periods that the tick is infected with the virus of louping-ill and becomes a carrier of infection, transmitting the virus once more back to the sheep when it again feeds in its next stage. Ticks can therefore become infected during their larval, nymphal, or adult stages, but as the virus probably does not pass through the egg, only the nymphal and adult female forms can infect sheep. It is fortunate that a number of factors tend to limit the dissemination of louping-ill in our sheep flocks.

(1) The sheep tick is itself limited in its distribution, since favourable conditions for its survival and multiplication occur only in the moss and rough vegetation over peat or acid soils. Ticks occur in such areas of Britain as the Border hills, Highlands and Western Isles of Scotland, Wales, the Pennine range, Cornwall, Devon, and Ireland. Owing to their strict environmental demands, ticks which are frequently scattered on low ground and good marginal pasture when hill sheep are brought on to them, usually fail to become established, although they may persist for a short time.

(2) As ticks, to become infected, must take in the virus of louping-ill from sheep sometime during their 3-year cycle, large tracts of tick-infested country remain free of the disease since the ticks never come into contact with virus-carrying sheep.

(3) Owing to the farming practice in many hill areas, sheep stocks are acclimatised to the ground, meaning that with the exception of a few rams, no new sheep are brought into the area for generation after generation. The risk therefore of introducing to louping-ill-free areas sheep carrying the virus of louping-ill is minimal.

(4) In the areas where louping-ill is present the acclimatised stocks build up a considerable immunity to the disease.

Louping-ill occurs only in the spring and early summer and again in the autumn in those areas where there is tick activity at this period. After the bite of the infected tick the majority of infections go undetected but a minority of animals, 6–18 days later, become dull and have a high temperature due to the virus multiplying in the bloodstream. Frequently, sheep recover from this phase without showing any further symptoms but, if not, the virus attacks the brain cells, causing symptoms of excitability, trembling, muscular spasm, and irregularity of walk, followed by paralysis, coma, and death. The factors which determine whether infection of the brain occurs have not been clearly determined but it is generally thought that simultaneous multiplication of the agent of tick-borne fever in the blood plays a part. Environmental factors such as poor weather and poor nutrition are also important, as is the genetic make-up of the sheep. The course of the disease may vary from several days to several weeks.

The post-mortem findings are negative but characteristic lesions may be found in the brain by microscopic examination.

Diagnosis in the field depends on the clinical symptoms occurring in tick-infested sheep. Field diagnosis can be confirmed in the laboratory by demonstrating the virus and characteristic lesions in the brain of a dead sheep. Serological confirmation in living sheep can also be used.

Immunity to the disease is well developed in acclimatised sheep. This is because newly-born lambs acquire immunity to the disease when sucking their mother's colostrum or first milk which on louping-ill farms contains considerable antibody to the virus. This immunity, being due to immune bodies developed by the mother, is short-lived unless it is reinforced by active immunity in the lamb. As a rule such active immunity is developed since most of the lambs become tick-infested soon after birth and therefore quickly come into contact with the virus of louping-ill but, being protected against a fatal infection of virus by the colostral immunity, develop their own antibody. In this way immunity in an acclimatised flock remains constant and sufficient to prevent serious losses from the disease.

The incidence of louping-ill in any flock varies conversely with the immune state of the flock. Although the acclimatisation of a flock is often sufficient to prevent serious losses, in other instances the weight of virus infection may be sufficient to break down the immunity. On the other hand, the number of ticks and, hence, the weight of virus infection, may be insufficient to develop an overall acclimatisation immunity to the disease. It follows that serious outbreaks of louping-ill may occur either on very heavily tick-infested hills or on hills where infestation is light. The disease also causes considerable loss on farms where the virus of louping-ill has only recently been introduced to the tick population, usually by the purchase of virus-carrying sheep. Losses also arise on farms where infected ticks are moving in for the first time. In these last two instances the virus is active in a sheep population devoid of any acquired immunity. In addition, louping-ill may cause catastrophic loss when sheep from non-louping-ill farms (and these need not necessarily be tick-free) are taken to a hill infested with virus-infected ticks, while accidental introduction of non-immune sheep to infected ticks also occurs on marginal hill farms when the low-ground sheep are moved on to a tick-infested hill or to marginal fields. It can be readily understood that the management of sheep stocks on louping-ill-infected farms is by no means a simple matter and is full of pitfalls for the unwary. It would appear that the sheep-tick system alone has all the essential ingredients for the perpetuation of louping-ill virus but wild fauna which are potential hosts for the virus may on occasion be important. Virus has been demonstrated in voles, grouse, wood mice, and hedgehogs and it occurs also in red deer.

There is no successful treatment for louping-ill.

Control of louping-ill by vaccination is now much more effective since the formulation of a very good inactivated oily adjuvant vaccine, the louping-ill virus having been grown in tissue cultures in the laboratory. Sheep should be injected twice to bring about a high level of immunity which will protect for several years but recently introduced vaccines can be effective after one injection. Lambs born to ewes receiving vaccine in the last month of pregnancy acquire a protective immunity via the colostrum. To control the disease on farms where louping-ill is endemic all ewe lambs should be vaccinated in the following spring or in areas where the tick is active in the autumn in the following autumn. All purchased sheep should be vaccinated 28 days before going to tick-infested grazings or in flocks where the disease has newly appeared the whole flock should be vaccinated. Tick control which is described later is of little or no value for control of louping-ill. Such efforts may make the louping-ill problem worse since a reduction in tick numbers often lessens the overall flock immunity that occurs on acclimatised farms.

## Listeriosis (circling disease)

Listeriosis is an infection of the brain of adult sheep which is caused by the organism *Listeria monocytogenes* which is commonly in tissues and faeces of many species and usually does little harm. It can survive in soil, food and forage and faeces for many months.

A sheep suffering from circling disease, as the name suggests, walks in circles with uncoordinated steps. Other symptoms such as neck stiffness, drooping of eye or ear are generally confined to one side. The mouth becomes open with protruding tongue and saliva drools. The sheep eventually collapses and the legs begin to paddle. Death is usual in about 4 days to a week. Serious outbreaks of the disease occur at the end of winter. The predisposing cause which enhances the invasiveness of the causal organism is their rapid multiplication in silage being fed to sheep in winter.

Diagnosis is based on clinical signs but mainly by laboratory examination.

Treatment by antibiotics can cure the infection but the symptoms are permanent. No vaccines are available and control is by the careful feeding of silage.

### Johne's disease

This is a chronic wasting disease and, although it rarely assumes serious proportions in the sheep flock, it is nevertheless one of the commonest causes of pining among adult animals. The cause of the disease is a microbe called *Mycobacterium johnei*, an organism belonging to the same family as those bacilli which cause tuberculosis and leprosy. The sheep may be affected by several types of the organism but the commonest one is probably the same as that causing Johne's disease in cattle. It is very persistent under field conditions and can survive for a long time in a suitable environment.

The disease most commonly affects young adult animals but may be seen also in older sheep. Johne's disease is almost invariably introduced to a flock by subclinically·infected animals, either sheep or cattle, which are often brought on to a farm. Spread within a flock occurs by infective faeces contaminating pastures, water, or foodstuffs, and especially the milk of ewes with lambs at foot. Recently it has been shown that lambs may be infected before birth by the mother. After infection, which usually occurs in the young animal, the incubation period is very long so that symptoms do not occur until many months or even years after infection. All infected animals do not develop the disease and whether or not infected animals develop symptoms depends on one or more of several factors, such as pregnancy, poor nutrition, worm infestations, mineral deficiencies and the genetic make-up of the sheep. (Plate 19.2.)

The clinical symptoms are those of an ill-thriving animal but scouring is rarely present.

The post-mortem picture is very variable. Where the non-bovine coloured form of the microbe is involved, the most striking feature is the intense chrome-yellow appearance of the affected portion of the intestine. The colour is directly due to the presence of packed masses of the bacilli, yellow in colour, which cause the gut to be thickened. Greater variation in the lesions is seen where the bovine type of organism is involved, since little or

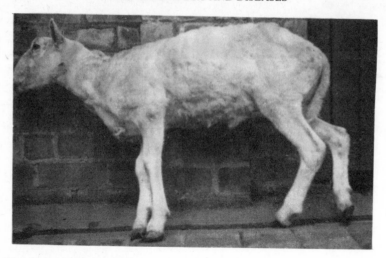

*Plate 19.2* This emaciated animal (piner) could be affected with stomach worms, Johne's disease, scrapie, liver fluke or undernutrition. Laboratory help is often required for differential diagnosis.

no thickening of the bowel may be seen. The lymph glands of the gut are invariably grossly enlarged and project obviously from the mesenteric fat.

Infection can be diagnosed satisfactorily only by post-mortem examination and by laboratory culture of the microbe. Diagnostic tests such as the johnin test and the blood test used commonly for clinical recognition in cattle are of little value in the sheep.

The condition can be best controlled by slaughtering affected sheep and their offspring in the case of ewes, as soon as symptoms are recognised. A vaccine similar to that used in cattle is well worth trying in flocks where serious losses occur but the vaccine does not eliminate the infection so that sheep can still be carriers. Good feeding of the flock and an adequate lime content of the soil are also important.

**Foot rot**

The term 'foot rot' appears to have caused some confusion of thought within recent years but it is considered that the name should be confined to the following specific condition which affects sheep of all ages, although lambs do not show such serious lesions as do older animals.

The disease is commonest and most severe in areas of heavy rainfall and lush pasture or on badly drained land, although normal weather conditions in all areas of Britain give a suitable environment for foot rot to develop. The importance of these predisposing factors is such that they are often actually regarded as the real cause of the trouble, whereas the symptoms are due to a specific microbe and the disease is contagious.

Foot rot can be readily transmitted to healthy sheep by inoculating them with material from the feet of affected animals and, following work on the causation of foot rot in Australian sheep, it is now known that the causal organism is *Bacteroides nodosus* associated with *Fusiformis necrophorus*. The former microbe will not multiply in the soil nor will it survive outside a sheep's foot for periods much over a week so that the real and only important source of infection is a sheep suffering from foot rot or by sheep carrying the organism but showing no lesions or symptoms.

The first sign of foot rot is the occurrence of lameness in a number of sheep, one or more limbs being affected. Close examination of the foot shows that the horn of the hoof is overgrown and under this horn and under the sole is a suppurative decay giving rise to an evil-smelling material which gradually spreads, separating the horn from its sensitive corium. If neglected the foot becomes grossly malformed and the horn wall may become partly detached. The pain may be so intense that the animal walks and grazes on its knees rather than use the affected feet. A more benign form of the disease occurs commonly in which only the skin between the claws and the skin-horn junction is affected.

Diagnosis is relatively easy since the characteristic lesions are not readily confused with other conditions of the feet, apart possibly from foot-and-mouth disease. Demonstration of the causal organism in the laboratory confirms the diagnosis.

Treatment of early individual cases of foot rot is straightforward, although the paring of the feet is not an easy matter since it requires considerable experience to know how much horn to remove without doing unnecessary harm to the underlying sensitive matrix. The use of a special hoof-clipper for this purpose is recommended. The cleaning of the hoof is followed by the application of 5 per cent formalin or a suitable antibiotic being painted on to the foot or by injection. The caustic chemicals so often favoured by shepherds should not be used since they cause considerable damage, inflict pain, retard healing, and do no better than formalin.

When foot rot affects large numbers of sheep in a flock a foot-bath containing 5 per cent formalin or 10 per cent copper sulphate may be used instead of individually dressing the feet which must, of course, be pared in both instances.

Foot rot may be eradicated from a flock or completely prevented from entering by adequate management. The method employed depends on two factors: firstly the isolation, segregation, and treatment of all infected animals, and secondly on the fact that the causal organism has a survival period away from the sheep of not more than 2 weeks. Healthy sheep are separated from infected animals by careful examination of the feet of all sheep. Those animals free from the disease are passed through a foot-bath and then taken to clean pasture (pasture which has not carried sheep for 14 days) after an hour's rest on a concrete floor. The infected sheep are treated as already described, made to stand for a time on concrete and then taken to another clean pasture. This procedure is repeated in a few days' time and the cured sheep are then allowed back to the clean flock. It may be necessary to treat a number of times in this way before all sheep are clean; frequently a few resist all efforts at cure and these animals should be culled. The success of the method depends on complete segregation of all infected animals and this in turn depends on thorough examination of the sheep's feet so that all animals carrying *F. nodosus* will be segregated from healthy sheep. The method has drawbacks because the diagnosis of carrier animals may not be easy under field conditions and the application of the segregation rules are often not feasible under many of our traditional methods of sheep husbandry but the procedure is well worth trying on farms where the disease is troublesome. Vaccines for the prevention of foot rot also have a place in controlling the disease but they are expensive and require veterinary advice before use.

### Foot abscess

Foot abscess is sometimes called foot rot but it is a distinct clinical entity and should not readily be confused. The infection may take one of two forms. In both types the area above the coronet becomes swollen, the digits are spread and the animal is very lame. In the less severe form a discharging sinus appears on the coronet after which relief from pain occurs. On the other hand, the sinus may penetrate downwards to the sole, and set up new sites of infection, the condition frequently becoming chronic. The characteristic unpleasant smell of foot rot is absent. In the more severe type, commonly seen in rams, the sinus does not discharge to the surface until the deep-seated infection has involved the ligaments, bones, and joints of the affected foot, causing abscess formation in these tissues. The leg above the affected area usually swells and the pain is intense. Post-mortem examination of the foot shows death of tissues and a small quantity of evil-smelling, black watery fluid, the whole being surrounded by pus and a thick encapsulating wall. The animal frequently shows secondary complications, such as pneumonia, which in many cases is the actual cause of death.

It is thought that *Fusiformis necrophorus* (a microbe normally present in soil) often in association with *Corynebacterium pyogenes* and other organisms, is the causal organism of foot abscess.

Treatment of the less severe form is usually satisfactory, especially when commenced in the early stages of the infection. Slight surgical interference of the sinus and abscess should be carried out and the cavity packed with a suitable antibiotic. Excessive proud flesh can be treated with 10 per cent copper sulphate solution. In the more severe form, if seen early, full doses of a sulphonamide along with an antibiotic are fairly satisfactory. Otherwise, radical surgery with removal of the digit is the best method of treatment but even this is not very successful unless veterinary advice is sought early.

There is no specific method of prevention but there are several predisposing factors which can be avoided. Abscess formation is often associated with excessive horn growth so that regular examination of feet and attention to those that are overgrown is an essential feature of good shepherding. The overgrowth of horn in certain sheep and not in others often has a genetic basis so that rams should be selected with good foot conformation. Constant exposure of sheep to wet conditions also leads to poor overgrown horn which in turn will lead to foot abscess unless the feet have constant attention. When sheep are overwintered indoors the type of flooring often causes foot problems and abscess formation so that advice as to type of flooring to be used should be obtained. Penetration of the sole of the foot with a foreign body such as nails, barbed wire, etc. is a relatively common predisposing cause of foot abscess and requires immediate surgical treatment to prevent serious consequences.

### Scald

This is a condition usually seen in young sheep on lush pastures. The skin of the interdigital space becomes inflamed, moist, and painful, and this causes some degree of lameness. It has been variously ascribed to the action of irritant juices from certain plants (clovers and buttercup) or to mass invasion by skin-penetrating worm larvae but recent work suggests that *Fusiformis necrophorus* is the cause. It is considered that these lesions may predispose to foot rot or foot abscess.

The lesions frequently heal spontaneously, but regular use of the foot-bath is probably advisable, as well as a change of pasture. Simple treatment with the foot-rot dressings already mentioned is effective, while chloromy-cetin tincture is very efficient for valuable animals.

### Soil balling

Sheep on heavy clay soils, such as on the turnip break, often simulate the symptoms of foot rot due to the soil balling between the digits. Examination of the affected feet quickly shows the real cause of the lameness.

### Strawberry foot rot

This is an ill-defined condition which has recently assumed some prominence in certain sheep-rearing areas. As the majority of outbreaks are, in fact, caused by the virus of orf, the term 'strawberry foot rot' should be discouraged, since it is more confusing than useful.

### Redfoot

This is seen in newly-born lambs in which the horn of the hooves becomes loosened or detached, leaving the sensitive horn matrix exposed. Although widespread in distribution it never affects more than a few lambs on any farm. The causal agent or factor has not been found and experiments to transmit the disease from lamb to lamb have always failed. Lameness is so severe that the lamb is unable to suck, and quickly dies from starvation. Methods of prevention are unknown and treatment is unsatisfactory.

### Foot-and-mouth

Foot-and-mouth is a notifiable disease which is highly contagious, although sheep are less susceptible than cattle or pigs and so few outbreaks of the disease originate in sheep.

The causal agent is a small virus of which there are a number of different immunological types. Although foot-and-mouth disease can cause rapid death in young lambs since the virus attacks the heart muscle with a very high incidence in an affected flock, typical symptoms occur much more frequently in older animals.

The major clinical sign is lameness caused by vesicles and ulcers of the skin in the interdigital space and on the bulbs of the heels. These lesions can readily be confused with foot rot or scald and since mouth lesions tend to be minimal in sheep the presence of the disease in a flock can readily be overlooked.

Control of the disease in Britain is by compulsory slaughter.

### Infectious opthalmia (pink eye, snow or heather blindness)

This is a highly contagious inflammatory disease of the eyes of all ages of sheep and is common in all sheep-rearing districts. The cause is an organism called *Mycoplasma conjunctivae* or more commonly the organism *Chlamydia psittacovis*. Other organisms may also be involved. In Scotland the names 'heather blindness' and 'snow blindness are given to the condition, but severe outbreaks can occur in sheep on ley pastures in summer and autumn.

The infection may cause lesions in one or both eyes. In the early stages of the disease there is a watery discharge from the eyes, an intense inflammation of the conjunctiva and of the lining membranes of the eye follows. An opacity develops and this may be partially over the eye or less commonly over the whole eye so that the cornea becomes completely opaque. In very severe cases ulceration of the surface of the eye occurs and pus may form within the eyeball. Fortunately in the majority of cases opacity is only partial and spontaneous healing begins within a few days and there is no serious upset in vision. In severe cases the duration of the disease may be many weeks during which time the animals lose condition and a few may even die from starvation. Even very badly affected eyes generally heal satisfactorily and permanent loss of sight is uncommon.

The source of infection is carrier animals, either recovered cases or normal sheep, which are for some reason not susceptible to the disease. It would seem that flies may transmit the infection from sheep to sheep, but this is not likely in 'snow blindness' when the disease may be truly contagious. Recovered sheep are immune to further infection for one to several years.

Specific diagnosis is by demonstration of the causal organisms in the laboratory. Although the majority of cases recover without treatment severe cases benefit by the local application of sulphonamides or antibiotics, etc. Eye ointments containing zinc sulphate or oxide of mercury are also useful. Veterinary advice as to which drug to use is advisable.

There is no method of preventing the disease except by decreasing stocking density and increasing trough space for each animal when sheep are housed.

It should be remembered that inflammation of the eye is one of the early clinical signs of several specific diseases of sheep including pneumonia.

**Sturdy (gid)**

Sturdy is a brain disease of adult sheep caused by the intermediate cysts of the intestinal tapeworm of the dog, *Taenia multiceps*. The dog infected with these tapeworms defaecates on grass, etc. and so deposits the eggs of the worm which are accidentally ingested by the grazing sheep. The eggs then develop into embryos within the sheep and after burrowing through the intestinal wall reach the brain and spinal cord of the sheep via the bloodstream. Once in the brain the tapeworm embryo develops into a cyst form which grows in size during several months and this causes symptoms due to pressure within the brain.

The clinical symptoms vary according to the location of the cysts in the brain but most commonly it causes uncoordination of gait, circling and apparent blindness. Sometimes it causes jerky movements and trembling. There is no raised temperature as occurs in brain infections. Most infected animals die within a few weeks if not treated.

Treatment which can be very successful is by surgical removal of the cyst.

Prevention is by the treatment of all farm dogs with anthelmintics to rid them of tapeworms. This should be done regularly.

## Cerebrocortical necrosis (CCN)

This mainly affects older lambs and reference should be made to the appropriate section in Chapter 17. However, it can on occasion occur in adult sheep.

## Urolithiasis

This is mainly seen in male sheep and in castrated males which have been housed for considerable periods of time.

This clinical features are those associated with uretheral obstruction which generally occurs at the vermiform appendage or sigmoid flexure. The obstruction is caused by numbers of small urinary calculi containing calcium, phosphorus, magnesium and ammonium compounds.

Affected animals are obviously very uncomfortable, kicking at the abdomen and straining to urinate. The little urine that is passed is often bloodstained and small chemical crystals may surround the end of the sheath. Often after a period of straining the urinary bladder bursts and the symptoms subside and this is followed by collapse and death.

Treatment of the obstruction is by surgical methods and frequently the simple removal of the vermiform appendage is all that is required.

Successful prevention of the formation of the calculi is achieved by feeding about 5 per cent common salt in the concentrate feed along with plenty of drinking water at all times.

## Bright blindness

This is a condition which occurs in adult hill sheep particularly those on the hills of Northern England and Scotland. Sheep eating bracken for extensive periods develop a degeneration of the retina of the eye.

Most cases occur in the late summer and autumn or perhaps they are noticed more readily as the days shorten and affected sheep have more difficulty in seeing in the decreasing daylight. Many sheep eventually become blind and cannot see at all. The main symptom is that of wasting because the sheep are unable to see to eat.

The name 'bright blindness' is given to the condition because if the eyes are examined with an opthalmoscope a very bright sheen is seen on part of the retina.

There is no specific treatment and prevention depends essentially upon management changes but these are not very easy to achieve in hill areas.

## Ulcerative vulvitis and balantitis

This is a venereal disease in which ulcers occur on the vulva, prepuce and penis of adult ewes and rams.

The clinical signs are seen soon after mating. The vulva becomes inflamed and bloody and as many as a third of the flock may be affected. Ulcers which can be severe follow and scab over. The vagina is not affected. Most rams in an affected flock show lesions which consist of inflammation and ulceration of the soft tissue of the glans penis with much haemorrhage. Despite the severity of the lesion there is little or no interference with mating and normal lambing follows.

The cause of the condition is not well established but it is now thought that it is not a form of contagious pustular dermatitis (orf) although there is still doubt about this.

Treatment is mainly by the use of local washes and emollients. Prevention is by good husbandry including isolation of affected animals early in an outbreak.

## Sheep tick

The only species which is important in Britain is *Ixodes ricinus* which transmits the diseases of louping-ill, tick-borne-fever and tick pyaemia which have already been described. Attachment of the tick to sheep is mainly in the region of the ears, face and the other wool-free areas of the body where they pierce the skin and suck blood; adults, nymphs and larvae all do this. It is a three-host tick but all three stages can occur on the same sheep or other host. The generation time from egg to adult can vary from 2–6 years, much of this time being spent on the ground. An average infestation is about 200 of all stages but the numbers may be much greater or very few. Unless the infestations are very heavy the sheep tick probably does little harm apart from the diseases it carries.

Control of the tick has been attempted in several ways, by pasture improvement, by keeping sheep and cattle off infested areas for a period, and by dipping. None of the methods has been successful, although spring dipping and spring spraying are widely used to reduce the number of ticks on the sheep. This, however, does little or nothing to control tick-borne disease. The persistence of the dips in the fleece is very short-lived so that frequent dipping would be necessary to completely control tick attachment. This, however, is not feasible both economically and management-wise. When sheep are purchased from tick-infested farms they should be dipped before being taken to tick-free pastures.

### Keds (melophagus ovinus)

Keds are dark brown in colour and resemble sheep ticks but they have only six legs whereas ticks have eight. The life-cycle is a very simple one, the female ked laying developed pupae which never leave the sheep and, in turn, become pupae-laying females.

The ked causes considerable irritation of the skin so that infested sheep bite and rub themselves on fences, stones, etc. Although the ked sucks blood its main economic effect is broken fleeces. Infestations are mainly seen in the autumn and winter months but are now uncommon in Britain.

Winter dipping with an organo-phosphorus dip is very successful in controlling the parasite.

### Lice

There are three varieties of sheep lice. The body louse or common louse which is a biting louse (*Damalinia ovis*) moves about on the skin under the wool. The foot louse (*Linognathus pedalis*) which is a sucking louse remains static on the legs and on occasion on the face and scrotum. The face louse (*Linognathus ovillus*), which is also a sucking louse, is found on the face, head and neck. All three varieties have a simple life-history confined to the sheep but they have a considerable capacity to multiply so that the speed of build-up can be very considerable. Infestations are heaviest in the winter and particularly when sheep are confined indoors in close contact.

Lightly infested sheep show few symptoms but when the number of body lice is large affected sheep become restless, constantly scratch and rub and even bite themselves. Wool damage and loss of weight follows. Neither foot lice nor face lice cause any appreciable symptoms and neither is commonly noticed. Infestations are probably not very common.

Lice infestations can be adequately controlled by winter dipping with organo-phosphorus dips.

### Sheep scab (mange)

The mange mites *Psoroptes communis* var. *ovis* and *Sarcoptes scabei* are the two designated in the United Kingdom Sheep Scab Order of 1977. Both were eradicated for many years but outbreaks of *P. communis* have returned in recent years. They are seen during the winter months but the mite, although not active in the summer, still remains on affected sheep to become active again during the following winter. There are a number of stages during the life-cycle but none leave the sheep except for another sheep. The condition is very contagious.

Psoroptic mites infest the skin of the wool-bearing areas of sheep and cause restlessness to be followed by intense rubbing and scratching against any suitable projection, such as fence stobs, etc. The wool is damaged, becomes lighter in colour and then begins to be shed and the skin becomes thickened with a crust or scab being formed. Secondary bacterial infection often follows with pus formation and a number of affected sheep die while others rapidly lose condition. The disease can cause enormous economic loss if not controlled.

Sarcoptic mange, on the other hand, affects those areas on the sheep free from wool, the head around the eyes and on the ears, but otherwise the condition is similar.

Specific diagnosis is by demonstrating the mites in skin scrapings which is easy unless the sheep have recently been dipped, but requires laboratory aids.

Treatment by double-dipping with an approved dip is mandatory. Preventive measures are based on compulsory double-dipping using approved dips and the control of movement of sheep. These are controlled by the local authority.

Chorioptic mange caused by *Chorioptes ovis* also occurs in various parts of the world including Britain; it is not a designated disease. The lesions are mainly seen on the legs and feet, sometimes on the scrotum and occasionally around the eyes. The lesions and symptoms are broadly similar to the other types of mange and treatment and prevention also depends on proper dipping being carried out.

### Blowfly (strike)

Blowfly is a condition which affects all breeds of sheep in nearly all environments but clipped sheep are less liable to strike. The term 'strike' describes the attack of the blowfly, described generally as the greenfly (*Lucilia cuprina*), or *Phormiaterrae novae* (blackfly), upon the faecal and urine-soiled areas of the fleece, usually the breech and tail but sometimes the body also. The flies lay eggs on the sheep and from these develop the maggots which cause the lesions of 'strike' by damage to the skin. Scabs quickly form over the raw areas of skin. The flies are not obligate parasites.

Affected sheep are restless and move around with a characteristic 'hangdog' appearance and they endeavour to kick and bite at the 'strike' area which has a most unpleasant foetid smell. Secondary infection often causes sheep to collapse and die but modern treatment rarely permits this to occur. The maggots can readily be seen in the skin wounds. Diagnosis is by the observation of the obvious lesions.

The disease is encouraged by humid weather conditions which encourage the rapid development of the flies mainly in June to September in Britain. Dirty unclipped sheep are much more susceptible.

The prevention and treatment of blowfly strike is now very efficient and all sheep should be dipped in appropriate dips in midsummer. This is very essential when the weather at this time is warm and humid. Antibiotic therapy may be necessary if secondary infection is serious.

### Headfly

Headfly which attacks the head mainly around the base of the horn is called *Hydrotaea irritans*. Its main operational area appears to be the English–Scottish Border.

The life-cycle of the headfly involves four stages, egg, larva, pupa and adult, the later stages of pupa and adult appearing in the year following the laying of eggs, which is usually around the months of July, August and September. Flies are only active during the day and feed on many types of animals other than sheep. Woodland cover increases the number of flies in the environment.

All classes and ages of sheep are affected but lambs suffer most. Open wounds of the head, particularly around the base of the horns, attract enormous numbers of headfly which, although not biting, irritate and enlarge the wounds allowing secondary invasion by bacteria which cause fever and loss of condition in the sheep. Scarring and disfigurement follow. Antibiotic therapy may be necessary.

Until recently, control depended mainly on fitting protective hoods to the sheep, but there are now fairly efficient fly-repellents available which are sprayed on to the head and neck of sheep at frequent intervals. At the moment, permethain is giving good results. Housing of sheep is also helpful. There is no method of controlling the fly population in the environment.

### Mycotic dermatitis (lumpy wool)

Mycotic dermatitis is caused by a bacterial infection called *Dermatophilus congolensis* which attacks the skin under the wool. It is much more common after heavy rain and outbreaks are severe in wet years. It is the cause of much down-grading of pelts and wool as well as actual wool loss. All ages of sheep can be affected. Other species including man can be affected.

The skin of affected sheep becomes inflamed and reddened and this is followed by exudation of fluid which becomes dry and thickened into a crust formation. If this is removed it leaves a raw bleeding surface. The presence of lumpy wool is often not recognised until the wool on the sheep is handled as at shearing time. The bases of the wool staples are hard and firmly adherent to the crusts of the skin, hence the name 'lumpy wool'. Affected animals show no clinical signs. When the infection occurs on the legs it is often known as strawberry foot rot.

Diagnosis is relatively simple by recognition of the lesions but specific diagnosis depends upon laboratory demonstration of the causal organisms.

The infection is so widespread in nature that eradication is not feasible. Treatment by the use of dips containing zinc sulphate or potassium aluminium sulphate is reasonably effective. Antibiotic therapy is also indicated in severely affected sheep.

### Periorbital eczema (staphylococcal dermatitis)

This has already been described in the section on microbial diseases in Chapter 18, but it can also occur in sheep of most ages that are being confined and trough-fed.

## Clostridial wound infections (gas gangrene, black quarter)

One form of this disease called metritis or post-parturient gas gangrene has already been described in the section dealing with microbial diseases in Chapter 18. These diseases are all of a similar nature and result in sudden or quick death in sheep of various ages. Death is caused by infection of tissues with one or more of several clostridial organisms, *Clostridium perfringens*, *Cl. septique*, *Cl. chauvoei* and *Cl. oedematiens* which kill quickly due to their powerful toxins.

Infection, often called post-vaccination gas gangrene, is not uncommon when vaccination programmes of various types have been carried out, one of the above organisms being introduced into the tissues by a contaminated injection (other organisms may also be incriminated such as corynebacteria and streptococci). This is the main reason for carrying out all injections in as sterile a manner as possible. Danger also lies in the further use of a bottle of serum or vaccine which has been stored for any period of time after being first opened, since bacteria may be introduced into the partly used bottle and multiply during unsuitable storage. A precaution against this type of contamination is to use a separate needle inserted and held in the cap of the vaccine bottle through which the syringe is filled, different needles being used for the actual inoculation of each sheep. However, in many such injection infections the presence of the actual vaccine under the skin or in the muscle seems to be sufficient stimulus to cause a latent organism to multiply. Other predisposing conditions which give rise to infection are surgical operations such as castration, docking, and clipping. Dipping may also precipitate infection while the condition known as 'big head', due to fighting in rams, is of the same type. On occasion, it may be difficult to determine the actual predisposing cause of bacterial multiplication but since the site of activity in these cases is often in the muscles of either a fore or hind limb, the condition is frequently referred to as 'blackquarter'.

Post-mortem diagnosis is not difficult providing the dead animal is examined quickly. In all cases the findings are similar except for the location of the site of infection and reference should be made to the section on post-parturient gas gangrene.

General preventive measures against infection are those of good husbandry, such as well-chosen and disinfected sheds and pens, along with personal cleanliness of the shepherd. The choice of suitable dry days for carrying out vaccination programmes, along with adequate sterilisation of syringes and needles and the use of clean and mud-free pens for confining the sheep, all help in preventing post-vaccination infection, while similar precautions for castration and docking lessen the risks considerably. It would appear, however, that trouble does often arise despite all reasonable precautions and when this occurs in serious proportions specific inoculations can be carried out. Serum hyperimmunised against the type of infection involved in the outbreak can be very efficient. Treatment with antibiotics is also very effective provided the animals are seen in the early stages of their illness. In general, preventive vaccination is indicated, especially now that combined vaccines are available which give protection against many of the different clostridial microbes of the sheep.

## Cruels

Cruels in sheep is characterised by chronic abscess formation of the skin of the head and neck. It occurs in relatively low incidence but, since it not infrequently attacks rams, it may be of considerable economic importance. The cause is a bacterium known as *Actinobacillus lignieresi*, the microbe that causes wooden tongue in cattle. In both cattle and sheep the organism is commonly found on the skin and in the mouth of healthy animals but the conditions which cause it to become pathogenic are unknown, except that as it occurs frequently in rams it is possible that wounds caused by fighting may be one predisposing factor.

The earliest signs of the disease are small swellings of the skin in the region of the lips, cheeks, lower jaw, neck, and base of the horns. These swellings slowly increase in size until they burst and discharge a tenacious green-yellow pus. Later, healing of the abscess may occur but other abscesses form in adjacent tissue and spread to the nearest lymphatic glands. In untreated cases the infection frequently spreads to the lung, stomach, and liver causing slow emaciation of the animal followed by death.

Post-mortem lesions consist of varying sizes of abscess in the skin, lymph glands, and internal organs. They contain the characteristic greenish pus.

Treatment by one of several antibiotics is very successful even in advanced cases.

## Broken mouth

Broken mouth, or the loosening and loss of permanent incisor teeth, is very common in culled sheep and is thought to be one of the important reasons for culling. Hill ewes do not achieve their maximum reproductive potential before their fifth lamb crop and since culling for broken mouth is frequent in sheep as young as $3\frac{1}{2}$ years old, broken mouth must reduce productivity in affected flocks and increase ewe replacement costs. Whether such culling for broken mouth is justified or not is debatable since the real loss of productivity probably occurs in younger sheep which have a full mouth but whose teeth and gums are very sore long before the incisor teeth fall out. Grazing to such sheep is very much more difficult and painful than it is later on when the mouth is really broken.

The cause of broken mouth is not known despite years of research mainly in the fields of energy and mineral nutrition. It is now generally recognised that the development of broken mouth is associated with inflammation of the surrounding tissues of the incisor and premolar teeth which occurs in most sheep during eruption of the teeth, but on those farms where broken mouth occurs this inflammatory condition progresses and becomes both severe and chronic, so removing the supporting tissues around the teeth. The incisor teeth are pulled out of their weakened sockets by the tearing action of grazing and even more by the biting of turnips, etc. The molar teeth, although often showing similar inflammation of the surrounding tissues, do not commonly fall out. On the other hand, the molars tend to be forced into place by the chewing or cud action. Excessive tooth wear as a result of soil ingestion on poor grazings may be a complementary factor. However, the cause of the excessive and continuing inflammation around the teeth on 'broken mouth' farms is not clear.

Little or nothing can be done to prevent the basic chronic periodontal inflammation but improved grazing and fewer sheep to the hectare can overcome the consequences to some extent and prevent affected sheep becoming 'poor doers'. However, most shepherds and flockmasters firmly believe that culling of broken-mouthed ewes is necessary. Experimentally, it has been shown that extraction of loose incisor teeth and the filing down of other affected incisor teeth allows such sheep to graze adequately and perform well. The fitting of 'artificial' or 'false' teeth has been attempted on a number of affected farms but this is rarely justified on economic grounds.


## Copper poisoning

This is very common in all ages of sheep and is a problem on many sheep farms which feed hay and concentrates as the main diet because such feeds are often low in molybdenum and sulphur and therefore enhance the availability of copper. Copper is also widely used on farms as a food additive in mineral mixtures, worm drenches, etc. and in concentrated feeds. It is also used in therapy and prevention of disease as, for instance, in pregnant ewes for the prevention of swayback in their offspring. Sheep accumulate excess copper in their bodies, particularly in the liver, and since the size of the toxic dose of copper is small such additives, etc, if given in an uncontrolled way, frequently cause copper poisoning. Pastures dressed with copper-rich pig slurry can also be dangerous to sheep grazing them so such pastures are best grazed by cattle in the first place.

Although copper poisoning may not reveal itself for many weeks the clinical disease when it occurs is a very acute syndrome. This is brought about by the sudden release of copper from the liver into the blood where it causes a sudden breakdown of the red blood corpuscles. The reason for this sudden out-pouring of copper is not known. Affected animals are dull, go off their feed and have a very rapid heart beat and their temperature is often raised. All the membranes of the mouth, eyes, nose and vagina become very yellow as also does the skin under the wool. The urine may be red in colour. Diarrhoea can be profuse if a considerable excess of copper has been taken. Death follows quickly.

Post-mortem examination shows the whole carcass to be jaundiced and the liver very large, brown in colour and readily broken when handled. Laboratory diagnosis is based on the quantity of copper in the liver, the level of copper being in excess of a 1000 parts per million on a dry matter (DM) basis. Copper poisoning can readily be confused with kale poisoning.

Treatment and prevention consists of radically reducing the amount of copper in the diet and giving a mixture of ammonium molybdate and sodium sulphate. Care in the use of copper for therapy is also very important.

The occurrence of copper poisoning is influenced by the breed of sheep, Texel and Suffolk sheep absorb copper very readily, and even between sheep with differing genetic make-up within a flock. It has recently been shown that sheep that are bred in the Orkneys, where seaweed is grazed regularly, are very susceptible to copper poisoning when they are introduced to normal pastures. Such sheep have a high absorption rate for dietary copper.

# 20 Long Incubation Diseases of Adult Sheep

## Jaagsiekte (pulmonary adenomatosis)

With its strange Afrikaans name, jaagsiekte has caused considerable economic loss in various countries of the world including South Africa and Iceland. It has been known to occur in Britain for many years but only recently has it become a real problem on some farms, especially where intensive management is being introduced.

Sheep pulmonary adenomatosis is an infectious tumour of sheep and two viruses are associated, one a ribonucleic acid tumour virus. The disease can readily be transmitted to healthy sheep by inoculating the lung and nasal fluid but so far the viruses have not caused any abnormality when injected into sheep.

The disease is insidious in its onset but later runs a more rapid course, death occuring in 2–3 months. The incubation period is very long, often 6 months to several years. The first sign is coughing with rapid and forced breathing but generally the animal is not fevered. Later in the disease, clear fluid begins to run from the nose, especially when the head is lowered, and as much fluid as will fill a tumbler sometimes pours out when the animal's head is forced down to the ground. The affected individuals lose condition and may be very emaciated at the end.

The changes at post-mortem examination are usually limited to the lungs. The lesion is first observed as a small, grey nodule (or nodules) which gradually enlarges in size with new ones forming around it until the whole lung is very big and heavy. The condition is not a pneumonia but a type of cancerous growth of the lung. Secondary tumours sometimes occur in tissues other than the lung. Pleurisy may occur as also may lung abscesses.

The spread of the disease is somewhat peculiar, for it occurred as an epidemic in Iceland until it was eradicated, whereas in other countries apart from South Africa, the incidence remains low. Methods of husbandry in that they determine the weight of infection, are of obvious importance and close housing of sheep undoubtedly encourages cross-infection.

There are no specific methods of cure or prevention but early slaughter of affected animals reduces the weight of infection in a flock. There is no serological test to pick out latent cases.

## Scrapie

This is a fatal progressive disorder of the central nervous system of sheep and goats which occurs as a natural infection in many countries of the world but not apparently in Australia or New Zealand. It has been present in Europe for over 200 years and perhaps longer.

*Plate 20.1* Two sheep showing the postures often assumed by animals affected with scrapie.

In Suffolk sheep which have been examined more extensively than any other breed the symptoms of scrapie are frequently 'classical'; impaired social behaviour followed by locomotor incoordination and fine trembling of the head and neck. Biting of the limbs, rubbing against fences and loss of condition follow. The duration of symptoms is usually 1–2 months but can range from a week to many months. A practical problem in diagnosis based on clinical signs is that pruritus, debilitation and ataxia are not always seen together and breed differences are considerable. Death without any suspicious symptoms is not unusual. In other words, clinical signs are not always diagnostic by themselves. (Plate 20.1.)

The most prominent lesion in scrapie is vacuolation of nerve cells in the brain along with neuronal degeneration, etc. Also there is often but not always, interstitial spongy degeneration in the same areas. These lesions may be present in the brain for a month or so before clinical symptoms occur so that they may by chance be seen in a normal-looking sheep when it is killed. Death due to scrapie occurs most frequently in sheep between 2–5 years old with a few cases in slightly younger sheep. If sheep are culled at about 6 years of age it is probable that about 25 per cent of scrapie cases will be missed from older sheep (5–15 years).

This, of course, indicates that no quarantine system can be 100 per cent certain of picking-up the disease. It has also been suggested that some infected sheep may die of old age before developing the disease. There is considerable evidence that the majority of sheep developing scrapie have been infected since early life and the scrapie agent has replicated within the sheep during the whole of this period without causing any apparent symptoms. Hence, the reason for calling scrapie a slow virus infection.

A genetic basis was originally postulated for scrapie but it is now accepted that scrapie is due to horizontal (contagion) and vertical (maternal) transmission of an infectious agent with a host genetic factor controlling whether clinical disease develops or not in an infected sheep. The agent can withstand heat, formalin, ultraviolet light and ionising irradiation, etc. to an extent that one would not expect of a conventional animal virus.

Scrapie can be experimentally transmitted in series to sheep, goats, mice, rats, golden hamsters, moles, gerbils and New World apes, and there are a number of strains of the agent which preserve their identity through different hosts. In the experimental disease in inbred strains of mice, the incubation period and the distribution of brain lesions vary with the strain of agent and the type of mouse, but for obvious reasons in sheep these factors are not known and one has to resort to the mouse for agent typing.

There is a complete absence of any demonstrable specific immune process in scrapie disease and the agent has not been shown to be antigenic. In addition, standard tissue culture techniques in the laboratory have failed to demonstrate the agent. In consequence, there are no laboratory diagnostic tests available.

Sheep vary in their response to experimental infection, some developing the disease and others not. Such differences appear to be individually based, there being no clear breed difference. It also varies according to the strain of infecting scrapie agent. A genetic relationship for 'susceptibility' and 'resistance' to SSBP/1 pool of agents in Cheviot and Herdwick sheep has been succesfully determined. There are two difficulties which have to be appreciated. One is that the 'resistant' sheep need not be resistant to other strains of agent other than SSBP/1. The other difficulty is that the 'resistant' sheep may only be resistant to the clinical disease and not to the multiplication of the agent. It is widely assumed that sheep in the field may vary in their susceptibility to natural scrapie but whether this is so is unknown.

There is no doubt that the majority of commercial flocks in Britain are either free of the disease or the incidence is so low that it is of no significance. There are, however, a few areas in Britain where scrapie is an economic problem, even to the commercial flockmaster, and 10–15 per cent of a flock may die with scrapie before they reach culling age. It is in the ram-producing flocks that scrapie can cause real problems, and in some such flocks the incidence may be as high as 30 per cent or even more. Unfortunately it is not possible to give factual figures since scrapie is widely feared, for a breeder's reputation can be harmed so its occurrence is nearly always kept secret. Nevertheless it is fairly certain that scrapie is not as serious now as it was at the beginning of the century, the reason being that control (but not eradication) is now feasible based on familial incidence. Such control measures do, however, damage genetic improvement schemes by limiting selection potential.

The diagnosis of scrapie in a sheep is based upon clinical signs which may be typical of the disease but their absence or their uncharacteristic features should not deter one from histopathological examination which often shows typical lesions, and these are usually considered to be sufficient for diagnostic purposes.

It is now well established that maternal transmission of scrapie agents accounts for much of the familial pattern seen in outbreaks of natural scrapie. Naturally affected ewes often have several pregnancies before dying from scrapie and the present evidence is that at least the last two crops of lambs may be affected. There is no information as yet on whether maternal transmission depends upon infection of the egg, the embryo or of the lamb at the time of birth. Milk does not seem to be involved. It is now very clear that scrapie can also be spread by contagion to flockmates. The placenta is one source of infection but non-pregnant ewes and possibly rams can spread the disease but little work has been done as to the excretion of the agent from infected animals. There is good evidence that pasture contamination is an important source of infection but again little work has been done on the degree of persistence of infectivity of the pasture. What little evidence there is suggests the agent may persist for years. Whether mechanical transfer occurs from farm to farm, etc. is unsure, vector transmission has not been proved. Contamination can clearly be by more than one strain of agent and most certainly in Britain many sheep are infected by several strains. It is thought that lambs are more readily infected than adults but young sheep have been shown to be susceptible.

Scrapie is not the only slow virus infection for in recent years a number of other diseases of both man and animals have been shown to have similar features to those of scrapie and are also transmissible by inoculation to primates, hamsters, etc. Kuru of New Guinea, Creutzfeld Jakob disease, familial Alzheimer's disease and probably other senile and presenile dementias of man along with transmissible mink encephalopathy are presently being investigated. The demonstration of amyloid and senile plaques produced at will in the brains of certain mouse strains inoculated with certain scrapie agents could be a model for an even wider field of dementias in man. Concern about the similarity of scrapie to these disorders of man has recently caused the United States Department of Agriculture to ban the use for human food not only of sheep affected with scrapie but also of related or contact sheep or goats. Although there is no evidence at all that sheep meat causes these diseases in man, the fear of such a possibility could have serious consequences for the sheep industries of countries where scrapie is endemic.

There is no treatment for scrapie. Control and prevention can never be total in countries where the disease is endemic but it can be contained within economic limits by culling affected ewes and all their offspring. In-contact spread from affected rams does not appear to be a high risk. Genetic control is not very feasible since a sheep immune to one strain of scrapie need not to be resistant to other strains.

### Maedi-Visna

This is a progressive pneumonia of adult sheep and the lesion is thought to be a tumour rather than inflammation. In many countries there has been considerable confusion between maedi and jaagsiekte but the two diseases are quite distinct. In Iceland maedi caused enormous economic loss until it was slaughtered out. It is common in South Africa where it is known as Graaff–Reinet disease and in Kenya where it is called Laikipia disease. It also occurs in the United States and in a number of countries in Western Europe. It now occurs in Britain which was free of the disease.

Maedi is a highly infectious and contagious condition caused by a virus of the ribonucleic tumour virus group. After infection the incubation period is very long, at least 3–4 years, or as long as 7 years. The routes by which infection passes between sheep is probably multiple; contagion from sheep to sheep by the respiratory route, although likely, is probably not as common as that from ewe to lamb by the colostrum and milk but not in the uterus. Once infection takes place a persistent and long-lasting infection occurs in the lungs despite the development of a slow antibody rise and a multiplication of immune white blood cells. It is also known that the virus frequently mutates to give new strains of the agent during this phase. It is thought that the virus causing Acquired Immune Deficiency Syndrome (AIDS) of man may be similar but not identical with the maedi virus.

The clinical course is also long, sheep lasting for as long as 6 months after the first signs are seen. There is progressive loss of condition with quick and forced breathing being the predominant symptom. Less commonly sheep show nervous symptoms such as change in behaviour, paralysis and incoordination of movement when the disease is called Visna. Both forms can occur in the same sheep.

At post-mortem examination the lungs are heavy, enlarged overall and spongy to the touch, the local lymph glands are also very large. When examined under the microscope the lung is infiltrated with tumour-like cells which are very different from those seen in jaagsiekte. There are no gross post-mortem lesions in sheep infected with Visna but brain lesions can be seen under the microscope.

Diagnosis on clinical grounds is not sufficient and laboratory aids are necessary. Examination of lung and brain under the microscope is satisfactory or causal virus can be grown. Serological tests in the living affected animal are quicker and these can also be used to diagnose carrier and preclinical cases.

There is no method of treatment nor is there any specific vaccine for prevention.

Control of maedi-visna is by preventing spread of infection within a flock or hopefully by eliminating infection from flocks and countries. This is done by repeated blood testing of infected flocks and by the slaughter of all positive reactors. It would appear that positive sheep can be eliminated from a flock in a few years. As a result the Ministry of Agriculture in Britain has set up an Accredited Scheme to help farmers eradicate the disease from their farms.

Since it is known that lambs are free of infection when in the uterus some flockmasters have resorted to caesarian section to develop a maedi-free flock, but this is by no means easy for it demands very good husbandry and hygiene.

As with jaagsiekte, methods of husbandry determine the incidence of the disease. Close housing or close contact of sheep undoubtedly encourage the spread of maedi within a flock once it is introduced.

# Index